Advanced Research in Plant Science

Advanced Research in Plant Science

Edited by **Molly Ismay**

SYRAWOOD
PUBLISHING HOUSE

New York

Published by Syrawood Publishing House,
750 Third Avenue, 9th Floor,
New York, NY 10017, USA
www.syrawoodpublishinghouse.com

Advanced Research in Plant Science
Edited by Molly Ismay

International Standard Book Number: 978-1-68286-068-7 (Hardback)

Printed in the United States of America.

Contents

Preface

Plant science is a significant branch of biological sciences and the study of this discipline has aided many advances in a multitude of disciplines, ranging from genetic engineering to environmental science. The topics covered in this extensive book deal with the core subjects of plant science. It presents the complex subject of botany in the most comprehensible and easy to understand language. The topics covered in this book deal with the primary areas of botany and include plant structure, growth, reproduction, biochemistry, genetics, taxonomy, etc. This book will offer the readers new insights and will prove to be a vital tool for everyone who is researching and studying this field.

This book has been the outcome of endless efforts put in by authors and researchers on various issues and topics within the field. The book is a comprehensive collection of significant researches that are addressed in a variety of chapters. It will surely enhance the knowledge of the field among readers across the globe.

It gives us an immense pleasure to thank our researchers and authors for their efforts to submit their piece of writing before the deadlines. Finally in the end, I would like to thank my family and colleagues who have been a great source of inspiration and support.

Editor

Impact of Culture Filtrate of *Piriformospora indica* on Biomass and Biosynthesis of Active Ingredient Aristolochic Acid in *Aristolochia elegans* Mart

Uttamkumar S. Bagde[1,2], Ram Prasad[2] & Ajit Varma[2]

[1] Applied Microbiology Laboratory, Department of Life Sciences, University of Mumbai, India

[2] Amity Institute of Microbial Technology, Amity University-Uttar Pradesh, India

Correspondence: Uttamkumar S. Bagde, Department of Botany, School of Biological Sciences, Dr. Harising Gour Central University, Sagar-470003 (Madhya Pradesh) India. E-mail: bagdeu@yahoo.com; usbagde@gmail.com

Abstract

The mycorrhiza plant partnership is the basic, essential and integral part of plant survival and growth. In the present investigation we are reporting the effect of culture filtrate of *Piriformospora indica,* a growth promoter and bioprotector fungus on *Aristolochia elegans* Mart. The culture filtrate of the fungus increased overall growth, biomass, and active ingredient-aristolochic acid in the leaves of plants. In untreated control plants, the overall growth was reduced. *P. indica* culture filtrate application increased root number, root length, and root dry weightby 28%, 98%, and 123% respectivelyin plants of *Aristolochia*. Also stem height and shoot length was enhanced by 43% and 155% respectively.There was increase in number of leaves by 79% and length of leaves by 36%. The increase in total biomass was 136%. The improvement in content of aristolochic acid in leaves was between7.6% and 28.8%in treated plants as against untreated control plants

Keywords: *Aristolochia elegans*, aristolochic acid, *Piriformospora indica,* symbiotic fungus

1. Introduction

The fungus *Piriformospora indica*, is related to the Hymenomycetes of the Basidiomycota, which is a root endophyte that has capabilities of a typical, arbuscular mycorrhizal fungus, (Verma et al., 1998; Weiss et al., 2004; Prasad et al., 2008a; Bagde et al., 2010a, 2010b, 2010c, 2011), but unlike Arbuscular Mycorrhizal fungi it is cultivable in axenic conditionseasily. This mycorrhiza like fungus can form association with roots for enhanced growth and development of plants (Varma et al., 1999, 2001; Oelmüller et al., 2009; Sirrenberg et al., 2007; Prasad et al., 2008a; Bagde et al., 2011). *P. indica* interacted mutually with various plants including *Fabaceae* and *Rhamnaceae* species (Varma et al., 2001), *Arabidopsis* (Peškan-Berghöfer et al., 2004), tobacco (Barazani et al., 2005), *Poaceae* species (Waller et al., 2005).

P. indica enhanced nutrient uptake, helped plants to survive in extreme drought,temperature and salt conditions, exhibited systemic resistance to toxins, acted as biofertilizer, bioprotector, stimulator of growth, increased seed production, and played a key role in increasing the tolerance to insects (Varma et al., 1999; Waller et al., 2005, 2007; Serfling et al., 2007; Prasad et al., 2008a, 2008b). It helps in biological hardening to tissue culture raised plants, provides protection against 'shock of transplantation 'and pathogens of roots (Sahay & Varma, 1999; Hazarika, 2003; Prasad et al., 2008a, 2008b).

Aristolochia elegans plant belongs to Aristolochiaceae family. It is generally called pine vines or Dutchman's pipes. It is an annual plant with slender, woody stems, climber on support with heart shaped leaves and calico flower. It grows all over the world including in tropical climates such as India and other countries and is also called birthworts and commonly used to ease pain of childbirth, treat malaria and other diseases (Kimura & Kimura, 1981). Plant is cultivated and used in some medicinal preparations in China (Lopes et al., 2001). It contains important alkaloid aristolochic acid which is antimicrobial in nature (Imran & Bagde, 2007) and is useful for variety of ailments. The fruits and roots have been used by Chinese people in medicine as anodynes, antiphlogistics, expectorants and anti-asthmatic agents and is also used in treatment of snakebite, anti-tumor, anti-platelet aggregator agent and lung inflammation (Vila et al., 1997; Wu et al., 1999; Tian-Shung et al., 2000).

However, its certain harmful activities such as mutagenicity and carcinogenicity have also been reported (Arlt et al., 2002).

So far all the accounts are on the interaction of fungus propgules but in this communication we document that *P. indica* culture filtrate also enhanced the overall growth parameters of *Aristolochia elegans* and contents of aristolochic acid in leaves.

2. Materials and Methods

2.1 Mycobiont

P. indica culture for this study was procured from Amity University'sAmity Institute of Microbial Technology, India.

2.2 Photosymbiont

Aristolochia elegans Mart. (Aristolochiaceae) is the perennial shrub cultivated as ornamental plant in India. Species of *Aristolochia* are cultivated and used in medicinal preparations (Lopes et al., 2001). The plantlets were procured from Jijamata Udyan Byculla, Mumbai, India and were multiplied in environmentally controlled green house. Sterile substratum was used to conduct the experiments.

2.2.1 Culturing the Fungus *P. indica*

A. elegans was cultivated and maintained on modified synthetic media fortified with 1.2% agar (w/v) in dark at 28 ± 2 °C (Hill & Käfer, 2001; Prasad et al., 2005). pH of medium was kept at 6.5. For mass propagation, the fungus was also cultivated in liquid broth medium under constant shaking at 120 rpm in dark (GFL 3019, Germany). Media were sterilized in autoclave at 15 psi pressure for 15 minutes.

2.2.2 Separation of Culture Filtrate

The fungus was grown in liquid medium for 15days and was first filtered through sterile muslin cloth followed by bacterial filter (Millex-GV, 0.22 µm Filter Unit, Millipore) and kept at 4 °C if not used afresh.

2.2.3 Co-Cultivation Experiments

Plantlets grown for Fifteen days were transferred to sterile 10″ diameter plastic pots containing sterile unfertilized garden soil autoclaved on three consecutive days. Initially two plantlets were planted in each pot. Once they got acclimatized then one of the plantlet was removed, finally retaining only one plantlet in each pot. To each pot containing 1 kg of soil 15ml of freshly eluted culture filtrate to experimental pots and an equal volume of sterile nutrient medium were added to control pots one day before transfer of the plantlet into the pots. Again after a period of one month this treatment was repeated.

2.2.4 Growth Conditions

Pots were kept in green house at temperature of 26 ± 2 °C and 16 h light/8h dark and 60%-70% relative humidity and a light intensity of 20,000 lux. Growth of plants was measured after 90 days by use of centimeter scale. For estimation of dry biomass, plant was chopped and dried at 80 °C for 12 h in a Memmert oven and dry biomass was estimated after cooling at room temperature and weighing on electric- mono-pan balance.

2.3 Aristolochic Acid Analysis

Leaves of *Aristolochia* were used to extract and estimate aristolochic acid. For preparation of extract, leaf material was ground to fine powder by mechanical grinding using HPLC grade methanol and formic acid. 2 gm. of ground sample was taken in a bottle, thoroughly mixed with a mixture of 50 ml of methanol (80%) and 20 ml of 10% formic acid in water. The contents were stirred for 30 minutes at 500 rpm (Innova Model 2001 bench top platform shaker, New Brunswick, USA) and then centrifuged for 4 minutes at 4000 rpm. The supernatant was taken for determination of aristolochic acid (Gaudreault et al., 2001; Flurer et al., 2001). Estimation of aristolochic acid in the leaves extract of *A. elegans* was carried out by HPTLC using standard aristolochic acid as reference (Sigma, USA) in the range of 0-200 µg/ml. Stationary phase used for HPTLC contained Silica gel 60 (Merck) plates of 10x10 cm size. The mobile phase used for the chromatogram consisted of toluene, ethyl acetate, water and formic acid in the ratio of 20:10:1:1. The sample used was 10 µg. For developing the plate twin trough chamber was saturated for 20 minutes and the plate was dried with hair drier (cold air) for 5 minutes. The plate was evenly sprayed with tin (II) chloride reagent and further dried at 100 °C for a minute. Plates were observed under UV light at 366 nm and acid content was measured. This was determinedafter 15, 30, 45, 60, 75 and 90 days.

3. Results and Discussion

When morphological appearance of *P. indica* was observed on Käfer agar medium the pattern of growth of the fungus was marked by uniform rhythmic zonation (Figure 1a). The rapid growth on Käfer nutrient broth was observed after 15 days incubation at temperature of 28 ± 2 °C (Figure 1b). The colonies showed prominent crowded balls of coral morphology in conformity with previous studies by various workers (Varma et al., 2001; Singh et al., 2003). The important characteristics of this organism have been described earlier (Varma et al., 2001).

Figure 1(a). Growth of *P. indica* on solidified agar medium

Figure 1(b). Cultivation of *P. indica* in aspergillus broth medium

The observations made in this study indicated that culture filtrate of fungus exerted positive impact on various parameters of the plant as depicted in Figures 2 & 3 and Tables 1 & 2. When *A. elegans* plant was treated with culture filtrate *P. indica*, it enhanced the number, length and biomass of the root (Table 1). Pretreatment also resulted in an increase in root number, root length and root dry weight by 28%, 98%, and 123% respectively in *Aristolochia*. Increased root length and number can enhance absorption of more nutrients due to increased absorbing area resulting in improved plant growth (Marschner & Dell, 1994). Similar observations were made by other workers (Mugnier & Mosse, 1987; Varma et al., 2001). Inoculation of culture filtrate in case of grasses, trees and herbaceous sp. also showed enhancement of plant growth (Varma et al., 2001).

Figure 2. Effects of *P. indica* culture filtrate inoculation on *Aristolochia elegans*(A treated and B untreated)

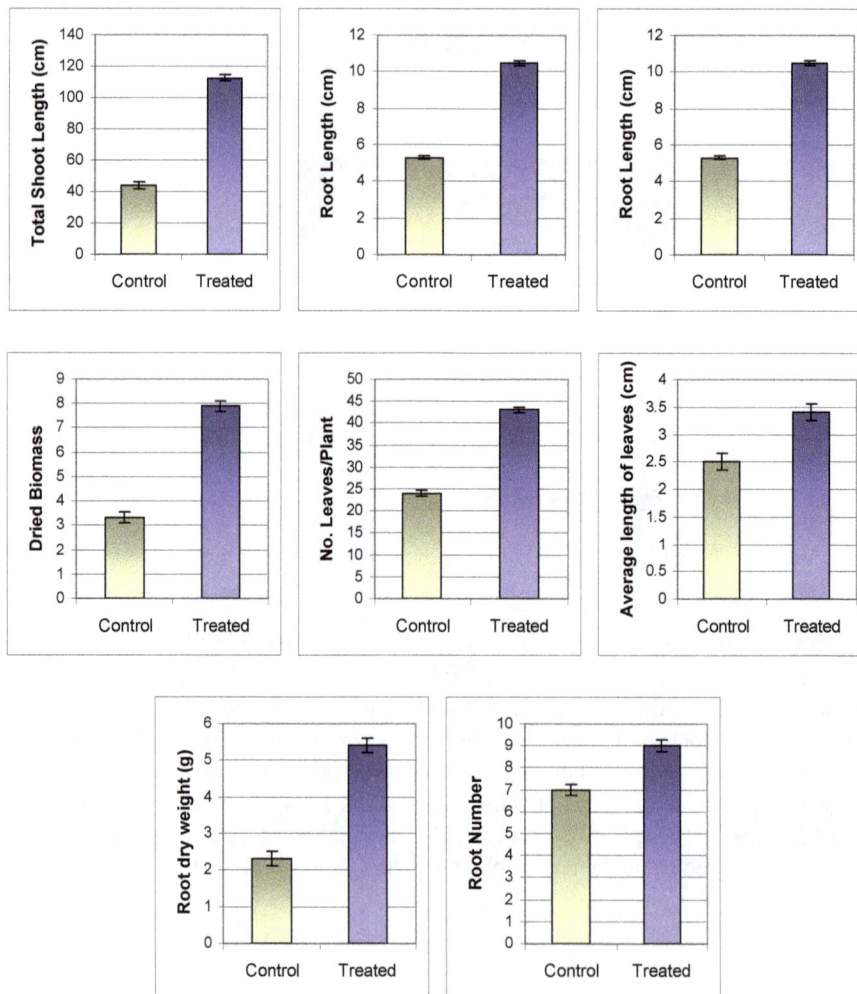

Figure 3. Effect of *P. indica* culture filtrate on *Aristolochia elegans*

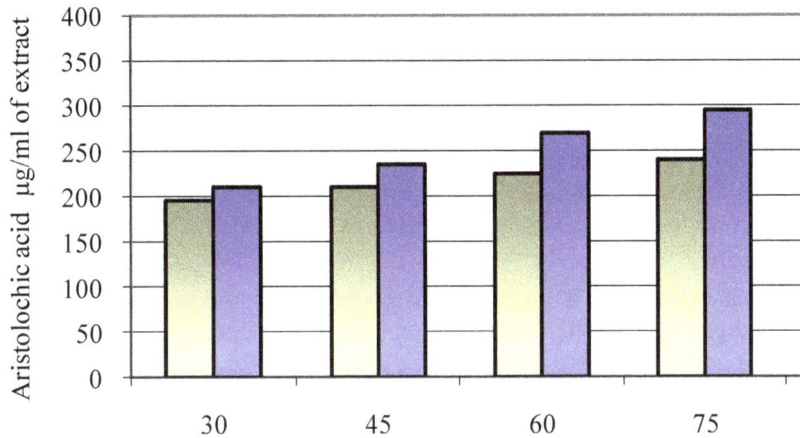

Figure 4. Effects of *P. indica* inoculation on Concentrations of Aristolochic acid in leaves of *Aristolochia elegans*

Table 1. Effect of *P. indica* culture filtrate on growth performance of *Aristolochia elegans*. Average values are for five replicates

	Characteristics	Control (untreated)	Experimental (*P. indica* treated)	S. E.	S. D.	Percent increase over control
Root	Number	7.00	9.00	± 0.26	0.2	28
	Length(cm)	5.30	10.5	± 0.12	0.12	98
	Dry weight (g)	2.30	5.4	± 0.20	0.20	123
Stem	Height(cm)	7.2 0	10.5	± 0.02	0.02	43
	Shoot length(cm)	44.00	112.5	± 2.17	2.17	155
Leaves	Number	24.00	43.0	± 0.68	0.68	79
	Length(cm)	2.50	3.4	± 0.15	0.15	36
Total Biomass (g)	Roots, Stems, Leaves,	3.32	7.9	± 0.22	0.22	136

S. E = Standard Error; S. D. = Standard Deviation.

Table 2. Effects of *P. indica* inoculation on Concentrations of Aristolochic acid in leaves of *Aristolochia elegans*

Days After planting	Aristolochic acid µg/g of extract			Percent increase over the Control
	Un-treated (Control)	Treatment with *P. indica*	S. D.	
30	195 ± 1.76	210 ± 1.76	1.76	7.6
45	210 ± 2.10	235 ± 2.10	2.10	11.9
60	225 ± 5.64	270 ± 5.64	5.64	20.0
75	240 ± 4.85	295 ± 4.85	4.85	22.8
90	260 ± 6.47	335 ± 6.47	6.47	28.8

Enhanced root growth, and root length was observed after application of the fungus in several plant species studied earlier (Varma et al., 2001). Not only the mycelium but even culture filtrate enhanced growth of the plants. Earlier it was reported that plant root cells can be killed by colonization of this fungus (Deshmukh et al., 2006), however it also increased root growth, weight and branching (Varma et al., 1999; Waller et al., 2005).

Increased rooting of calli of *N. tabacum* and cuttings of other plants was also noticed (Varma et al., 1999; Drudge et al., 2007). When culture filtrate of *P. indica* was applied, a diffusible factor from it enhanced root growth of *Arabidopsis*. There was stunted but highly branched roots in treated plants (Sirrenberg, 2007). The overall increment in the plant growth reported in this study may be due to increased nutrients uptake by the roots. This may also be due to application of culture filtrate that contained many growth promoters that exerted desirable effect on plant.

When culture filtrate of *P. indica* was applied, Increments in stem height and shoot length of plants were observed in plants in the present study (Table 1). Stem height increased by 43% and shoot length by 155%. This is in conformity to observations made in earlier investigations (Varma et al., 2001; Nautiyal et al., 2010; Bagde et al., 2010a, 2011).

The number of leaves increased by 79% and length by 36% in *P. indica* culture filtrate treated plants in comparison to untreated plants (Table 1) Increase in number and length of leaves was also reported in other plants using fungal mass or fungal culture filtrate (Varma et al., 2001; Fakhro et al., 2010; Bagde et al., 2011). A greater number of leaves, with increased length produced in treated plants could have contributed to increased rate of photosynthesis (Kungu, 2004).

P. indica culture filtrate treatment enhanced growth as well as total biomass of plants in comparison to untreated control plants in present study (Table 1). Similarly there was reported increase in total biomass by 136% as against treated control plants in herbaceous species (Varma et al., 2001) and *Helianthus annus* (Bagde et al., 2011), winter wheat plants (Serfling et al., 2007).

When six strains of *Sebacina vermifera* were tested on *Panicum virgatum* roots, it was noticed that there was positive effects on plant height and biomass production. It was also observed that culture filtrates from some strains of *S. vermifera* increased seed germination in *P. virgatum* by 52% over the control. In spring barley *P. indica* increased plant biomass and grain yield by 11% (Waller et al., 2005). Serfling et al. (2007) observed that fungus *P. Indica* colonization increased plant biomass in winter wheat plant.

When fungal culture filtrate was applied to the soil before planting, it increased total content of aristolochic acid in leaves between 7.6% to 28.8% (Table 2). The quantity of leaves and content therein were augmented as compared to control plants when treated plantlets were transferred to the pots. This positive influence in promoting the plant growth and yield in terms of biomass and medicinal ingredients may be due to positive effect of stimulatory factors or components present in the culture filtrate.

Besides several reports pertaining to the association of cells of *P. indica* with plants that enhanced growth, present study reports positive effect of even culture filtrate of fungus on plant growth. This is due to special characteristics of culture filtrate that was used. Culture filtrate is a complex growth enhancer of which all ingredients are not known (Bagde et al., 2010b). Culture filtrate contains fungal exudates, hormones, enzymes, proteins etc. that increased root number, length, root dry weight, stem height, shoot length, number and length of leaves, total biomass and aristolochic acid content of leaves in culture filtrate treated plants. Similar observations were made in case of maize, *Bacopa monniera,* and tobacco (Varma et al., 2001), neem and maize (Kumari, 2002; Singh et al., 2003). In *Helianthus annus*, treatment with *P. indica* culture filtrate promoted overall growth of the plant in terms of increased, root collar diameter, number of secondary roots, root length, root weight, stem diameter, stem height, number of leaves, length and width of leaf, flower number, flower diameter, flower dry weight, number of seeds, weight of seeds and total biomass as compared to untreated control plants. Seed oil content considerably increased in treated plants. Seed oil content increased by 51.13 per cent in sun gold variety and 70.33 per cent in treated Japanese gold variety of *H. annus* plants (Bagde et al., 2011).

Varma et al. (2001) also reported that application of culture filtrate of *P. indica* led to increase in root length, shoot length and plant biomass in treated plants. In present study treatment of *Aristolochia elegans* increased growth of roots, stems, leaves, total biomass as well as aristolochic acid over untreated plants (Table 1). These observations are in conformity to observations of Singh et al. (2003) wherein treatment resulted in considerable increase in growth and development in *Azadiracta indica* and *Zea mays* plants. Similarly when *Helianthus annus* plants were treated with culture filtrate of *P. indica,* root number, length, root collar diameter and dry weight of root increased considerably (Bagde et al., 2011). Observations like these were also made by other investigators, who reported luxurious and elaborate root growth and biomass when treated with mycelia of fungus (Varma et al., 2001; Kungu, 2004).

According to Sirrenberg et al. (2007) actual mode of action of *P. indica* in enhancing the growth of plants was not yet clear. But it is suggested that effect was due to diffusible factor that could be IAA, as *P. indica* was found to produce IAA in culture filtrate in sufficient quantities and hence it must have contributed to the beneficial effect on its host plants.The fungus may in addition induce auxin production in the plant (Peškan-Berghöfer et al., 2004). *Plants* colonized with *P. indica* can tolerate physical stress, nutrient deficiency, biotic and abiotic stresses and can fight pathogens including invaders of insects and facilitated increase in seeds and early flowering in medicinal plants (Oelmüller et al., 2009).

References

Arlt, V. M., Stiborova, M., & Schmeiser, H. H. (2002). Aristolochic acid as a probable human cancer hazard in herbal remedies: A review. *Mutagenesis, 17*(4), 265-27. http://dx.doi.org/10.1093/mutage/17.4.265

Bagde, U. S. (2010a). *Impact of Bio inoculants on economically important plants: Development and formulations*. D. Sc. thesis submitted to Amity University Uttar Pradesh, India.

Bagde, U. S., Prasad, R., & Varma, A. K. (2010b). Characterization of culture filtrates of *Piriformospora indica. Asian J. Microbiol Biotech. Env. Sc., 12*, 805-809.

Bagde, U. S., Prasad, R., &Varma, A. K. (2010c). Interaction of *P. indica* with medicinal plants and plants of economic importance. *African Journal of Biotechnology, 9*(54), 9214-9226.

Bagde, U. S., Prasad, R., & Varma, A. K. (2011). Influence of culture filtrate of *Piriformospora indica* on growth and yield of seed oil in *Helianthus annus. Symbiosis (USA), 53*, 83-88. http://dx.doi.org/10.1007/s13199-011-0114-6

Barazani, O., Benderoth, M., Groten, K., Kuhlemeier, C., & Baldwin, I. T. (2005). *Piriformospora indica* and *Sebacina vermifera* increase growth performance at the expense of herbivore resistance in *Nicotiana attenuata. Oecologia, 146*, 234-243. http://dx.doi.org/10.1007/s00442-005-0193-2

Deshmukh, S., Hückelhoven, R., Schäfer, P., & Imani, J. (2006). The root endophytic fungus *Piriformospora indica* requires host cell death for proliferation during mutualistic symbiosis with barley. *Proc. Natl. Acad. Sci. USA, 103*, 18450-18457. http://dx.doi.org/10.1073/pnas.0605697103

Drudge, U., Baltruschat, H., & Franken, P. (2007). *Piriformospora indica* promotes adventitious root formation in cuttings. *Science and Horticulture, 112*, 422-426. http://dx.doi.org/10.1016/j.scienta.2007.01.018

Fakhro, A., Andrade-Linares, R. D., Bargen, S., & Bandte, M. (2010). Impact of *Piriformospora indica* on tomato growth and on interaction with fungal and viral pathogens. *Mycorrhiza, 20*, 191-200. http://dx.doi.org/10.1007/s00572-009-0279-5

Flurer, R. A., Jones, M. B., Vela, N., Ciolino, L. A., & Wolnik, K. A. (2001). Determination of aristolochic acid in traditional Chinese medicines and dietary supplements. *FDA Laboratory Information Bulletin 4212, 16*(7), 13.

Gaudereault, F., Richad, B., & Jean, F. P. (2001). Determination of aristolochic acid in natural health products. *Health Canada,* 10-12.

Ghimire, S. R., Charlton, N. D., & Craven, K. D. (2009). The mycorrhizal fungus, *Sebacina vermifera*, enhances seed germination and biomass production in switch grass (*Panicum virgatum* L). *BioEnergy Research 2*, 51-58. http://dx.doi.org/10.1007/s12155-009-9033-2

Hazarika, B. N. (2003). Acclimatization of tissue-cultured plants. *Current Science, 85*, 1704-1712.

Hill, T. W., & Käfer, E. (2001). Improved protocols for Aspergillus medium: trace elements and minimum medium salt stock solutions. *Fungal Genetics Newsletter, 48*, 20-21.

Imran, M. S., & Bagde, U. S. (2007). Screening of *Aristolochia elegans*for antibacterial activity. *Natl. J. of Life Sciences, 4*(2), 219-222.

Kimura, K., & Kimura, T. (1981). *Medicinal Plants of Japan in Color* (p. 63). Tokyo: Koikushe Publishing Co..

Kumari, M. (2002). *Mycorrhizae for better establishment of neem plantlets (biological hardening) and enhancement of therapeutic properties*. Thesis submitted to B. R. Ambedkar Bihar University, Muzaffarpur for the award of the degree of Doctor of Philosophy in Botany.

Kungu, J. B. (2004). Effect of Vesicular Arbuscular Mycorrhiza (V. A. M.) inoculation on growth performance of *Senna spectabilis*. In A. Bationo (Ed.), *Managing nutrient cycles to sustain soil fertility in sub Saharan Africa* (pp. 433-446). Nairobi, Kenya: Academy Science Publishers.

Lopes, X., Nascimento, I. R., Silva, T., & Da, L. M. (2001). Phytochemistry of the aristolochiaceae family. In R. M. M. Mohan (Ed.), *Research Advances in Phytochemistry* (Vol. 2, pp. 19-108). Global Research Network Kerala.

Marschner, H., & Dell, B. (1994). Nutrient uptake in mycorrhizal symbiosis. *Plant and Soil, 159*, 89-102.

Mugnier, J. & Mosse, B. (1987). V. A. M. infection in transformed root inducing T-DNA root grown axenically. *Phytopathology, 77*, 1045-1050. http://dx.doi.org/10.1094/Phyto-77-1045

Nautiyal, C. S., Chauhan, P. S., Mehta, D. S., Seem, K., Varma A., & Staddon, W. J. (2010). Tripartite interactions among *Paenibacillus lentimorbus* NRRL B-30488, *Piriformospora indica* DSM 11827, and *Cicer arietinum* L. *World Journal of Microbiology and Biotechnology, 26*, 1393-1399. http://dx.doi.org/10.1007/s11274-010-0312-z

Oelmüller, R., Sherameti, I., Tripathi, S., & Varma, A. (2009). *Piriformospora indica*, a cultivable root endophyte with multiple biotechnological applications. *Symbiosis, 49*, 1-18. http://dx.doi.org/10.1007/s13199-009-0009-y

Peškan-Berghöfer, T., Shahollari, B., Pham, H. G., Hehl, S., Markent, C., Blank, V., ... Oelmüller R. (2004). Association of *Piriformospora indica* with *Arabidopsis thaliana* roots represent a novel system to study beneficial plant-microbe interactions and involve in early plant protein modifications in the endocytoplasmic reticulum and in the plasma membrane. *Plant Physiology, 122*, 465-471. http://dx.doi.org/10.1111/j.1399-3054.2004.00424.x

Prasad, R., Bagde, U. S., Pushpangadan, P., & Varma, A. (2008a). *Bacopa monniera* L: Pharmacological aspects and case study involving *Piriformospora indica*. *International Journal of Integrative Biology, Singapore, 3*, 100-110.

Prasad, R., Sharma, M., Chatterjee, S., Chauhan, G., Tripathi, S., Das, K. S., ... Varma, A. (2008b). Interactions of *Piriformospora indica* with medicinal plants. In A. Varma, & B. Hock (Eds.), *Mycorrhizae 3rd Edition* (pp. 655-678). Germany: Springer-Verlag.

Rai, M. K., Varma, A., & Pandey, A. K. (2004). Enhancement of antimycotic potential in *Spilanthes calva* after inoculation of *Piriformospora indica*, a new growth promoter. *Mycoses, Germany, 47*, 479-481. http://dx.doi.org/10.1111/j.1439-0507.2004.01045.x

Sahay, N. S., & Varma, A. (1999). *Piriformospora indica*: a new biological hardening tool for micropropagated plants. *FEMS Microbiology Letters, 181*, 297-302. http://dx.doi.org/10.1111/j.1574-6968.1999.tb08858.x

Serfling, A., Wirsel, S. G. R., Lind, V., & Deising, H. B. (2007). Performance of the biocontrol fungus *Piriformospora indica* on wheat under greenhouse and field conditions. *Phytopathology, 97*, 523-531. http://dx.doi.org/10.1094/PHYTO-97-4-0523

Singh, A. N., Singh, A. R., Kumari, M., Rai, M. K., & Varma, A. (2003). Biotechnological importance of *Piriformospora indica*. A novel symbiotic mycorrhiza-like fungus: An Overview. *Indian Journal of Biotechnology, 2*, 65-75.

Sirrenberg, A., Göbel, C., Grond, S., Czempinski, N., Ratzinger, A., Karlovsky, P., Santos, P., ... Pawlowski, K. (2007). *Piriformospora indica* affects plant growth by auxin production. *Physiologia Plantarum, 131*, 581-589. http://dx.doi.org/10.1111/j.1399-3054.2007.00983.x

Tian-Shung, W., Yu-Yi, C, & Yann-Li. (2000). The constituents of the root and stem of *Aristolochiacucurbitifolia* Hayata and their biological activity. *Chem Pharm Bull, 48*, 1006-1009. http://dx.doi.org/10.1248/cpb.48.1006

Varma, A., Singh, A., Sudha, S. N., Sharma, J., Roy, A., Kumari, ... Kranner, I. (2001). *Piriformospora indica*: A cultivable mycorrhiza-like endosymbiotic fungus. In K. Esser, & P. A, Lemke (Eds.), *Mycota IX* (pp. 123-150). Germany: Springer-Verlag.

Varma, A., Verma, S., Sudha, S. N., Britta, B., & Franken, P. (1999). *Piriformospora indica*-a cultivable plant growth promoting root endophyte with similarities to arbuscular mycorrhizal fungi. *Applied and Environmental Microbiology, 65*, 2741-2744.

Verma, S., Varma, A., Rexer, K. H., Hassel, A., Kost, G., Sarbhoy, A., ... Franken, P. (1998). *Piriformospora indica* gen. nov., a new root-colonizing fungus. *Mycologia (USA), 90*, 896-909. http://dx.doi.org/10.2307/3761331

Vila, R., Mundina, M., Muschietti, L., Priestap, H., Bandoni, A. L., Priestap, H., ... Bahigueral, S. (1997). Volatile constituents of leaves, roots and stems from *Aristolochia elegans*. *Phytochemistry, 46*(6), 1127-1129. http://dx.doi.org/10.1016/S0031-9422(97)00400-7

Waller, F., Achatz, B., & Kogel, K. H. (2007). Analysis of plant protective potential of the root endophytic fungus *Piriformospora indica* in cereals. In A. Varma, & R. Oelmüller (Eds.), *Advanced Techniques in Soil Microbiology* (pp. 343-354). Heidelberg, Berlin: Springer. http://dx.doi.org/10.1007/978-3-540-70865-0_22

Waller, F., Baltruschat, H., Achatz, B., Becker, K., Fischer, M., Fodor, J., ... Kogel, K. H. (2005). The endophytic fungus Piriformospora indica reprograms barley to salt stress tolerance, disease resistance and higher yield. *Proceedings of the National Academy of Sciences (USA), 102,* 13386-13391. http://dx.doi.org/10.1073/pnas.0504423102

Weiss, M., Selosse, M. A., Rexer, K. H., Urban, A., & Ober W. F. (2004). Sebacinales: a hitherto overlooked cosm of heterobasidiomycetes with a broad mycorrhizal potential. *Mycological Research, 108,* 1003-1010. http://dx.doi.org/10.1017/S0953756204000772

Wu, T., Chan, Y., Leu, Y., Wu, P., Li, C., & Mori, Y. (1999). Four Aristolochic acid esters of rearranged ent. Elemone Sesquiterpenes from *Aristolochia heterophylla. J. Nat. Prod., 62,* 348-351. http://dx.doi.org/10.1021/np980212y

Detection for Salt Tolerance Character in Two Selected Genotypes of Wheat

I. H. AL-Mishhadani Ibrahim[1], Bilal F. Zakariya[2], N. Ismail Eman[1] & M. Dawood Wisam[2]

[1] Biotechnology Research Center, AL-Nahrain University, P. O. Box 64074, Jadriah, Baghdad, Iraq

[2] Biology Department, AL-Razi College of Education, University of Diyala, Iraq

Correspondence: I. H. AL-Mishhadani Ibrahim, Biotechnology Research Center, AL-Nahrain University, P. O. Box 64074, Jadriah, Baghdad, Iraq. E-mail: hassanir1955@yahoo.com

Abstract

Using the genetic variation to improve salt tolerance in some wheat genotypes is a very important achievement to increase the production of salinized soils. The aim of this study is to determine the realized improvement in salt tolerance in some selected genotypes of wheat, which induced through plant breeding programs. The selected genotypes were derived from F2 populations after exposed to high salinity condition for six cycles of screening and selection. Salt tolerance of these selected genotypes (2H, N3) was tested during germination, early seedling and tillering stages in salinilized soils with three salinity levels (2, 8 and 15 ds/m) as compared with the local cultivar (Tamooze-2). Results showed that all selected genotypes were significantly superior in shoots and roots growth at 8 and 15 ds/m to those of the local cultivar. At all salt levels, the highest reduction in shoots and roots growth was in the local cultivar. Results also indicated that the highest values of K^+/Na^+ and ca^{+2}/Na^+ ratios were in the selected genotypes (2H, N3) at 8 and 15 ds/m. therefore, significant improvement in salt tolerance was achieved in the selected genotypes through plant breeding programs. The conclusion of these results is the salt tolerance of the selected genotypes correlated with the highest values of K^+/Na^+ and ca^{+2}/Na^+ ratios in their upper leaves, then the selected genotypes were more salt tolerance than local cultivar, which had the lowest K^+/Na^+ and ca^{+2}/Na^+ ratios in upper leaves.

Keywords: plant breeding, genotypes, salinity, salt tolerance

1. Intrduction

Salinity is a major factor limiting plant growth and then leads to lower agriculture production in arid and semi-arid regions. Salinity in Iraq especially in middle and south caused a big reduction in the yield production of the most field crops cultivars. Damaging effects of salt accumulation in agriculture soils have influenced ancient and modern civilizations. Therefore, so tolerance to salinity stresses is a key topic to consider for crop improvement (Zohary, 1973). In irrigated agriculture, improved salt tolerance of crops can lessen the leaching requirement, and so lessen the costs of an irrigation scheme, both in the need to import fresh water and to dispose of saline water (Pitman & Lauchli, 2002). Salt tolerant crops have a much lower leaching requirement than salt sensitive crops. In dry land agricultures improved salt tolerance can increase yield production under saline condition. When the rainfall is low and the salt remains in the subsoil, increased salt tolerance will allow plants to extract more water (Munns et al., 2006). The introduction of deep-rooted perennial species necessary to lower the water table, but salt tolerance will be required for the de-watering species and for the annual crops that follow, salt as will be left in the soil when the water – table is lowered.

Wheat is the most important crop in Iraq. One the most efficient ways to increase wheat yield under saline condition is to improve the salt tolerance in some wheat genotypes. Salt tolerance of crops may vary according to their growth stage (Mass & Poss, 1989). In general, there are variations between growth stages of wheat plants in their salt tolerance, the most sensitive to salinity during the vegetative stage and less sensitive during the flowering and grain – filling stages (Mass & Tiller, 1994). Therefore, the plant exposure and selection for salt tolerance under saline condition during the germination and early seedling stage are very important for improve salt tolerance in wheat plant, provided there is significant correlation between these stages and the later growth stages in their salt tolerance (AL-Mishhadani, 2012). The efficient breeding programs to overcome salinity problem by improving salt tolerant cultivars, those need information on the genetic basis of salt tolerance, genetic variation in this character, mode of inheritance, magnitude of gene effect, their mode of action, and Na$^+$ concentration in shoos [Gorham et al., 1990; Munns & James, 2003; Munns et al., 2006). Also, increased salt tolerances requires new genetic sources of this character, more efficient techniques for identifying salt tolerance germplasm, and determine salt level of exposure (AL-Mishhadani, 2012). He also reported that some salt tolerant genotypes of wheat were identified through plant breeding programs. The aim of the this work is detection for salt tolerance in two wheat genotypes which selected through plant breeding programs under salinized soil condition as compared with local cultivar (Tmooze-2).

2. Material and Methods

The growth of the two selected wheat genotypes was examined in salinized soils experiment having treatment of 2, 8, 15 ds/m as compared with local cultivar. These selected genotypes were derived from plant breeding programs after 6 cycles of exposure and selection under high salinity level (30 ds/m). Salinity levels were prepared by mixed two kind of soil that only differed in their saline degree (Ec). The saline degree of the above levels was determined by using the electrical conductivity meter. The experiment was carried out in plastic posts, which filled with salinized soils. These pots were set up in a glasshouse in a completely randomized design with three blocks. Each pot contained one observation, the total of pots were 27 ($3\times3\times3\times$), 7 seeds of each genotype were sown on each pot. All the agriculture treatments were carried out. Seeds and plants were watered with tap water (300 mill /pot) according to the field capacity. Plant of all selected genotypes and local cultivars were harvested after 4 month from the sowing date and the following characters were measured for each pot after oven drying for 3 days at 80°C:

1) Shoot dry weight (g)

2) Root dry weight (g)

3) K$^+$/Na$^+$ ratio in upper leaves

4) Ca$^+$/ Na$^+$ ratio in upper leaves

The data were subjected to analysis of variance according to the experimental design.

3. Results

3.1 Shoots and Roots Dry Weight (g/plant)

There were significant differences in dry weight reduction of shoots and roots with increasing salinity level (Figures 1&2). The dry weights of shoots and roots were more affected at 8 &15 ds/m than those at 2 ds/m, but the reduction at 15 ds/m was higher than at 8 ds/m. The growth of roots was more affected by salinity than shoot growth (Figures 1&2) and (Plates 1&2). The results also showed that there were significant differences between the genotypes and local cultivar in their responses to salinity. At all salinity levels, shoots and roots dry weights of the selected genotypes were less affect by salinity as compared with those of local cultivars (Figures 1&2) and (Plates 1&2). Also there was difference between the selected genotypes in their shoots and roots dry weights but not significant. This mean the growth of shoots and roots of the selected genotypes was proximatlly similler under salinity condition and were both higher than those of the local cultivar (Plates 1&2).

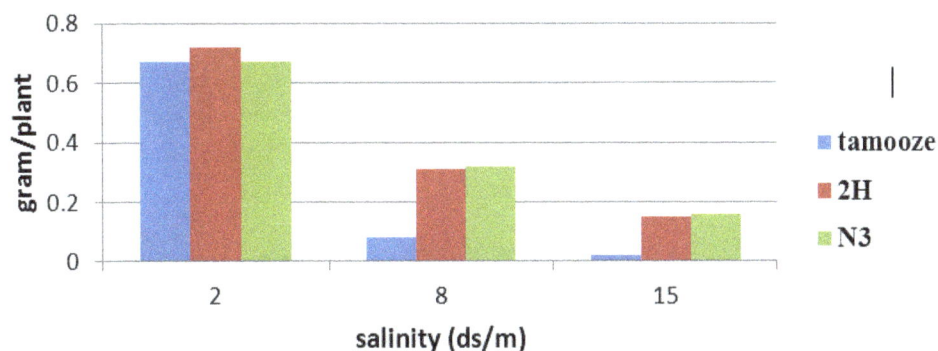

Figure 1. Shoot dry wight of selected genotypes of wheat under salinity condition

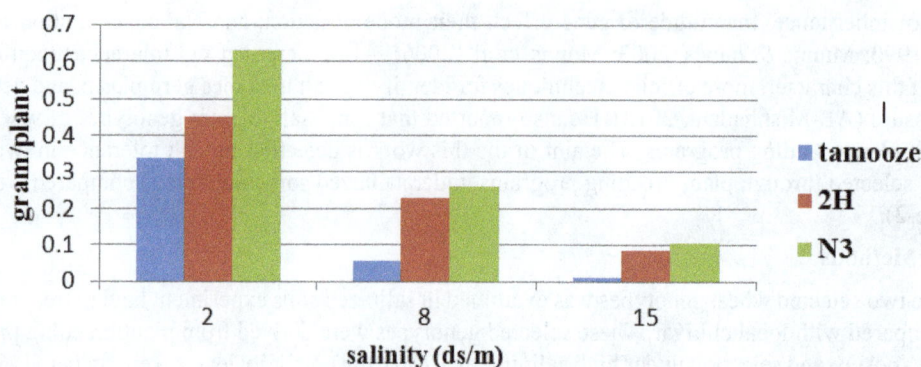

Figure 2. Root dry wight of selected genotypes of wheat under salinity condition

Results of analysis of variance for these data revealed that the interaction (salinity × genotypes) was significant, reflecting there were differences between the selected genotypes and local cultivar in their response to the salinity at all salt levels (Figures 1&2) and (Plates 1&2). The dry weight of these genotypes significantly decreased at 8 and 15 ds/m, but there are significant differences between the selected genotypes (2H &N3) and local cultivar (Tamooze- 2) in their dry weights reduction at each salt level. At 8, 15 ds/m levels, the dry weights of shoots and roots of the selected genotypes were proximatilly similler, but they differed at 2 ds/m. Also the results showed that the biggest reduction in dry weight was at 15 ds/m as compared with the 2 and 8 ds/m, especially in the local cultivar (Tamooze-2). The results of shoots growth (Plate 1) revealed that the two selected genotypes were superior in germination percentage, in tillering, and leaves numbers especially at 8, 15 ds/m.

Plate 1. shoot growth of the selected wheat genotypes under salinity condition as compared with the local cultivar

Plate 2. Root growth of the selected wheat genotypes under salinity condition as compared with the local cultivar

3.2 K^+/Na^+ Ratio

The data for K^+/Na^+ ratio in upper leaves of all genotypes under salinity condition is presented in Figure 3. The results show that there are significant differences between the selected genotypes and local cultivar in their K^+/Na^+ ratio at all salinity levels. K^+/Na^+ ratio in upper leaves of these genotypes and local cultivar was significant decreased with increasing salinity, but there are significant variation between the selected genotypes and local cultivar in their K^+/Na^+ ratio reduction at each salinity level (Figure 3). At 8&15 ds/m levels, the selected genotypes had much higher K^+/Na^+ ratio in their upper leaves than this of the local cultivar. Also the results showed that there are differences between the two selected genotypes in the K^+/Na^+ ratio of their upper leaves at all salinity levels. At each salinity level the N3 genotype had higher K^+/Na^+ ratio than the 2H genotype and local cultivar. At the highest salinity level, the K^+/Na^+ in upper leaves was greater in the two selected genotypes than in the local cultivar (Figure 3).

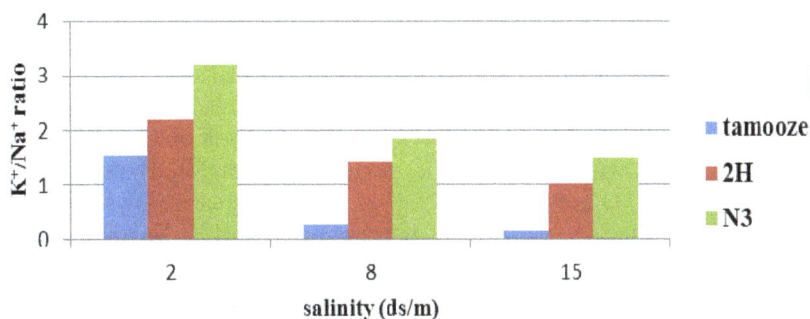

Figure 3. K^+/Na^+ ratio in upper leaves of selected genotypes of wheat under salinity condition

3.3 Ca+/ Na+ Ratio in Upper Leaves

The data for Ca^+/ Na^+ ratio in upper leaves is presented in Figure 4, which showed that there are significant reductions in Ca^+/ Na^+ ratio in upper leaves under salinity conditions as compared with the control condition. All the genotypes had reduction in their Ca^+/ Na^+ ratio under salinity condition, but the reduction was more clear in local cultivar (Tamooze-2) as compared with the selected genotypes. At 8 ds/m, the two selected genotypes were not differed in their Ca^+/ Na^+ ratio in upper leaves, while they differed significantly with the local cultivar. At 15 ds/m, there are significant variations between the genotypes and local cultivar in their Ca^+/ Na^+ ratio (Figure 4), but the selected genotypes had much higher Ca^+/ Na^+ ration than the local cultivar. This result reflect that maximum Na^+ accumulation was observed in local cultivar, whereas, it was lowest in the two selected genotypes under salinity conditions. Reverse, at the same salinity conditions, maximum Ca^+ contents in upper leaves were recorded in the two selected genotypes (2H & N3). Whilst, the lowest contents were recorded in the local cultivar, therefore, maximum Ca^+/ Na^+ ratio was observed in these selected genotypes, especially at 15 ds/m.

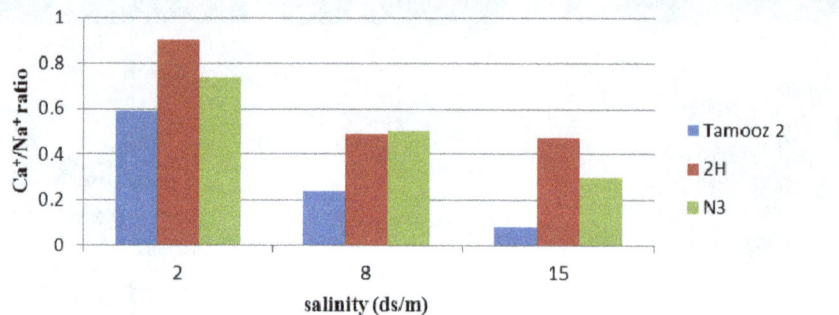

Figure 4. Ca^+/Na^+ ratio in upper leaves of selected genotypes of wheat under salinity condition

4. Discussion

Under saline condition, due to excessive amounts of exchangeable Na^+, low K^+/Na^+ ratio occur in the soil. Plants subjected to such environment, take up high amounts of Na^+, whereas the uptake of K^+ is considerably reduced, and these increased with increasing salinity levels. These conditions were found in local cultivar (Figure 3), which showed that K^+/Na^+ decreased with increasing salinity level. This revealed that Na^+ concentration in upper leaves highly increased and K^+ concentration highly decreased in the local cultivar with increasing salinity level as compared with those of the two selected genotypes, which their Na^+ concentration less increased and K^+ concentration less decreased at each salinity level. There for the K^+/Na^+ ratio of the two selected genotypes higher than this of the local cultivar (Figure 3). These conditions improved the growth of shoots and roots of the two selected genotypes under salinity condition as compared with the growth in local cultivar (Figure 1, 2 & Plate 1, 2). At each salt concentration the growth of the both characters of the two selected genotypes much higher than those of the local cultivar, this improvement more correlated with their K^+/Na^+ ratio in upper leaves.

This result reflecting the fact that the high K^+/Na^+ ratio in plant more correlated with the salt tolerance in this plant under salinity condition. The same conclusion are reported by Marschner (1995), Wenxue et al., (2003) and Munns et al. (2000) that they revealed reasonable amounts of both K^+ and Ca^+ are required to maintain the integrity and functioning of cell membranes, which are very important for plant growth under saline condition. This result showed that the salt tolerance of the selected genotypes associated with the maintenance of adequate K^+ in plant leaves tissue under salt stress which seems to be dependent upon selective K^+ uptake, selective cellular K^+, and Na^+ compartmentation and distribution in the shoots (Munns et al., 2000; Carden et al., 2003). Therefore, high K^+/Na^+ selectivity in plants under saline conditions has been suggested as an important selection criterion for tolerance (Corham et al., 1997; Ashraf, 2002; Wenxue et al., 2003). On the other hand, kafkafi (1984) concluded that roots of the salt tolerant Beta vulgaris had a greater affinity for K^+ relative to Na^+ than did the salt sensitive phaseolus vulgaris, nd this increased K^+/Na^+ ration in the shoots. This conclusion agree with the result of this work, which showed that each salt concentration the K^+/Na^+ ration in upper leaves of the selected genotypes (N3,2H) much higher than those of the local cultivar. Also Munns et al. (2000) reported that selection for salt tolerant genotypes of durm wheat were screened for low Na^+ concentration in their leaves and the associated enhanced K^+/Na^+ discrimination. Also they reported there was wide genetic variation was found in Na^+ accumulation and K^+/Na^+ discrimination, according to this, there are wide genetic variations also was found in salt tolerance among durm wheat genotypes. The results of this study were agreed with the above conclusion, which showed that selected genotypes with high K^+/Na^+ ration were more salt tolerance than the local cultivar with low K^+/Na^+ (Figures 1,2,3 & Plates 1,2). These result indicated that Na^+ exclusion from upper leaves is important at high salinity level (Munns, 2005). Because this caused reduction in Na^+ and increased in important elements such as K^+, Ca^+ and Mg^+ concentrations in upper leaves to prevent leaf injury and enhancing growth and yield production (Husain et al., 2003).

Also Munns (2005) reported in addition of lowering Na^+ accumulation in upper leaves, K^+ and organic solutes should accumulate in the cytoplasm and vaculles to balance the osmotic pressure in the leaves cells. They would be important for turgor maintenance of cells and for a metabolic protective, and so maintain growth at high salinity levels. Similarly, in soybean, lauchlia and wieneke (1979) found that the salt tolerant CV.'Lee "accumulated more k^+ in its leaves than did the salt sensitive CV.'Jackson". Asimilar mechanism of ion uptake has also been observed in barley (Wenxue et al., 2003).

Calcium is also known to play a crucial role in maintaining the structural and functional integrity of plant membranes in addition to its considerable roles in cell wall stabilization, regulation of ion transport, selectivity, and activation of cell wall enzymes (Rengel, 1992; Marschner, 1995). These effects are very important for physiological processes in the leaves cells and also plant growth. However, the differences between the two selected genotypes and local cultivar in their shoots and roots growth (Figures 1, 2 & Plates 1, 2) under salinity conditions refer to their difference in Ca^+ concentration in upper leaves. Therefore these differences caused variation between them in their responses to the salinity, which more correlated with Ca^{+2}/Na^+ ratio of the upper leaves (Figure 4), and then their salt tolerance also more correlated with the high Ca^{+2}/Na^+ ratio in their upper leaves. The low Ca^{+2}/Na^+ ratio of a saline medium and also in plant leaves plays a significant role in growth inhibition in addition to causing significant changes in morphology and anatomy of plants (Cramer, 1992). On the other hand, Soussi et al. (2001) and Unno et al. (2002) reported that the maintenance of calcium transport and accumulation in plant tissue under salt stress is an important determinant of salinity tolerance. In most cases salt tolerance of a crop cultivar can be increased by an increasing in the Ca^{+2} concentrations in the upper leaves, which depend on its concentration in the saline growth medium and plant selectivity mechanism that controlled by genetic.

Ca^{+2} maintenance and Na^+ exclusion, which are related to salinity tolerance in most crop cultivars, were genetically controlled with additive major genetic components (Foolad, 1997) and the inherent genetic capability to maintain Ca^{+2} in tissue and to exclude Na^+ from shoots were highly heritable traits, suggesting Ca^{+2}/Na^+ and K^+/Na^+ ratios might be promising indicators for discriminating between salt tolerant and salt sensitive plants of many crops.

These results of the shoots and roots growth reflect the large improvement in salt tolerance obtained in the selected genotypes through plant breeding programs after exposure to 30 ds/m drainage water for six cycles of screening and selection. Similar improvement was obtained in some selected genotypes of wheat through these plant breeding programs (AL-Mishhadani, 2012; AL-Mishhadani et al., 2014). This improvement in salt tolerance in these selected genotypes may due to these genotypes were selected from F2-F7, which generations generally contain much wider range of genetic variation in salt tolerance. Very small numbers of plants are still surviving after exposure these genetic materials to the 30ds/m drainage water for 6-8 weeks. This refer to salt tolerant genes are segregated in these plants, which exhibited salt tolerance character in these survived plants under saline conditions (gene × environment interaction) (Al-Mishhadani, 2012). Therefore, the salt tolerance is correlated with the number and kind of segregated salt tolerant genes that control the salt tolerance mechanisms and determine the degree of tolerance in plant (Munns, 2005). The superiority of selected genotypes in salt tolerance of the local cultivar (Plates 1&2) may due to exposure the genetic materials to high salt concentration (30ds/m). Similarly, AL-Mishhadani (2012) reported that selection at high salt concentration may be useful for identifying high salt tolerant progeny. Also the superiority of shoots growth (Plate 1 & Figure 1) of these selected genotypes as compared with those of local cultivar may due to their superiority in roots growth (Plate 2 & Figure 2) and in K^+, Ca^{+2} concentrations in upper leaves.

Generally, these results concluded that the selected genotypes more salt tolerant than the local cultivar, the plant breeding programs were affective in improvement of salt tolerance in these selected genotypes, and Ca^{+2}/Na^+, K^+/Na^+ ratios more correlated with the salt tolerance of these selected genotypes.

Reference

AL-Mishhadani, I. I. H. (2012). Breeding and selection of same lines of bread wheat for salt tolerance. *Journal of Agricultural science and Technology, B2,* 934-939.

AL-Mishhadani, I. I. H. (2014). Estimation of new wheat genotypes for salt tolerance which induced through plant breeding programs. *J. of Agricultural Sci. and Technology, B4.*

Ashraf, M. (2002). Salt tolerance of cotton: some new advances. *Critical Reviews in Plant Sciences, 21*(1), 1-30. http://dx.doi.org/10.1016/S0735-2689(02)80036-3

Carden, D. E., Walker, D. J., Flowers, T. J., & Miller, A. J. (2003). Single-cell measurements of the contributions of cytosolic Na^+ and K^+ to salt tolerance. *Plant Physiology, 131*(2), 676-683. http://dx.doi.org/10.1104/pp.011445

CRAMER, G. R. (1992). Kinetics of maize leaf elongation II. Responses of a Na-excluding cultivar and a Na-including cultivar to varying Na/Ca salinities. *Journal of Experimental Botany, 43*(6), 857-864.

Foolad, M. R. (1997). Genetic basis of physiological traits related to salt tolerance in tomato, Lycopersicon esculentum Mill. *Plant Breeding, 116*(1), 53-58. http://dx.doi.org/10.1111/j.1439-0523.1997.tb00974.x

Gorham, J., Bridges, J., Dubcovsky, J., Dvorak, J., Hollington, P. A., LUO, M. C., & Khan, J. A. (1997). Genetic analysis and physiology of a trait for enhanced K+/Na+ discrimination in wheat. *New Phytologist, 137*(1), 109-116. http://dx.doi.org/10.1046/j.1469-8137.1997.00825.x

Gorham, J., Jones, R. W., & Bristol, A. (1990). Partial characterization of the trait for enhanced K+— Na+ discrimination in the D genome of wheat. *Planta, 180*(4), 590-597. http://dx.doi.org/10.1007/BF02411458

Husain, S., Munns, R., & Condon, A. T. (2003). Effect of sodium exclusion trait on chlorophyll retention and growth of durum wheat in saline soil. *Crop and Pasture Science, 54*(6), 589-597. http://dx.doi.org/10.1071/AR03032

Kafkafi, U. (1984). Plant nutrition under saline conditions. In J. Shainberg (Ed.), *Soil Salinity Under Irrigation, Processes and Management* (pp. 319-338). Berlin: Springer.

Lauchli, A., & Wieneke, J. (1979). Studies on growth and distribution of Na^+, K^+ and Cl^{-1} in soybean varieties differing in salt tolerance. *Z. Pflanzenen, Bodenk, 142*, 3-13. http://dx.doi.org/10.1002/jpln.19791420103

Marschner, H. (1995). *Mineral Nutrition of Higher Plants*. Acad. Pr., London (1995).

Mass, E. V., & Poss J. A. (1989). salt sensitivity of cowpea at various growth stages. *Irri Sci. 10*, 313-320.

Mass, E. V., & Tiller, C. M. (1994). Development in salt stressed wheat. *Crop Sci., 34*, 1594-1603. http://dx.doi.org/10.2135/cropsci1994.0011183X003400060032x

Munns, R. (2005). Gene and salt tolerance: bringing them to gather. *New Phytologist, 167*, 645-665. http://dx.doi.org/10.1111/j.1469-8137.2005.01487.x

Munns, R., & James, R. A. (2003). Screening methods for salt tolerance: a case study with tetraploid wheat. *Plant and Soil, 253*, 201-218. http://dx.doi.org/10.1023/A:1024553303144

Munns, R., Hare, R. A., James, R. A., & Rebetzke, G. J. (2000). Genetic variation for salt tolerance of durm wheat. *Aust. J. Agri. Res., 51*, 69-74. http://dx.doi.org/10.1071/AR99057

Munns, R., James, R. A., & Läuchli, A. (2006). Approaches to increasing the salt tolerance of wheat and other cereals. *Journal of Experimental Botany, 57*(5), 1025-1043. http://dx.doi.org/10.1093/jxb/erj100

Pitman, M. G., & Lauchli, A. (2002). Global impact of salinity and agricultural ecosysteme. In A. Lauchli, & U. Luttge (Eds.), *Salinity: Environment-plant-molecules*. the Netherlands: Dordrechr.

Rengel, Z. (1992). The role of calcium in salt toxicity. *Plant Cell Environ, 15*, 625-632. http://dx.doi.org/10.1111/j.1365-3040.1992.tb01004.x

Soussi, M., Ocana, A., & Lluch, C. G. (2001). Nitrogen fixation and ion accumulation in two chickpea cultivars under salt stress. *Agricoltura mediterranea, 131*, 1-8.

Unno, H., Maeda, Y., Yamamoto, S., Okamoto, M., & Takenaga, H. (2002). Relationship between salt tolerance and Ca2+ retention among plant species. *Japanese Journal of Soil Science and Plant Nutrition*, 715-718.

Wenxue, W., Bilsborrow, P. E., Hooley, P., Fincham, D., Lombi, E., & Forster, B. P. (2003). Salinity induced difference in growth, ion distribution and partitioning in barley between the cultivar may Thorpe and its derived mutant golden promise. *Plant Soil, 250*, 183-191. http://dx.doi.org/10.1023/A:1022832107999

Zohary, M. (1973). Geobotanical foundation of the Middle East, 2 Vols Amsterdam. *Stuttgart*, 33-44.

3

Cadmium-Induced Changes in Germination, Seedlings Growth, and DNA Fingerprinting of *in vitro* Grown *Cichorium pumilum* Jacq.

Wesam Al Khateeb[1]

[1] Department of Biological Sciences, Yarmouk University, Jordan

Correspondence: Wesam Al Khateeb, Department of Biological Sciences, Yarmouk University, Irbid, Jordan.
E-mail: wesamyu@gmail.com

Abstract

The aim of this study was to assess the effect of Cd^{2+} on germination, growth, proline content, lipid peroxidation, and DNA fingerprinting of *in vitro* grown *Cichorium pumilum*. Results showed that seed germination was highly inhibited by cadmium (down to 47% at 1600 µM $CdCl_2$). In addition, root and shoot growth showed significant decreases in response to $CdCl_2$ level. Analysis of proline content and lipid peroxidation showed that with increasing $CdCl_2$ levels in the growing medium, the amount of proline accumulation and lipid peroxidation increased gradually. Total chlorophyll content was found to increase only at higher tested levels of Cd^{2+} (800 and 1600 µM). The results also show that Cd^{2+} inhibits callus growth at different levels starting from 50 µM $CdCl_2$ compared with the control in the callus growth experiment. Callus growth ceased completely at 200 µM $CdCl_2$ and above. Random amplified polymorphic DNA (RAPD) analysis showed DNA alterations in Cd^{2+}-treated *C. pumilum* microshoots compared with the control. The results of this experiment showed that Cd^{2+} stress affects several physiological, biochemical, and molecular processes in *C. pumilum*.

Keywords: Cadmium, proline, lipid peroxidation, fingerprinting, *Cichorium pumilum*

1. Introduction

Chicory (*Cichorium pumilum* Jacq., Asteraceae), is a bushy perennial herb with blue or lavender flowers which grows as a wild plant on roadsides. Chicory is also known as blue sailors, endive, radicchio, French endive, red endive, sugarloaf, witloof, elit, and coffeeweed. It is a culinary and medicinal plant grown worldwide. In the Middle East, its leaves are widely used in salads after being blanched, as the unblanched leaves taste bitter. In Europe, the root is eaten like a vegetable after being boiled, or it can be roasted then ground for use as a coffee substitute (Robert et al., 2008). Al Khateeb et al. (2012) showed that *C. pumilum* methanolic extracts have high levels of phenolic compounds and showed very strong antioxidant properties. Moreover, they found that methanol and ethanol extracts obtained from *C. pumilum* have antimicrobial effects on 10 different bacterial species.

In the last few decades, a dramatic increase in the contamination of the environment, (including soil, air, and water) has been observed. It appears that anthropogenic activities are the main source of the pollution that is causing the environment contamination (Gratao et al., 2005). Recently, it has been shown that large areas of land have been contaminated with heavy metals as a result of urban activities, agricultural practices, and industry.

Heavy metals are defined as the group of elements that have specific weights higher than about 5 $g \times cm^{-3}$. A number of them (Co, Fe, Mn, Mo, Ni, Zn, Cu) are essential micronutrients which are required for normal growth and for many metabolic processes in plants. Metals which are considered nonessential (Pb, Cd, Cr, Hg, etc.) are potentially highly toxic for plants (Sebastiani et al., 2004). Contamination of soil by heavy metals is a global ecological problem because heavy metals are included in the main category of environmental pollutants which can remain in the environment for long periods. Their accumulation is potentially hazardous to humans, animals, and plants (Benavides et al., 2005).

Agricultural soil contamination can severely affect humans, both directly (through the food web) and indirectly (by damaging environmental health) (Nriagu, 1990). For plants, heavy metals are phytotoxic, causing growth inhibition and eventually plant death through mechanisms that are still not completely understood (Romero-Puertas et al., 1999). The toxic effect of increasing cadmium (Cd) concentration in the environment has become a major environmental concern (Shriarastava & Singh, 1989). Cadmium accumulation in soils may come from various sources: from air pollutants or through applications of commercial fertilizers, sewage sludge, manure, and lime (Kidd et al., 2007). Also, industrial effluents may contain a wide variety of pollutants depending on the industries involved (Iribar et al., 2000). Cd is generally present in soil as free ions or in different soluble forms, and its mobility is affected by pH and the presence of chelating substances and other cations (Hardiman & Jacoby, 1984). In plants, Cd is accumulated mainly in the edible parts, thus making crop yield a potential hazard for human and animal health. It has been suggested that Cadmium may cause damage even at very low concentrations, and healthy plants may contain Cd levels that are toxic for mammals (Chen et al., 2007). Pinot et al. (2000) showed that the main source of Cd accumulation in food is the Cd uptake by plants.

It has been shown that Cadmium can inhibit plant growth and photosynthesis, reduce chlorophyll content, and induce oxidative stress (Schill et al., 2003). In addition, the genotoxicity of Cd is directly related to its effect on the structure and function of DNA. Therefore, Cd toxicity not increases the mortality rate in the exposed organisms, but may also result in the modification of the population dynamics and the species biological diversity (Theodorakis et al., 2006). Furthermore, it has been shown that cadmium generates oxidative stress through the formation of reactive oxygen species (ROS). The formation of ROS by cadmium suggests that DNA can also be taken into account as a potential target of this metal (Błasiak, 2001).

It has been suggested that proline plays a role in protecting plants from heavy metal toxicity. Siripornadulsil et al. (2002) reported that proline reduces cadmium stress not by sequestering cadmium but by reducing cadmium-induced free radical damage, thus maintaining a more reducing environment in the cell. Xu et al. (2009) found that proline pre-treatment of *Solanum nigrum* reduces the reactive oxygen species levels and protects the plasma membrane integrity of callus under cadmium stress.

The objectives of this study were to study the effects of Cd on germination, seedling and callus growth, biochemical properties and DNA fingerprint of *in vitro* grown *C. pumilum*.

2. Materials and Methods

2.1 Plant Material, Seed Germination and Proliferation

Ripe fruits of *Cichorium pumilum* Jacq were collected from Irbid/Jordan during the summer of 2011. Seeds were surface sterilized with 2% sodium hypochlorite solution for 10 min, then washed with 70% ethanol for 30 s, followed by three rinses in sterile distilled water. Seeds were inoculated into Petri dishes containing germination medium of half strength Murashige and Skoog (MS) salts, 2% sucrose, and 0.8% Difco-Bacto agar in addition to different levels of $CdCl_2$. Plates were incubated in the dark at 24 ± 2 °C for 7 days, and then germination percentages, hypocotyl and root lengths were recorded.

Shoot tips were excised from in vitro grown seedlings and cultured on MS medium supplemented with 1 µM *Benzyl adenine* (BA) and 0.5 µM naphthaleneacetic acid (NAA) for shoot proliferation (Al Khateeb et al., 2012). The new microshoots were subcultured after 6 weeks on fresh medium containing different levels of $CdCl_2$ (50, 100, 200, 400, 800, and 1600 µM). Cultures were placed in a growth chamber (24 ± 2 °C and 16 h light in cool white fluorescent light) for further growth. After 6 weeks, the fresh weights of the shoots were recorded.

2.2 Callus Growth

Al Khateeb et al. (2012) protocol was used for callus induction. Three-week-old callus was divided into parts of 0.5 g. Then, these parts were subcultured onto the same medium supplemented with different $CdCl_2$ levels. Cultures were maintained at 24 ± 2 °C and 16 h light in cool white fluorescent light. Fresh weights were taken every week for a period of 6 weeks.

2.3 Chlorophyll Analysis

The effect of different concentrations of Cd^{2+} on chlorophyll content was tested. Microshoots grown on MS medium supplemented with different levels of $CdCl_2$ were extracted with 80% acetone overnight, the A_{645} and A_{663} were determined using spectrophotometer and chlorophyll content was calculated according to the method of Mackinney (1941).

2.4 Proline Analysis

500 mg of plant tissues from microshoots grown on MS medium supplemented with different levels of $CdCl_2$ were homogenized in 10 mL of aqueous solution of sulfosalicylic acid. The solution was then filtered rapidly through a Buchner funnel using Whatman filter paper N° 2. 2 mL of the filtrate was transferred to a test tube in addition to 2 mL of ninhydric acid and 2 mL of glacial acetic acid, followed by one hour incubation at 100 °C. The reaction was then stopped in an ice bath. Afterwards, 4 mL of toluene was added and the contents of the tube

were inverted for 20 seconds. After this, the toluene phase was separated by centrifugation at 13,000 g for 10 minutes. Finally, the absorbance was measured at 520 nm with a visible light spectrophotometer. The concentration of proline was determined from the calibration curve.

2.5 Lipid Peroxidation

Lipid peroxidation was estimated based on measuring the malondialdehyde (MDA) content. The MDA content in microshoots grown on MS medium supplemented with different levels of $CdCl_2$ was analyzed following Heath and Packer (1968). This assay is based on the reaction with thiobarbituric acid. Fresh microshoots (0.5 g) were ground in 20 mL of 0.1% tri-chloroacetic acid (w/v) then centrifuged for 15 min at 13,000 g. One mL of the supernatant was reacted with 5 mL of 20% TCA solution containing 0.5% thiobarbituric acid (w/v). After that, the mixture was heated for 45 min. at 95 °C and then cooled immediately in an ice bath. Next, the mixture was centrifuged for 5 min at 13,000 g, and finally the absorbance of the supernatant was measured using spectrophotometer at 532 and 600 nm. MDA content was calculated using the extinction coefficient of 155/(mM/cm) (Soltani et al., 2006).

2.6 DNA Extraction and RAPD Analysis

DNA was extracted from C. pumilum microshoots using modified CTAB (cetyltrimethylammonium bromide) method (Porebski et al., 1997). DNA concentration was measured spectrophotometrically at 260 nm. The RAPD reaction was performed in a total volume of 50 µL containing 5 µL template DNA, 10 × PCR buffer, 5 mM $MgCl_2$, 250 µM deoxynucleoside triphosphates, 1.5 U of Taq DNA polymerase and 1 µM of each primer. Primers were obtained from commercially available kits (OPA, OPC, and OPG) (Operon Technologies, CA and USA) (Table 1). DNA amplification program in the thermal cycler was as follows; 40 cycles of 94 °C for 1 min, 52 °C for 45 sec and 72 °C for 30 s. A final extension step was also used at 72 °C for 5 min. PCR products were loaded on agarose gel (1.5% agarose) and run with Tris-borate-EDTA (TBE) buffer for 90 min.

Table 1. Sequence information of RAPD primers used for C. pumilum fingerprinting

RAPD Primers	Sequence Information
OPC	GTCCCGACGA
OPG03	GAGCCCTCCA
OPG05	CTGAGACGGA
PM5	CGACGCCCTG
PM6	GCGTCGAGGG
OPAB 14	AAGTGCGACC
OPO 08	GCTCCAGTGT
OPAH 15	CTACAGCGAG
OPAO 01	AAGACGACGG
OPAP 20	CCCGGATACA
CRC	GCGAACCTCG
CRA22	CCGCAGCCAA

2.7 Statistical Analysis

Statistical significance was confirmed by analysis of variance (ANOVA) using SPSS for Windows (version 16.0). Results were expressed as mean ± standard error. Means were separated by using Tukey t-test at 0.05 level of probability. All experiments were repeated at least three times.

3. Results

Analysis of variance (ANOVA) showed that cadmium affects germination, hypocotyl and root length, fresh weights of shoots, chlorophyll and proline content, lipid peroxidation level, and callus growth of Cichorium pumilum significantly at 0.05 level of probability.

3.1 Effect of Cadmium on Germination Percentage

In general, the germination percentage of *Cichorium pumilum* decreased as Cd^{2+} level increased (Figure 1. A). The highest germination percentage (100%) was found in the control. A sharp reduction in germination percentage was observed when the seeds were treated with 50 μM $CdCl_2$ (80%). No significant differences were observed between 100, 200 and 400 μM $CdCl_2$ levels. The lowest germination percentage was observed when *C. pumilum* seeds were treated with the highest $CdCl_2$ concentration (1600 μM) which resulted in only 50% germination.

3.2 Effect of Cadmium on Hypocotyl and Root Length

Results show that the hypocotyl length of *C. pumilum* decreased with increasing $CdCl_2$ concentration (Figure 1. B). No significant difference was observed for hypocotyls length between control and the lowest levels of $CdCl_2$. No significant differences in hypocotyls length was observed between 100, 200, 400 and 800 μM $CdCl_2$ levels. Treating *C. pumilum* seedlings with 1600 μM $CdCl_2$ resulted in the highest reduction of the hypocotyls length.

A clear trend for the effect of $CdCl_2$ on root length was observed: root length of *C. pumilum* decreased when increasing $CdCl_2$ concentration (Figure 1. C&D). Root length was moderately affected by 50 μM $CdCl_2$, while higher levels (100 and 200 μM) decreased root length significantly to 50% of the control. Furthermore, seedlings grown on 1600 μM $CdCl_2$ had the shortest roots.

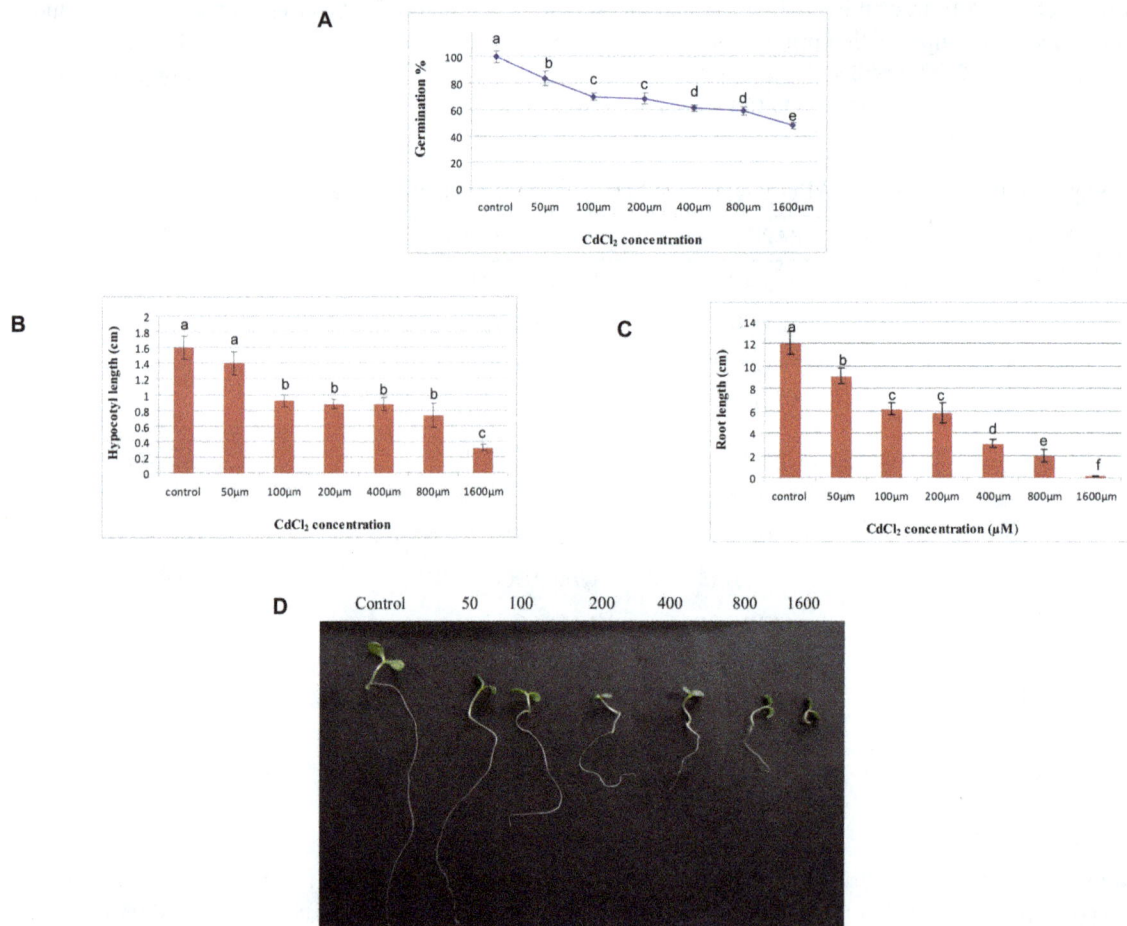

Figure 1. Germination percentage, hypocotyls length and root length of *C. pumilum* grown under different levels of $CdCl_2$. Data represents mean values ± standard error of ten replicates and the whole experiment was repeated three times. Means followed by the same letter are not statistically different at $p \leq 0.05$

3.3 Effect of Cadmium on Shoot Fresh Weight

Shoot fresh weight of *C. pumilum* was measured after 6 weeks of *in vitro* growth under different levels of $CdCl_2$.

Shoot fresh weight was affected adversely with increasing $CdCl_2$ concentration (Figure 2. A). Microshoots grown on MS medium supplemented with 50 μM $CdCl_2$ resulted in a significant reduction in shoot fresh weight with 7.6 g compared to 8.9 g in the control. Higher levels of $CdCl_2$ reduced shoot fresh weight severely. No significant difference for shoot fresh weight was observed between 100 and 200 μM $CdCl_2$. Sharp significant decreases in shoot fresh weights were observed for *C. pumilum* exposed to 800 and 1600 μM $CdCl_2$ which resulted in more than tenfold reduction compared to the control.

3.4 Effect of Cadmium on Chlorophyll Content

Chlorophyll content for *C. pumilum* microshoots at 50 and 100 μM $CdCl_2$ was found to be similar to control microshoots (Figure 2. B). In contrast, microshoots grown in MS medium supplemented with 200, 400, 800 and 1600 μM $CdCl_2$ showed higher levels of chlorophyll content than those grown in the control medium and lower levels of $CdCl_2$. No significant differences in chlorophyll content were observed between shoots grown on medium supplemented with 400, 800 and 1600 μM $CdCl_2$.

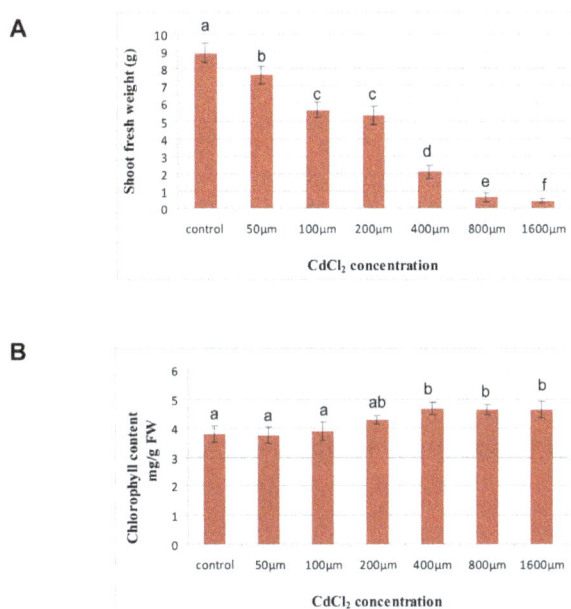

Figure 2. Shoot fresh weight and chlorophyll content of *in vitro* grown *C. pumilum* under different levels of $CdCl_2$. Data represents mean values ± standard error of eight replicates and the whole experiment was repeated three times. Means followed by the same letter are not statistically different at $p \leq 0.05$

3.5 Effect of Cadmium on Callus Growth

Different callus growth rates were observed based on fresh weight gain between the different levels of $CdCl_2$ (Figure 3). Calli grown on control medium showed the highest growth rate during the six weeks reaching a final fresh weight of 11.3 g. On the other hand, calli grown on MS medium supplemented with 50 μM $CdCl_2$ showed growth inhibition compared with the control and resulted in a final fresh weight of only 3.2 g. Calli grown on medium supplemented with 100 μM $CdCl_2$ showed more inhibition than that grown on 50 μM $CdCl_2$. Higher levels of $CdCl_2$ appear to be lethal for callus growth.

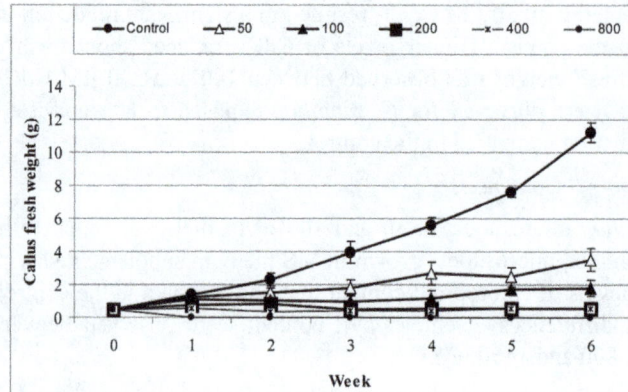

Figure 3. Callus growth curve of of *C. pumilum* grown under different levels of $CdCl_2$. Data represents mean values ± standard error of ten replicates and the whole experiment was repeated three times

3.6 Effect of Cadmium on Proline and Lipid Peroxidation Level

Proline content for *C. pumilum* microshoots was examined under different levels of $CdCl_2$. Results showed that proline content increased gradually and significantly as $CdCl_2$ level increased in the growth medium (Figure 4. A). The highest proline content was obtained from microshoots grown on the MS medium supplemented with the highest level of $CdCl_2$ (400 μM). Results showed that growing microshoots on 400 μM $CdCl_2$ increased proline content by more than tenfold.

The influence of Cd^{2+} on the lipid peroxidation rate of *C. pumilum* shoots was estimated by measuring MDA content, which is the product of lipid peroxidation. Lipid peroxidation rate in *C. pumilum* microshoots increased with increasing $CdCl_2$ level (Figure 4. B). Growing microshoots on MS medium supplemented with 50 μM $CdCl_2$ enhanced lipid peroxidation rate by more than twofold (compared with control). The highest lipid peroxidation rate was achieved in microshoots grown in MS medium supplemented with 400 μM $CdCl_2$.

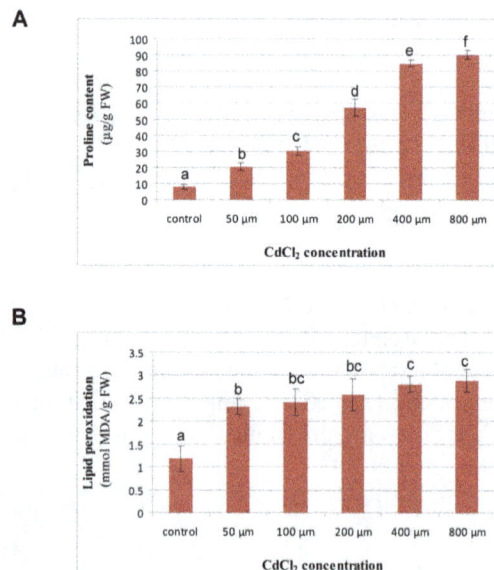

Figure 4. Proline content and lipid peroxidation rate of *C. pumilum* microshoots grown under different levels of $CdCl_2$ Data represents mean values ± standard error of eight replicates and the whole experiment was repeated three times. Means followed by the same letter are not statistically different at $p \leq 0.05$

3.7 DNA Fingerprinting Using RAPD Analysis

Genomic DNA was extracted from plants grown for 4 weeks on MS medium supplemented with different levels of Cd^{2+}. Twelve primers were used in this experiment (Table 1). Amplified profiles resulting from these primers showed variation between untreated and treated plants in terms of number and size of DNA bands. Figureure 5 shows RAPD profiles of treated and untreated samples of *C. pumilum* microshoots obtained from primer OPAP 20 as a representative for the other primers. The RAPD profiles obtained showed bands between 300 and 1800 bp in length. A total of 184 bands scored, only 36 were found to be polymorphic. Figure 5 shows the appearance or absence of bands at 200 and 400 µM $CdCl_2$.

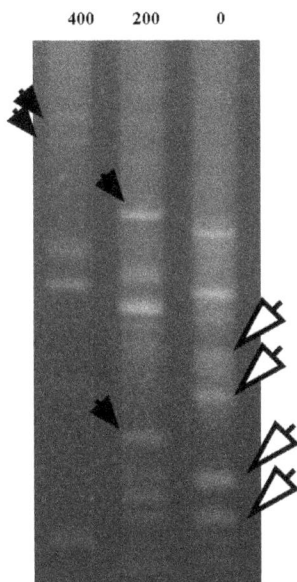

Figure 5. RAPD profiles (using primer OPAP 20) of genomic DNA extracted from *C. pumilum* microshoots grown under 0, 200 and 400 µM $CdCl_2$. Black arrows represent appearance of new bands and white arrows represent absence of bands relative to control (0)

4. Discussion

Plant growth and development under stress conditions are generally negatively affected. One of these stress conditions that affect plants is heavy metals. Recently, heavy metals have become a hot topic of research for many researchers around the world, mostly due to their detrimental effects on many organisms including plants.

Much research has been conducted on the effect of Cd^{2+} on crops and other agricultural plants. However, little information is available on the toxicity of Cd^{2+} on medicinal plants. Thus, the aim of this study was to assess the effect of cadmium on germination, growth, proline content, lipid peroxidation, and DNA fingerprinting of *in vitro* grown *C. pumilum*. Here, *in vitro* culture was used which is a convenient system for the study of mechanism of metal toxicity, as it eliminates the interfering processes of translocation and organ-specific trapping of metal ions.

The results of this study showed that Cd^{2+} levels affected all of the studied parameters in *C. pumilum* by different magnitudes. A significant reduction in percentages of seed germination, hypocotyl and root lengths of *C. pumilum* was observed here. This is in agreement with many published reports that studied the effects of Cd^{2+} on other plant species. Mathur et al. (1987) found that higher levels of cadmium inhibited germination percentage and the growth of the early seedlings of *Allium cepa*. Similarly, He et al. (2008) found that cadmium stress significantly inhibits germination index and shoot and root growth of rice. Pasquale et al. (1995) studied the influence of cadmium on the growth and biological activity of the medicinal plant *Coriandrum sativum* L. They found that growing plants under cadmium stress significantly reduced shoot and root lengths, resulting in leaf yellowing and in a major alteration in the essential oil quality and quantity. It has been shown that cadmium stress causes many problems in plants including growth and photosynthesis inhibition, alteration in nutrients and formation of free radicals (Sahu et al., 2007). Seed germination reduction due to heavy metals stress could be attributed to higher levels of seeds stored nutrients breakdown and/or change in permeability characteristics of the cell membrane (Shafiq et al., 2008). In peanuts, it has been shown that root and shoot growth and the initiation of lateral roots decreased with the increase in cadmium levels (Renjini & Janardhanan, 1989). The reason for reduced seedling length in metal treatments could be due to the reduction in meristematic cells present in this region and of some enzymes contained in the cotyledon and endosperms.

It has been shown that another species of Chicory (*Cichorium intybus* L.) showed a potential to be used as heavy metal bioindicator, *C. intybus* plants grown in nutrient solution supplemented with 0.5-50 µM cadmium showed high levels of Cd, in their shoots and roots (Simon et al., 1996). Another study (Kostantinos et al., 2008) showed that the fresh and dry weights of *Cichorium endivia* L. were not affected when grown on soil supplemented with different levels of Cd. Furthermore, they found that no toxicity symptoms were observed on *Cichorium endivia* plants.

Moreover, the result of this study showed that Cd^{2+} treatment significantly increased proline accumulation in *C. pumilum*. Proline accumulation is used as an indicator of stress conditions, including heavy metals. It has been shown that proline acts as a Cd^{2+} chelator in plants and forms a non-toxic complex with Cd^{2+} (Sharma et al., 1998). Similarly, Dinakar et al. (2009) found that proline content increased under cadmium stress in *Arachis hypogaea* L. It has been shown that plants subjected to $CdSO_4$ stress in the presence of proline showed a lower amount of reactive oxygen species compared to plants without proline. (Xu et al., 2009).

In addition to proline accumulation, the amount of lipid peroxidation also increased in response to Cd^{2+} stress. This is in agreement with Shah et al. (2001) who found an increase in malondialdehyde (MDA) levels (enhancement of lipid peroxidation) in rice seedlings after $Cd(NO_3)_2$ exposure. Similarly, Soltani et al. (2006) found that cadmium stress increased lipid peroxidation levels in *Brassica napus* plants. Lipid peroxidation is the main sign of free radical elevation. Plants may have two classes of antioxidative systems against the perceived oxidative stress: enzymatic antioxidants (such as superoxide dismutase (SOD)) and non-enzymatic low molecular weight antioxidants (such as proline, ascorbic acid, and glutathione) that can directly detoxify free oxygen radicals.

Different classical genotoxic assays have been used to examine the effect of heavy metals on plants including the comet assay and the micronucleus assay (Cambier et al., 2010). Recently, DNA fingerprinting has been successfully applied to test the effect of such stresses at the molecular level in different species (Korpe & Aras, 2011; Liu et al., 2012). The ability of cadmium to induce DNA mutations and/or damage has been shown previously (Gichner et al., 2004; Liu et al., 2012). Insertions and deletions, point mutations, base substitutions, single/double-strand breaks are examples of the effect of cadmium stress on DNA (Castano & Becerril, 2004). Here, RAPD profile shows different changes in the DNA fingerprint indicating that Cd^{2+} affects the genome integrity. Not all primers showed variations in the RAPD profile between treated and non-treated plants, which could be explained by the variation in genome sensitivity to heavy metals stress between different regions. Also, some genome areas could be protected from external damage (Liu et al., 2012).

In conclusion, the results of this study showed that Cd^{2+} had a toxic effect on germination, growth, proline content, lipid peroxidation, and DNA fingerprinting of *in vitro* grown *C. pumilum*. A reduction in hypocotyl and root length and shoot fresh weight was observed in seedlings grown under Cd^{2+} stress. A gradual increase in proline content and lipid peroxidation along with increasing Cd^{2+} concentration was also observed. The variation that occurred in the RAPD profiles of microshoots following Cd^{2+} treatment can be efficiently used as a sensitive tool to detect DNA damage and genotoxicity.

Acknowledgements

This work was supported by the deanship of research at Yarmouk University (Project #22/2009).

References

Al Khateeb, W., Hussein, E., Qouta, L., Alu'datt, M., Al-Shara, B., & Abu-zaiton, A. (2012). In vitro propagation and characterization of phenolic content along with antioxidant and antimicrobial activities of *Cichorium pumilum* Jacq. *Plant Cell, Tissue and Organ Culture, 110*(1), 103-110. http://dx.doi.org/10.1007/s11240-012-0134-9

Benavides, M., Gallego, S., & Tomaro, M. (2005). Cadmium toxicity in plants. *Braz. J. Plant Physiol, 17*, 21-34. http://dx.doi.org/10.1590/S1677-04202005000100003

Błasiak, J. (2001). DNA-Damaging effect of Cadmium and protective action of Quercetin. *Polish Journal of Environmental Studies, 10*, 437-442.

Castano, A., & Becerril, C. (2004). In vitro assessment of DNA damage after short- and long-term exposure to benzo(a)pyrene using RAPD and the RTG-2 fish cell line. *Mutat. Res., 552*, 141-151. http://dx.doi.org/10.1016/j.mrfmmm.2004.06.010

Chen, F., Dong, J., Wang, F., Wu, F., Zhang, G., Li, G., ... Wei, K. (2007). Identification of barley genotypes with low grain Cd accumulation and its interaction with four microelements. *Chemosphere., 67*, 2082-2088. http://dx.doi.org/10.1016/j.chemosphere.2006.10.014

Dinakar, N., Nagajyothi, P. C., Suresh, S., Damodharan, T., & Suresh, C. (2009). Cadmium induced changes on proline, antioxidant enzymes, nitrate and nitrite reductases in Arachis hypogaea L. *J. Environ. Biol., 30*(2), 289-294.

Gichner, T., Patková, Z., Száková, J., & Demnerová, K. (2004). Cadmium induces DNA damage in tobacco roots, but no DNA damage, somatic mutations or homologous recombination in tobacco leaves. *Mutat. Res., 559*, 49-57. http://dx.doi.org/10.1016/j.mrgentox.2003.12.008

Gratão, P., Polle, A., Lea, P., & Azevedo, R. (2005). Making the life of heavy metal-stressed plants a little easier. *Funct. Plant Biol., 32*, 481-494. http://dx.doi.org/10.1071/FP05016

Hardiman, R., & Jacoby, B. (1984). Absorption and translocation of Cd in bush beans in the environment: sources, mechanisms of biotoxicity, and biomarkers. *Rev. Environ. Health., 15*, 299-323.

He, J., Yan-fang, R., Cheng, Z., & De-an, J. (2008). Effects of Cadmium Stress on Seed Germination, Seedling Growth and Seed Amylase Activities in Rice (*Oryza sativa*). *Rice Science., 15*(4), 319-325. http://dx.doi.org/10.1016/S1672-6308(09)60010-X

Heath, R., & Packer, L. (1968). Photoperoxidation in isolated chloroplasts. I. Kinetics and stoichiometry of fatty acid peroxidation. *Arch. Biochem. Biophys., 125*, 189-198. http://dx.doi.org/10.1016/0003-9861(68)90654-1

Iribar, V., Izco, F., Tames, P., Antiguedad, I., & da Silva, A. (2000). Water contamination and remedial measures at the Troya abandoned Pb-Zn mine (The Basque Country, Northern Spain). *Environ. Geol., 39*, 800-806. http://dx.doi.org/10.1007/s002540050496

Kidd, S., Domínguez-Rodríguez, J., Díez, J., & Monterroso, C. (2007). Bioavailability and plant accumulation of heavy metals and phosphorus in agricultural soils amended by long-term application of sewage sludge. *Chemosphere., 66*, 1458-1467. http://dx.doi.org/10.1016/j.chemosphere.2006.09.007

Korpe, A., & Aras, S. (2011). Evaluation of copper-induced stress on eggplant (*Solanum melongena* L.) seedlings at the molecular and population levels by use of various biomarkers. *Mutat. Res., 719*, 29-34. http://dx.doi.org/10.1016/j.mrgentox.2010.10.003

Kostantinos, A., Akoumianakis, & Harold, C. (2008). Effect of cadmium on yield and cadmium concentration in the edible tissues of endive (*Cichorium endivia* L.) and rocket (*Eruca sativa* Mill.). *Food, Agriculture and Environment, 6*, 206-209. http://dx.doi.org/10.1016/j.mrgentox.2010.10.003

Liu, W., Sun, L., Zhong, M., Zhou, Q., Gong, Z., Li, P., ... Li, X. (2012). Cadmium-induced DNA damage and mutations in Arabidopsis plantlet shoots identified by DNA fingerprinting. *Chemosphere., 89*, 1048-1055. http://dx.doi.org/10.1016/j.chemosphere.2012.05.068

Mackinney, G. (1941). Absorption of light by chlorophyll solution. *J. Biol. Chem., 140*, 315-322.

Mathur, K., Srivastava, R., & Chaudhary, K. (1987). Effect of Cd and Cr metals on germination and early growth performance of *Allium cepa* seeds. *Proc. Nat. Acad. Sci. India. Sect. B (Biol. Sci.)., 57*, 191-196.

Nriagu, J. (1990). Global metal pollution. Poisoning the biosphere. *Environment., 32*, 28-33. http://dx.doi.org/10.1080/00139157.1990.9929037

Pasquale, R., & Rapisarda, A. (1995). Effects of cadmium on growth and pharmacologically active constituents of the medicinal plant Coriandrum sativum L. *Air. Soil. Pollut., 84*, 147-157. http://dx.doi.org/10.1007/BF00479594

Pinot, F., Kreps, S., Bachelet, M., Hainaut, P., Bakonyi, M., & Polla, B. (2000). Cadmium in the environment: sources, mechanisms of biotoxicity, and biomarkers. *Rev. Environ. Health., 15*, 299-323. http://dx.doi.org/10.1515/REVEH.2000.15.3.299

Porebski, S., Bailey, G., & Baum, B. (1997). Modification of a CTAB DNA extraction protocol for plants containing high polysaccharide and polyphenol components. *Plant Molecular Biology Reporter, 15*, 8-15. http://dx.doi.org/10.1007/BF02772108

Renjini, M. B. J., & Janardhanan, K. (1989). Effect of heavy metals on seed germination and early seedlings growth of groundnut, sunflower and gingerly. *Geobios., 16*, 164-170.

Robert, C., Happi Emaga, T., Wathelet, B., & Paquot, M. (2008). Effect of variety and harvest date on pectin extracted from chicory roots (*Cichorium intybus* L.). *Food Chemistry., 108*, 1008-1018. http://dx.doi.org/10.1016/j.foodchem.2007.12.013

Romero-Puertas, M., McCarthy, I., Sandalio, M., Palma, M., Corpas, J., Gómez, M., & del Río, A. (1999). Cadmium toxicity and oxidative metabolism of pea leaf peroxisomes. *Free Radical. Res., 31*, 25-31. http://dx.doi.org/10.1080/10715769900301281

Sahu, R. K., Katiyar, S., Tiwari, J., & Kisku, G. (2007). Assessment of drain water receiving effluent from tanneries and its impact on soil and plants with particular emphasis on bioaccumulation of heavy metals. *J. Environ. Biol., 28*, 685-690.

Schill, O., Gorlitz, H., & Kohler, H. (2003). Laboratory simulation of a mining accident: acute toxicity, hsc/hsp70 response, and recovery from stress in Gammarus fossarum (*Crustacea smphipoda*) exposed to a pulse of cadmium. *BioMetals., 16*, 391-401. http://dx.doi.org/10.1023/A:1022534326034

Sebastiani, L., Scebba, F., & Tognetti, R. (2004). Heavy metal accumulation and growth responses in poplar clones Eridano (*Populus deltoides* × *maximowiczii*) and I-214 (*P.* × *euramericana*) exposed to industrial waste. *Env. Exp. Bot., 52*, 79-84. http://dx.doi.org/10.1016/j.envexpbot.2004.01.003

Shafiq, M., Iqbal, M. Z., & Athar, M. (2008). Effect of lead and cadmium on germination and seedling growth of *Leucaena leucocephala. J. Appl. Sci. Environ. Manage., 12*(2), 61- 66.

Shah, K., Ritambhara, G., Verma, S., & Dubey, R. (2001). Effect of cadmium on lipid peroxidation, superoxide anion generation and activities of antioxidant enzymes in growing rice seedlings. *Plant Science., 161*(6), 1135-1144. http://dx.doi.org/10.1016/S0168-9452(01)00517-9

Sharma, S., Schat, H., & Vooijs, R. (1998). In vitro alleviation of heavy metal-induced enzyme inhibition by proline. *Phytochemistry., 46*, 1531-1535. http://dx.doi.org/10.1016/S0031-9422(98)00282-9

Shrivastava, G., & Singh, P. (1989). Uptake, accumulation and translocation of cadmium and zink in Abelmoschus. *Plant Physiol Biochem., 16*, 17-22.

Simon, L., Martin, H., & Adriano, D. (1996). Chicory (*Cichorium intybus* L.) and dandelion (*Taraxacum officinale* Web.) as phytoindicators of cadmium contamination. *Water, Air, and Soil Pollution. 91*, 351-362. http://dx.doi.org/10.1007/BF00666269

Siripornadulsil, S., Traina, S., Verma, D., & Sayre, R. (2002). Molecular mechanisms of proline-mediated tolerance to toxic heavy metals in transgenic microalgae. *Plant Cell., 14*, 2837-2847. http://dx.doi.org/10.1105/tpc.004853

Soltani, F., Lagha, G., & Kalantari, M. (2006). Effect of cadmium on photosynthetic pigments, sugars and malondealdehyde content in (*Brassica napus* L.). *Iranian Journal of Biology, 19*(2), 136-145.

Theodorakis, C., Lee, K., Adams, S., & Law, C. (2006). Evidence of altered gene flow, mutation rate, and genetic diversity in redbreast sunfish from a pulpmill-contaminated river. *Environ. Sci. Technol., 40*, 377-386. http://dx.doi.org/10.1021/es052095g

Xu, J., Yin, H., & Li, X. (2009). Protective effects of proline against cadmium toxicity in micropropagated hyperaccumulator, *Solanum nigrum* L. *Plant Cell Rep., 28*(2), 325-333. http://dx.doi.org/10.1007/s00299-008-0643-5

4

Antimicrobial Action of Epidermal Mucus Extract of *Clarias gariepinus* (Burchell, 1822) Juveniles-Fed Ginger Inclusion in Diet

A. A. Nwabueze[1]

[1] Department of Fisheries, Delta State University, Asaba Campus, Nigeria

Correspondence: A. A. Nwabueze, Department of Fisheries, Delta State University, Asaba Campus, Asaba, Nigeria. E-mail: aanwabueze@gmail.com

Abstract

The antimicrobial activity of epidermal mucus extract of *C. gariepinus* juveniles-fed ginger inclusion in diet was investigated and compared with the activity of epidermal mucus extract of *C. gariepinus* juveniles (control) without ginger in diet. This study demonstrates the antimicrobial role of ginger in improving protection of fish against bacterial infection as shown by the higher zones of inhibition observed for epidermal mucus extract of fish-fed ginger in diet as compared with control. Zones of inhibition for epidermal mucus of treatment fish were 30.7 mm, 29.8 mm, 26.3 mm and 19.3 mm for *Bacillus*, *Escherichia*, *Staphylococcus* and *Streptococcus* species respectively. Though these values were not significantly ($P > 0.05$) higher than those obtained for the control fish with zones of inhibition of 25 mm, 11.2 mm, 9.0 mm and 7.3 mm for *Bacillus*, *Escherichia*, *Staphylococcus* and *Streptococcus* species respectively, the higher values recorded for the treatment fish shows that ginger inclusion in fish diet had an antibiotic effect against isolates of bacteria in fish samples from cultured ponds. The addition of ginger in *C. gariepinus* diet is encouraged as its action is indicative of the potentials of ginger in preventing emergence of resistant bacteria and improving the antimicrobial role of fish mucus and therefore the quality of *C. gariepinus*.

Keywords: antimicrobial, *Clarias gariepinus*, epidermal mucus, ginger

1. Introduction

1.1 The Problem

All fish live in microbe-rich environment and are vulnerable to invasion by pathogenic and opportunistic micro-organisms. The environment of fish being aquatic is very challenging with fish in constant interaction with a wide range of pathogenic and non-pathogenic micro-organisms (Subramanian, Mackinnon, & Ross, 2007). Environmental degradation due to pollution of natural water bodies and poor culture conditions of some culture fish ponds have elicited the presence of these pathogenic agents (Nwabueze, 2011, 2012) with bacteria being one of the most common micro-organisms of farmed catfishes.

1.2 Importance of the Problem

Fish is sometimes overwhelmed by these pathogenic agents and sometimes succumb to infections. In culture ponds, especially in the tropics, a number of antimicrobial agents have been used to combat several diseases of fish. In recent times though, an increase in the antibiotic resistant strains of some of these micro-organisms have been observed. Antimicrobial activity has been demonstrated in fish mucus of several fish species, yet this activity seems to vary from fish species to species such as rock fish (*Sebastes schlegelii*), rainbow trout (*Oncorhynchs mykiss*) and tilapia (*Tilapia hornorum*) and can be specific toward certain bacteria (Noya, Magarinos, Toranzo, & Lamas, 1995). *Clarias gariepinus* is a valuable farmed food fish in Nigeria. Its high nutrition and importance in aquaculture have encouraged a lot of researches in the improvement of the quality of the fish species.

1.3 Relevant Scholarship

Increase in antibiotic resistance of some microbial strains has lead to investigations on the use of traditional plants for their antibacterial and medicinal values (Bhalodia & Shukla, 2011). Fishes produce mucus from the epidermal cell layer for protection against some of these pathogenic agents in the aquatic medium. Fish epidermal mucus has been reported to be a component of innate immunity playing an important role in the prevention of colonization by parasites, bacteria and fungi (Muroga, Higashi, & Keitoku, 1987; Kanno, Nakai, & Maroga, 1989; Ebran, Julien, Orange, Auperin, & Molle, 2000). Fish epidermal mucus has also been known to contain a variety of biologically active compounds and antimicrobial peptides that are constitutively expressed to provide immediate protection of fish from potential pathogenic microbes and parasites (Hjelmeland, Christe, & Raa, 1983). Kasai, Ishikawa, Komata, Fukuchi, Chiba, Nozaka, Nakamura, Sato and Miura (2009) reported an antibacterial protein l-amino acid oxidase (LAAO) from the epidermal mucus of flounder *Platichthys stellatus* which exerted antibacterial activity against *Staphylococcus epidermids*, *S. aureus* and methicillin-resistant *S. aureus*. It has also been noted that this antibacterial protein exerted antibacterial activity in a variety of animal fluids such as snake venom (Du & Clemetson, 2002), fish epidermal mucus and extract (Jung, Mai, Iwamoto, Arizono, Fujimoto, Sakamaki, & Yonehara, 2000; Kitani et al., 2007; Kitani, Kikuchi, Zhang, Ishizaki, Shimakura, Shiomi, & Nagashima, 2008) body surface mucus of the giant African snail (Obara, Otsuka-Fuchino, Sattayasai, Nonomura, Tsuchiya, & Tamiya, 1992). It is therefore important to consider ways of improving fish health by using biological agents to prevent infections rather than using drugs to provide cure. Ginger plant is a spice that has been noted for its medicinal values as an analgesic, sedative, antipyretic and antibacterial agent (Patrick-Iwuanyanwu, Wegwu, & Ayalogu, 2007). Ginger also has anti-microbial, anti-oxidative and seasoning qualities.

1.4 Hypothesis

This study examines the antimicrobial activity of epidermal mucus extract of *C. gariepinus*-fed ginger inclusion in diet, to know if ginger can enhance protection in fish as expressed by the epidermal mucus extract of cultured *C. gariepinus*.

2. Materials and Methods

2.1 Study Area

The antibiotic activity of epidermal mucus extract of *C. gariepinus* (average weight of 40.6 ± 0.1 g and total length of 17.2 ± 0.4 cm) juveniles-fed ginger inclusion in diet was investigated from January to May, 2013 at the Department of Fisheries and Faculty of Agriculture Research Laboratories of Delta State University, Asaba Campus, Asaba, Nigeria.

2.2 Acclimation and Experimental Tanks

Sixty juvenile fish were obtained from the Faculty of Agriculture Research Farm and acclimated for 7 days in stock tank during which time fish were fed commercially available feed at 4% body weight. Stock tank (45 cm × 45 cm × 90 cm) was well aerated and contained 50 litres of borehole water, tank water temperature was 26.4 °C and tank water was changed twice. Ten apparently healthy fish samples each distributed into two smaller tanks (40 cm × 40 cm × 60 cm) containing 25 litres of water and labelled, A (control) and B (Treatment) with two replicates of A_1, A_2 and B_1, B_2 making up a total of 60 *C. gariepinus* fish were used.

2.3 Preparation of Experimental Diet

Ginger root was bought from a local market in Asaba, back was peeled off and 100 g ginger was cut into bits and sun-dried for 5 days and later ground into powder form using an electric blender, sieved and stored in container away from moisture and direct sun light according to Stoll (2000) until further use. Fish in experimental tanks were fed formulated diet with fish in the treatment tanks B, B_1 and B_2 having an addition of 0.5% (0.025 kg) inclusion of ginger in fish diet (Table 1). Fish were fed for 4 months after which the epidermal mucus extract of fish in control and treatment tanks were collected and examined for their antimicrobial activities.

Table 1. Composition of experimental diets

Ingredients	Control diet composition (kg)	Treatment diet composition (kg)	Concentration of diets (%)
Groundnut cake	0.78	0.75	15.0
Soya bean	0.78	0.75	15.0
Fish meal	0.78	0.75	15.0
Vitamin C	0.026	0.025	0.50
Bone meal	0.21	0.20	4.0
Premix	0.10	0.10	2.0
White maize	0.52	0.50	10.0
Wheat (rice bran)	0.52	0.50	10.0
Palm oil	1.46	1.4	28
*Ginger	-	0.025	0.50

*Added only for fish in treatment tanks.

2.4 Isolation and Identification of Bacteria Isolates

Bacterial isolates were obtained from epidermal mucus of apparently healthy 6 months old live *C. gariepinus* fish weighing about 480 g, harvested from a culture pond in Asaba. Mucus samples were carefully and aseptically scraped from the dorsal surface of fish skin using a sterile soft rubber spatula. The aqueous mucus scrapings were pooled and cultured on Sabourand Dextrose agar at 37 °C for 24 h. Distinct growth colonies observed were sub cultured into sterile Petri dishes with Nutrient agar by streaking method to obtain pure isolates which were incubated aerobically at 37 °C for 24 h. A growing edge of distinct isolates was picked from pure cultures and streaked with inoculating loop onto sterile Petri dish with Nutrient agar as the growth medium for each of the isolates. Isolates were identified using routine Gram Staining Techniques and Biochemical Characterization according to MacFaddin (1980). Bacterial isolates obtained were *Staphylococcus*, *Streptococcus* , *Escherichia* and *Bacillus* species. Each pure colony was suspended in physiological saline (0.85% NaCl) and standardized to correspond to 1.5×10^8 CFU/ml of each bacterial isolates.

2.5 Preparation of Plate Inoculums

Fresh sterile cotton-tipped swap was dipped into suspension of pure bacterial isolates and was used to inoculate sterile Mueller-Hinton Agar plates by streaking and ensuring even distribution of the inoculums. This was done for each bacterial isolates.

2.6 Collection and Sterilization of Fish Epidermal Mucus

Mucus from the control and experimental fish samples was carefully scraped from the anterior to the posterior of the dorsal body using a sterile soft rubber spatula. Mucus scrapings were pooled from the ten experimental and also pooled from the ten control fish separately. Mucus samples were then thoroughly mixed with equal quantity of sterilized physiological saline (0.85% NaCl) and centrifuged at 5000 rpm for 15 minutes according to kuppulakshmi, Prakash, Gunasekaran, Manimealai and Sarojini (2008). The supernatant was collected and stored at 4 °C and were used to prepare the antibiotic disc for the antimicrobial studies.

2.7 Preparation of Antibiotic Disc

Sensitivity disc was prepared employing pipette delivery to impregnate 20 µl of the epidermal mucus extract of control and treatment *C. gariepinus* fish onto a disc using disc diffusion method (Cavalieri et al., 2005). Discs were firmly pressed down to ensure complete level contact with agar.

2.8 Sensitivity Testing

After the introduction of the disc, the plate was allowed to incubate at 37 °C for 24 h. Zones of inhibitions were measured to the nearest millimetres. Sensitivity testing was carried out on epidermal mucus extract of fish from control and treatment tanks on the bacterial lawn for the different isolates. A clear zone of inhibition (plaque) indicates absence of bacterial growth while no discernable plaque around the disc, means that the bacteria are growing normally. The presence of a plaque means sensitivity while the absence of a plaque means resistance.

2.9 Statistical Analysis

Student 't' test statistic was employed to analyze data collected for zones of inhibition of the four different strains of bacteria isolated from the fish epidermal mucus from both the control and treatment tanks. Differences between means were considered significant when $P < 0.05$.

3. Results

Bacterial isolates used for this study were *Staphylococcus*, *Streptococcus*, *Escherichia* and *Bacillus* species. All bacteria isolates were sensitive to epidermal mucus extract of fish from both control and treatment experiments. Zones of inhibition of epidermal mucus extract of control and treatment fish against bacterial isolates are presented in Table 2. The resulting zones of inhibition for epidermal mucus of treatment fish were 30.7 mm, 29.8 mm, 26.3 mm and 19.3 mm for *Bacillus*, *Escherichia*, *Staphylococcus* and *Streptococcus* species respectively. While for the control fish values of 25 mm, 11.2 mm, 9.0 mm and 7.3 mm were zones of inhibition for *Bacillus*, *Escherichia*, *Staphylococcus* and *Streptococcus* species respectively. *Bacillus* species was observed to be more sensitive to the activity of ginger than the other isolates. *Streptococcus* species was the least sensitive. The same trend was also observed for epidermal mucus of the control fish.

Zones of inhibition were observed to be higher in the epidermal mucus of fish-fed ginger inclusion in diet as compared with epidermal mucus of control fish without ginger in diet. Zones of inhibition of the four different strains of bacteria isolated from fish epidermal mucus from both the control and treatment tanks were analyzed separately. Results of analysis however show that there were no significant difference ($P > 0.05$) in the zones of inhibition of epidermal mucus of control fish as compared with the epidermal mucus of treatment fish (Table 3). In this study, the mucus extracted from *C. gariepinus* (control fish) showed inhibitory effect on selected bacteria isolates. However, *C. gariepinus*-fed ginger in diet had more inhibitory effect on the selected bacteria isolates.

Table 2. Zones of inhibition of epidermal mucus of *C. gariepinus* against bacteria isolates

S/n	Bacterial Isolates	Zones of Inhibition for Control and Experimental fish mucus (mm)			
		Experimental Tanks (Control)	Epidermal mucus of Control fish	Experimental Tanks (Treatment)	Epidermal mucus of Treatment fish
1	*Staphylococcus*	Tank A	8.8	Tank B	26.4
		Tank A_1	9.2	Tank B_1	25.7
		Tank A_2	9.1	Tank B_2	26.1
2	*Streptococcus*	Tank A	7.8	Tank B	18.5
		Tank A_1	7.1	Tank B_1	20.2
		Tank A_2	6.9	Tank B_2	19.3
3	*Escherichia*	Tank A	11.0	Tank B	30.3
		Tank A_1	11.1	Tank B_1	28.7
		Tank A_2	11.4	Tank B_2	30.0
4	*Bacillus*	Tank A	24.7	Tank B	31.3
		Tank A_1	26.1	Tank B_1	30.8
		Tank A_2	24.3	Tank B_2	29.9

Table 3. Results of analysis of zones of inhibition of epidermal mucus of control and treatment fish

S/n	Bacteria isolates	T Statistics	T Critical
1	*Staphylococcus*	- 53.5674	4.302653
2	*Streptococcus*	- 16.9336	4.302653
3	*Escherichia*	- 37.5034	4.302653
4	*Bacillus*	- 10.266	4.302653

T test: Paired Two Sample Means (two-tailed).

4. Discussion

The role of epidermal mucus as an important component of fish innate immunity was demonstrated in this study with fish epidermal mucus being a potential source of antimicrobial activity for specific fish pathogens. This fact is evidenced by the clear zones of inhibition observed for the epidermal mucus of the control fish signifying absence of bacterial growth thereby indicating that fish mucus has antimicrobial properties which prevented bacterial growth. This fact could be attributed to a complex system of innate defence mechanisms enabling fish epidermal mucus to have a potential broad spectrum-antimicrobial activity. Balasubramanian, Baby, Arul, Prakash, Senthilraja and Gunasekaran (2012) noted that despite an intimate contact with high concentrations of pathogens (bacteria and viruses) in their environment, the fish can still maintain a healthy system under normal conditions. Fish skin is a complex limiting structure providing mechanical, chemical and immune protection against injury and pathogenic micro-organisms (Fontenot & Neiffer, 2004). Many researchers have proved that mucus exhibits good resistance to invading pathogens (Fletcher, 1978; Ingram, 1980; Austin & McIntosh, 1988; Fouz, Devaja, Gravningen, Barija, & Tranzo, 1990). Fish mucus layer confers an innate immune protection against pathogen entry. Mucus covering fish surfaces exposed to water acts as an innate and adaptive first line of defence against pathogen entry (Shephard, 1994).

In fish, the epidermal mucus is considered a key component of innate immunity. The composition and rate of mucus secretion has been observed to change in response to microbial exposure or to environmental fluctuations (Ellis, 2001). Raj et al. (2011) reported that skin mucus removal and epidermal lesions in *Cyprinus carpio* in invitro experiments enhanced the entry of CyHV-3 virus while the presence of skin mucus of *Cyprinus carpio* conferred protection against the entry of the virus. Numerous studies on innate immunity in fish have shown that fish epidermal mucus can inhibit the growth of some bacteria and therefore may have a potential source of novel antimicrobial components in it (Wei, Xavier, & Marimuthu, 2010). Kasai et al. (2009) observed inhibition of growth of *Staphylococcus epidermidis* and *S. aureus* and noted that the proliferation of *S. epidermidis* in particular was strongly suppressed, the effect being most marked among all the bacteria strains studied.

The antibacterial activity of fish mucus may be due to the presence of antibacterial glycoproteins which are able to kill bacteria by forming large pores in the target membranes (Ebran, Julien, Orange, Saglio, Lemaitre, & Molle, 1999). Fish mucus is believed to play an important role in the prevention of colonization by parasites, bacteria and fungi and thus act as a chemical defence barrier (Gobinath & Ravichandran, 2011).

Results show an antibiotic effect of ginger against isolates of bacteria in fish samples from cultured ponds. The antimicrobial role of ginger in improving protection of fish against bacterial infection was shown by the higher zones of inhibition observed for epidermal mucus extract of fish-fed ginger in diet when compared with epidermal mucus extract of control fish. Dugenci, Arda and Candan (2003) have shown that the rainbow trout fish fed with diets containing aqueous extracts of mistletoe (*Viscum album*), nettle (*Urtica dioica*), and ginger (*Zingiber officinale*) exhibited significant non-specific immune responses. Pandy (2013) also reported that all medicinal plants are able to stimulate only non-specific immune responses and suggested that vaccines might be a better way to prevent deadly diseases and as such the plants could be used as vaccine adjuvant. In addition, Idris, Omojowo, Omojasola, Adetunji and Ngwu (2010) while working on the effect of different concentrations of ginger on smoked-dried *C. gariepinus*, found that ginger reduced the free fatty acid values, trimethylamine values as well as reduced the fungi load of processed fish. The antimicrobial action of ginger has also been reported by Patel, Thaker and Patel (2011) in invitro studies using ginger in combination with honey against *Staphylococcus* isolates. Ginger definitely has antimicrobial properties and its use in the prevention of emergence of resistant bacteria is of high benefit.

5. Conclusion

This study has demonstrated the antibacterial role of fish mucus as shown by the higher zones of inhibition observed for mucus extract of fish-fed ginger in diet as against inhibition zones of mucus extract of fish without ginger in diet. The addition of ginger in *C. gariepinus* diet is encouraged as its action is indicative of the potentials of ginger in preventing emergence of resistant bacteria. Ginger inclusion in fish diet is beneficial in improving the antimicrobial role of fish mucus and hence the quality of *C. gariepinus*.

References

Austin, B., & McIntosh, D. (1988). Natural antibacterial compounds on the surface of rainbow trout. *Journal of Fish Diseases, 11*(3), 275-277. http://dx.doi.org/10.1111/j.1365-2761.1988.tb00550.x

Balasubramanian, S., Baby, R. P., Arul, P. A., Prakash, M., Senthilraja, P., & Gunasekaran, G. (2012). Antimicrobial properties of skin mucus from four freshwater cultivable fishes (*Catla catla,*

Hypophthalmichthys molitrix, Labeo rohita and *Ctenopharyngodon idella*). *African Journal of Microbiology Research, 6*(24), 5110-5120. http://10.1016/s1995-7645(11)60091-6

Bhalodia, N. R., & Shukla, V. J. (2011). Antibacterial and antifungal activities from leaf extracts of *Cassia fistula* l.: An ethnomedicinal plant. *Journal of Advanced Pharmaceutical Technology & Research, 2*(2), 104-109. http://dx.doi.org/10.4103/2231-4040.82956

Cavalieri, S. J., Harbeck, R. J., McCarter, Y. S., Ortez, J. H., Rankin, I. D., Sautter, R. L., … Spiegel, C. A. (2005). Manual of Antimicrobial Susceptibility Testing. Coordinating Editor, M. B. Coyle. *American Society for Microbiology (ASM)*.

Du, X. Y., & Clemetson, K. J. (2002). Snake venom L-amino acid oxidases. *Toxicon, 40*(6), 659-665. http://dx.doi.org/10.1016/S0041-0101(02)00102-2

Dugenci, S. K., Arda, N., & Candan, A. (2003). Some medicinal plants as immune-stimulant for fish. *Journal of Ethnopharmacology, 80*, 99-106. http://dx.doi.org/10.1016/S0378-8741(03)00182-X

Ebran, N., Julien, S., Orange, N., Auperin, B., & Molle, G. (2000). Isolation and characterization of novel glycoproteins from fish epidermal mucus: correlation between their poreforming properties and their antibacterial activities. *Biochim Biophys Acta, 1467*, 271-280. http://dx.doi.org/10.1016/S0005-2736(00)00225-X

Ebran, N., Julien, S., Orange, N., Saglio, P., Lemaitre, C., & Molle, G. (1999). Pore forming properties and antibacterial activity of protein extracted from epidermal mucus of fish. *Comparative Biochemical Physiology A, 122*, 181-189. http://dx.doi.org/10.1016/S1095-6433(98)10165-4

Ellis, A. E. (2001). Innate host defence mechanisms of fish against viruses and bacteria. *Dev. Comp. Immunol., 25*, 827-839. http://dx.doi.org/10.1016/S0145-305X(01)00038-6

Fletcher, T. (1978). Defense mechanisms in fish. In D. Malins & J. Sargent (Eds.), *Biochemical and Biophysical perspertives in marine biology* (pp. 189-222). London: Academic Press.

Fontenot, D. K., & Neiffer, D. L. (2004). Wound management in teleost fish: biology of the healing process, evaluation and treatment. *Vet. Clin. North Am. Exot. Anim. Pract., 7*, 57-86. http://dx.doi.org/10.1016/j.cvex.2003.08.007

Fouz, B., Devaja, S., Gravningen, K., Barija, J. L., & Tranzo, A. E. (1990). Antibacterial action of the mucus of the turbot. *Bulletin European Association of Fish Pathology, 10*, 56-59.

Gobinath, R. A. C., & Ravichandran, S. (2011). Antimicrobial peptide from the epidermal mucus of some estuarine cat fishes. *World Applied Sciences Journal, 12*(3), 256-260.

Hjelmeland, K., Christe, M., & Raa, J. (1983). Skin mucus protease from rainbow trout, *Salmo gairdlneri*, Richardson and it's biological significance. *Journal of Fish Biology, 23*(1), 13-22. http://dx.doi.org/10.1111/j.1095-8649.1983.tb02878.x

Idris, G. L., Omojowo, F. S., Omojasola, P. F., Adetunji, C. O., & Ngwu, E. O. (2010). The effect of different concentrations of ginger on the quality of smoked-dried catfish (*Clarias gariepinus*). *Nature and Science, 8*(4),59-63.

Ingram, G. A. (1980). Substances involved in the natural resistance of fish to infection - A Review. *Journal of Fish Biology, 16*, 23-60. http://dx.doi.org/10.1111/j.1095-8649.1980.tb03685.x

Jung, S. K., Mai, A., Iwamoto, M., Arizono, N., Fujimoto, D., Sakamaki, K., & Yonehara, S. (2000). Purification and cloning of an apoptosis-inducing protein derived from fish infected with Anisakis simplex, a causative nematode of human anisakiasis. *Journal of Immunology, 165*, 1491-1497.

Kanno, T., Nakai, T., & Maroga, K. (1989). Mode of Transmission of vibrios among ayu *Plecoglossus activelis*. *Journal of Aquatic Animal Health, 1*(1), 2-6. http://dx.doi.org/10.1577/1548-8667(1989)001%3C0002:MOTOVA%3E2.3.CO;2

Kasai, K., Ishikawa, T., Komata, T., Fukuchi, K., Chiba, M., Nozaka, H., … Miura, T. (2009). Novel l-amino acid oxidase with antibacterial activity against methicillin-resistant Staphylococcus aureus isolated from epidermal mucus of the flounder Platichthys stellatus. *Biochemistry FEBS Journal, 277*(2), 453-465. http://dx.doi.org/10.1111/j.1742-4658.2009.07497.x

Kitani, Y., Kikuchi, N., Zhang, G., Ishizaki, S., Shimakura, K., Shiomi, K., & Nagashima, Y. (2008). Antibacterial action of L-amino acid oxidase from the skin mucus of rockfish *Sebastes schlegelii*.

Comparative Biochemical Physiology B: Biochemistry and Molecular Biology, 149, 394-400. http://dx.doi.org/10.1016/j.cbpb.2007.10.013

Kitani, Y., Tsukamoto, C., Zhang, G., Nagai, H., Ishida, M., Ishizaki, S., ... Nagashima, Y. (2007). Identification of an antibacterial protein as L-amino acid oxidase in the skin mucus of rockfish *Sebastes schlegeli. Biochemistry FEBS Journal, 274*(1), 125-136. http://dx.doi.org/10.1111/j.1742-4658.2006.05570.x

Kuppulakshmi, C., Prakash, M., Gunasekaran, M. G., Manimealai, G., & Sarojini, S. (2008). Antibacterial properties of fish mucus from *Channa punctatus* and *Cirrhinus mrigala. European Review for Medical and Pharmacological Sciences, 12*, 149-153

MacFaddin, J. F. (1980). *Biochemical Tests for Identification of Medical Bacteria* (2nd edition) (p. 313). Baltimore: The Williams and Wilkins Co..

Muroga, K., Higashi, M., & Keitoku, H. (1987). The isolation of intestinal microflora of farmed red seabream (*Pagrus major*) and black seabream (*Acanthopagrus schlegeli*) at larval and juvenile stages. *Aquaculture, 65*, 79-88. http://dx.doi.org/10.1016/0044-8486(87)90272-9

Noya, M., Magarinos, B., Toranzo, A. E., & Lamas, J. (1995). Sequential pathology of experimental pasteurellosis in gilthead seabream sparus aurata a light microscopic and electron microscopic study. *Diseases of Aquatic Organisms, 21*,177-186. http://dx.doi.org/10.3354/dao021177

Nwabueze, A. A. (2011). Public Health Implications of Aquatic Snails around Fish Ponds in Okwe, Delta State. *International Journals of Agriculture and Rural Development (IJARD), 14*(2), 652-656.

Nwabueze, A. A. (2012). Diseases Status of *Clarias gariepinus* (Burchell, 1822) and Some Fish Ponds in Asaba, Nigeria. *International Journal of Agriculture and Rural Development, 15*(3), 1216-1222.

Obara, K., Otsuka-Fuchino, H., Sattayasai, N., Nonomura, Y., Tsuchiya, T., & Tamiya, T. (1992). Molecular cloning of the antibacterial protein of the giant African snail, Achatina fulica Férussac. *European Journal of Biochemistry, 209*, 1-6. http://dx.doi.org/10.1111/j.1432-1033.1992.tb17254.x

Pandy, G. (2013). *Some medicinal plants to treat fish ectoparasitic infections. International Journal of Pharmaceutical and Research Sciences (IJPRS), 2*(2), 532-538.

Patel, R. V., Thaker, V. T., & Patel, V. K. (2011). Antimicrobial activity of ginger and honey on isolates of extracted carious teeth during orthodontic treatment. *Asian Journal of Tropical Biomedicine*, 558-561.

Patrick-Iwuanyanwu, K. C., Wegwu, M. O., Ayalogu, E. O. (2007). The protective nature of garlic, ginger and vitamine C on CCl4-induced hepatotoxicity in rats. *Asian Journal of Biochemistry, 2*(6), 409-414. http://dx.doi.org/10.3923/ajb.2007.409.414

Raj, S. T., Fournier, G., Rakus, K., Ronsmana, M., Ouyang, P., Michel, B., ... Vanderplasschen, A. (2011). Skin mucus of *Cyprinus carpio* inhibits cyprinid herpesvirus 3 binding to epidermal cells. *Veterinary Research, 42*, 92. http://dx.doi.org/10.1186/1297-9716-42-92

Shephard, K. L. (1994). Functions for fish mucus. *Reviews in Fish Biology & Fisheries, 4*, 401-429. http://dx.doi.org/10.1007/BF00042888

Stoll, G. (2000). *Natural crop protection in the tropics*. Margraf Verlag, Weikersheim.

Subramanian, S., Mackinnon, S. L., & Ross, N. W. (2007). A comparative study on innate immune parameters in the epidermal mucus of various fish species. *Comparative Biochemical Physiology, 148B*, 256-263. http://dx.doi.org/10.1016/j.cbpb.2007.06.003

Wei, O. Y., Xavier, R., & Marimuthu, K. (2010). Screening of antimicrobial activity of mucus of Snakehead fish, *Channa striatus* (Bloch). *European Review for Medical and Pharmacological Sciences, 14*, 675-681.

5

Variability in Compensatory Ability and Relative Invasive Potential in Ornamental Cleomes

Nadilia N. Gómez Raboteaux[1] & Neil O. Anderson[2]

[1] Research Analyst – RIM, Pioneer Hi-Bred Seed Co., Johnston, IA, USA

[2] Department of Horticultural Science, University of Minnesota, Saint Paul, MN, USA

Correspondence: Neil O. Anderson, 305 Alderman Hall, 1970 Folwell Avenue, Saint Paul, MN 55106, USA. E-mail: ander044@umn.edu

Abstract

Tolerance to herbivory is an important trait influencing invasive potential of exotic plant species. However, though many invasive plant species have been introduced as ornamentals, tolerance to herbivory has not been evaluated among cultivars of ornamental crops to assess relative invasive potential. A greenhouse study was performed to compare tolerance to simulated herbivory in five *Cleome* cultivars (Sparkler White, Sparkler Rose, Queen White, Queen Rose and Solo). The herbivory treatments simulated deer, rabbit and invertebrate herbivore damage by clipping the main stem, removing leaves, or punching out leaf pieces, respectively. Data were collected for flowering time, vegetative and reproductive biomass, ratio of reproductive: vegetative biomass (reproductive effort), number of flowering and vegetative shoots, ratio of number of flowering: total shoots (reproductive allocation), and number of ovules/flower. Cultivars showed different norms of tolerance ranging from under compensation to overcompensation with differences among cultivars within series. The response differed among patterns of simulated herbivory with stem clipping having the most dramatic and negative effect on plant growth and reproduction relative to whole leaf and partial leaf defoliation. The response also varied depending on cultivar and trait. For example, compensation in vegetative, but not reproductive, biomass, was observed across most cultivars after clipping. Significant interactions of herbivory treatment x cultivar were detected for total shoot number and the ratio of flowering: total shoots in the stem clipping experiment, indicating shifts in relative cultivar ranks. The implication of variation in tolerance to herbivory is discussed in relation to ornamental crop development and invasive species risk assessment.

Keywords: biomass allocation, *Cleome hassleriana, Cleome serrulata,* plant-herbivore interactions, *Polanisia dodecandra,* simulated herbivory

1. Introduction

Plant-herbivore interactions play a significant role in determining whether an exotic species becomes invasive (Bigger & Marvier, 1998; Crawley, 1989). When an exotic species is introduced to a new region, a shift in the composition of herbivore is likely. The exotic species may be released from the herbivores present in its native range and possibly exposed to a different set of herbivores in the area of introduction. Therefore, response to different patterns of herbivory is likely to affect the probability of population establishment and growth.

Several hypotheses on plant-herbivore interactions have been proposed to explain rapid population growth in invasive species. On one hand, the enemy release hypothesis (ERH) proposes that population size of exotic species will increase because they are no longer constrained by the herbivores in their native range which would normally limit population size expansion (Keane & Crawley, 2002). In contrast, herbivores may increase population size if they have a positive effect on plant fitness. This idea has been explored as the grazing optimization hypothesis (GOH), which suggests that plant productivity can increase with moderate levels of herbivory (de Mazancourt, Loreau, & Dieckmann, 2001). Another hypothesis, the evolution of improved competitive ability (EICA), suggests that invasive species have increased fitness because resources are reallocated from defense to growth and reproduction in response to reduced herbivory (Blossey & Nötzold, 1995; Joshi & Vrieling, 2005).

Whereas resistance through plant defense compounds acts primarily to deter specialist herbivores, tolerance enables a species to sustain herbivore damage from a broad range of herbivores (Müller-Schärer, Schaffner, & Steinger, 2004). Selection for tolerance to herbivory by means of an increase in leaf size, branching or tillering, increased growth rates and net photosynthetic rate, activation of dormant meristems, and changes in plant architecture and allocation patterns in response to herbivory (Strauss & Agrawal, 1999; van Kleunen & Schmid, 2003; Bossdorf, Schröder, Prati, & Harald, 2004) is expected to be an important evolutionary change leading to increased survival and establishment of exotic species. The degree to which plants tolerate herbivory is known as compensatory ability or compensation (Strauss & Agrawal, 1999).

Compensatory ability can be of value in estimating the relative invasive potential of ornamental cultivars before crops are commercially distributed (Belsky, 1986; Hochwender, Marquis, & Stowe, 2000). Many exotic species that have become invasive have been introduced as ornamentals (Reichard & White, 2001). Despite a lack of protocols to evaluate invasive potential during crop development and prior to market release of a new crop (Anderson & Gomez, 2004), cultivars are being advertised as less invasive than others based solely on performance in traditional breeder trials. Such trials are primarily designed to evaluate the suitability of cultivars to production regimes and, though they might provide information about life history traits, they are very limited in their ability to predict a crop's response to herbivores.

Breeder trials cannot provide information about response to herbivory because plants are routinely evaluated in cultivated habitats where herbivores are excluded. For example, *Cleome hassleriana* and *Polanisia dodecandra* (sold as *C. serrulata*) are annuals of the Capparaceae present in Northeastern US (Gleason & Cronquist, 1991) that are bred for ornamental purposes. *Cleome hassleriana* is native to southeastern Brazil and Argentina (Still, 1994), is commonly used as a garden ornamental plant and a cut flower (Nau, 1999), reseeds prolifically and has occasionally escaped from cultivation (Bailey, 1927; Kindscher, 1987; Gleason & Cronquist, 1991). *Polanisia dodecandra* is native to North America (Kindscher, 1987; Gleason & Cronquist, 1991). It grows in open prairies, sandy and rocky soils, open woodlands and disturbed sites (Whitson, 1991; Freeman, 1991). *Polanisia dodecandra* is distasteful to animals (Steffey, 1984). It produces glucosinolates that act as repellents against most herbivores, except those that have evolved tolerance to mustard-oils (Cronquist, 1988). *Cleome hassleriana* has spines along stems, petioles, and leaf mid-veins that provide mature plants with additional protection against herbivores. In non-cultivated habitats, however, mammals and invertebrates can cause removal of apical meristems, various degrees of defoliation, and seedling death despite these mechanisms that deter herbivores. In a preliminary study of growth and establishment of cleome cultivars in non-cultivated habitats significant mortality was observed in transplanted seedlings due to herbivory in prairies (94%) and roadside (76%) conditions but not in garden conditions (10%) (Gomez, unpublished data). This suggests that when cleomes are grown in gardens, they are rarely attacked by herbivores, but when grown in non-cultivated settings, herbivore damage, especially on seedlings, can dramatically reduce survival probabilities.

'Solo', on the other hand, is the product of a more recent domestication event. Its resemblance to the wild species *Polanisia dodecandra* suggests that the development of this cultivar is best described as a bottleneck of the indigenous species rather than a prolonged and intense selection for ornamental characteristics. Compensatory ability in *Cleome* cultivars is likely to contribute to invasive potential in non-cultivated environments. This study, using three types of herbivory experiments, aimed: (1) to determine whether *Cleome* cultivars exhibit tolerance to various patterns of simulated herbivory, (2) to compare the growth and reproduction response among *Cleome* cultivars in different patterns of herbivory, and (3) to determine the extent to which cultivars differ in their response to simulated herbivory. To our knowledge, this is the first study to compare tolerance to simulated herbivory among ornamental cultivars prior to naturalization in the context of assessing relative invasive potential.

2. Method

2.1 Study Species

Cultivars advertised as good alternatives for the more aggressive 'Queen' include 'Sparklers' and 'Solo'. The pedigree leading to the development and selection of Sparklers is proprietary information (T. Perkins, Goldsmith Seed Co., personal communication), but it is likely that they are derived from dwarf *Cleome* mutants, possibly a gibberellic acid (GA) mutant. In addition to their shorter stature relative to Queens, Sparklers have a greater percent of non-viable seeds than Queens. 'Solo' is advertised as a 'native' alternative to Queen. Despite its smaller flowers and unpleasant odor which make it less appealing as a garden plant (Whitson, 1991), 'Solo' is becoming more popular as more plant enthusiasts search for native annuals to incorporate into their gardens (Lee, 2000).

Four *C. hassleriana* cultivars ('Queen Rose'=QR, 'Queen White'=QW, 'Sparkler Rose'=SR, 'Sparkler White'=SW) and one *P. dodecandra* cultivar ('Solo'=S) were evaluated in this study (Table 1). Among the five cultivars studied are those that have naturalized outside cultivation and those that have been advertised as less aggressive alternatives. 'Queen' series cultivars have been marketed as ornamentals for almost a century as documented by references dating back to as early as 1912 in the Curtis' Botanical Magazine of London, the oldest current periodical devoted to garden plants (Bailey, 1927). Populations of 'Queen Pink' have been documented in the eastern half of the United States extending from as far west as Texas and ranging as far north into the upper Peninsula of Michigan (USDA Plants Database, 2005). Since colored flowers are most likely of dominant inheritance in cleomes, it is not possible to rule out the role of 'Queen White' in the origin of naturalized populations even though naturalized plants exhibit primarily, pink-colored flowers.

Table 1. *Cleome* cultivars and sources evaluated for tolerance to three different patterns of simulated herbivory (stem clipping, whole leaf defoliation, partial leaf defoliation) in greenhouse conditions

Cultivar	Abbreviation	Series[a]	*Cleome spp.*	Company
Solo	S	none[b]	*C. serrulata*[c]	Express Seeds
Sparkler Rose	SR	Sparkler	*C. hassleriana*	Ball Seed Co., Inc
Sparkler White	SW	Sparkler	*C. hassleriana*	Ball Seed Co., Inc
Queen Rose	QR	Queen	*C. hassleriana*	Ball Seed Co., Inc
Queen White	QW	Queen	*C. hassleriana*	Ball Seed Co., Inc

[a] A series is a group of cultivars that share most phenotypic traits of ornamental value but differ in flower color.

[b] 'Solo' is not grouped within a series because it is available only in one flower color.

[c] *C. serrulata* is the name under which this cultivar is being sold, even though based on anther number and viscid pubescence it is *Polanisia dodecandra*.

Seeds used in this study were purchased from Express Seed Co. (Oberlin, OH USA) and Ball Seed, Co. (West Chicago, IL USA). Seeds were sown on June 26, 2003 in the greenhouses of the University of Minnesota, St. Paul Campus (44° 59' 16" N, 93° 10' 54" W, St. Paul, MN) in 288-plug trays with germination mix (LP5 Germination Mix, Sun Gro Horticulture Sunshine, Pine Bluff, AR) and loosely covered with vermiculite (Medium Vermiculite Premium Grade, Sun Gro Horticulture Sunshine, Pine Bluff, AR). Trays were taken that same day to Wagner Greenhouses (44° 53' 37" N, 93° 18' 30" W, Minneapolis, MN) where seeds were germinated on greenhouse benches at approximately 26°C day /21°C night temperatures (Ball, 1965; Nau, 1999). Two weeks later, trays were brought back to the St. Paul Campus and placed in the greenhouses until August 18 when they were transplanted into 15.24 cm (6 inch) plastic pots filled with potting medium (SB300 Universal Professional Growing Mix, Sun Gro Horticulture Sunshine, Pine Bluff, AR). Plants were grown in the greenhouses of the University of Minnesota, St. Paul Campus for a month before being randomly assigned to one of the three simulated herbivory experiments described below.

2.2 Experimental Design

Three experiments differing in the pattern of simulated-herbivore damage were conducted, each of which were designed to answer the three aims of this paper (see above): (1) stem clipping simulated deer browsing by removing half of the main stem on treated plants including the apical and several lateral meristems; (2) whole leaf defoliation simulated rabbit herbivory by removing either all the main leaves or both main and axillary leaves at the base of the leaf blade where the leaf joins the petiole; (3) partial leaf defoliation simulates the effect of folivores that remove small portions of the leaf blade throughout the entire plant though allow damaged leaves to remain attached to the plant. Before initiating the experiments, the height of ten randomly selected plants from each cultivar population was recorded as an initial gauge of plant height differences and variation within and among cultivars.

Twenty plants of each cultivar were randomly selected for the stem clipping experiment. The distal half of the main stem was clipped off (Hester, Millard, Bailllie, & Wendler, 2004; Bergquist, Bergstrom, & Zakharenka, 2003) in ten randomly selected plants from each cultivar (reps). The remaining plants were untreated (control). All plants were set on a single bench in the greenhouse in a completely randomized design.

For the whole leaf defoliation experiment, twelve plants of each cultivar were randomly chosen. Four plants from each cultivar (reps) were randomly assigned to one of three treatments consisting of complete removal of all leaves from main and axillary shoots, removal of mature leaves on the main stem only, and a control treatment with no leaf removal. Leaves were removed by cutting off the petiole at the base of the leaf (Markkola, Kuikka, Rautio, Härmä, Roitto, & Tuomi, 2004). Plants were treated and distributed in a complete randomized block design across three greenhouse benches (blocks).

For the partial leaf defoliation experiment, four plants per cultivar (reps) were randomly assigned to each treatment level of leaf removal: low (L), 20% total leaf area; intermediate (I), 40% total leaf area in QR, QW and SR, 50% total leaf area in SW; and high (H), 80% total leaf area and one of three greenhouse benches (blocks) in a completely randomized block design. A calibrated leaf area meter was used to determine total leaf area in two plants per cultivar. Holes were punched using a 1/8" circular steel, paper punch to simulate three herbivory intensities. The number of holes punched per plant was determined for each cultivar based on the desired level of damage (L, I, H, control) and the average total leaf area per cultivar. Holes were haphazardly distributed on the leaf and plant, meaning that the holes were distributed throughout the plant avoiding damage to the leaf mid vein (Rogers & Siemann, 2003) and, in general, smaller leaves had fewer holes applied than larger leaves. 'Solo' was excluded from this experiment because leaflets were too small to be treated without being damaged completely.

Plants from all three experiments were grown simultaneously in the greenhouse under supplemental lighting (400 W high pressure sodium, high intensity discharge lamps, ~ 100-150 μmol^{-2} sec^{-1}, 0600-2200 HR) and fertilized weekly with a 20-20-20 N:P:K soluble fertilizer at 300 ppm N. One month after treatment, flowering phenology was scored for all plants. The scores ranged from 0 to 4 (0 = absence of flower structures of any kind, 1 = only flower buds present on plant, 2 = one or three open flowers present on plant, 3 = more than three open flowers, sometimes with more than one inflorescence with buds and a few open flowers, 4 = plants in full bloom, with more than one flowering inflorescences, sometimes with senescing flowers or seed pods).

Experiments were terminated 40 days after treatment. Axillary shoots were counted and recorded as flowering or non-flowering (flowering phenology). Inflorescences and vegetative tissues were separated, placed into paper bags, and dried in the oven at 75°C for at least four days until dry weights stabilized to obtain relative dry weights of reproductive and non-reproductive (vegetative) plant parts. Other traits measured, though not discussed in this paper, included number and length of axillary shoots, number of leaves, and two measurements of total height 1 month apart to estimate growth rate.

Before drying reproductive tissues, at least one, 1-cm long, ovary was harvested from every flowering plant and fixed in Farmer's solution (95% ethanol : glacial acetic acid solution, 3:1 v/v) (Chamberlain, 1932). This length was chosen because preliminary ovary dissections showed that ovules were well-developed and could be counted readily at this stage. Ovaries were transferred into 70% ethanol one week later and then into vials containing distilled, de-ionized water and stored in the lab, until they could be dissected. Gynoecia were measured from the base of the ovary to the tip of the stigma. They were dissected under a dissecting microscope, ovules were counted, and morphological abnormalities of the ovaries were recorded. The number of ovules per ovary was measured as a measure of reproductive output. It was not determined if the ovaries counted at this stage of development have been fertilized or not.

2.3 Statistical Analyses

Most traits were analyzed using Statistical Package for the Social Sciences (SPSS) version 11.0 unrestricted, univariate, fixed-effects, general linear model (SPSS Inc., 2001) with main factors and all possible interactions included in the model. Flowering phenology was analyzed using a two-way ANOVA and interaction by standard least squares (JMP IN 1989-2000). Post-hoc tests of Least Significant Difference (LSD) at α=0.05 were performed on initial plant heights. To determine differences among cultivars within the series of the same species, 'Solo' was dropped from the analysis and a model with cultivars nested within series was used. Ovary length was used as a covariate in the analysis of ovule counts because it differed across cultivars. Significant differences in the proportion of ovary abnormalities across treatments and cultivars were determined using a Likelihood Ratio test.

Prior to analysis, the data were tested for normality, heteroscedasticity, and outliers. Vegetative and reproductive biomasses in the stem clipping and whole leaf defoliation experiments were square-root transformed. Reciprocal square root transformations were used for total shoot number in all three experiments and flower shoot number in the stem clipping experiment. For number of total shoots, four outliers were removed from the whole leaf defoliation data. For the analyses of biomass, one outlier was removed from the whole leaf defoliation data. One outlier was dropped from the partial leaf defoliation data for the number of branches and four from the whole leaf defoliation data. This did not affect the overall outcome of the results. Back-transformed means and standard errors were used to make the graphs in the original scale, whenever a transformation was required for the analyses.

3. Results

Height of cultivars before treatment ranged from 12.7 cm in SR to 20.4 cm in QW (Figure 1). The average height for QW and QR plants before herbivory treatment was 20.4 and 18.1 cm, respectively. QW and QR plants were significantly taller than SR (12.7 cm), SW (14.5 cm), and S (13.2 cm) plants. Variation in height among individuals within cultivars (low S.E., Figure 1) was minor, however. Height before treatment might have had a residual effect on the response to simulated herbivory. However, size alone does not explain the differences in tolerance to herbivory since, as will be discussed below, differences among cultivars in tolerance to herbivory were not always consistent with initial plant height.

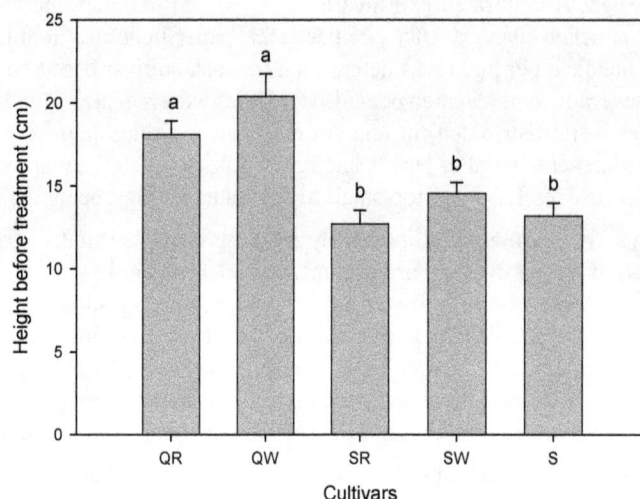

Figure 1. Mean (±S.E.) height (cm) of *Cleome hassleriana* and *Polanisia dodecandra* cultivars before simulated herbivory treatments. Height was measured on n = 10 randomly selected plants from each cultivar of *C. hassleriana* (QW = 'Queen White'; QR = 'Queen Rose'; SW = 'Sparkler White'; SR = 'Sparkler Rose'), and *P. dodecandra* (S = 'Solo'). Letters indicate groups with similar means according to Least Significant Difference post-hoc test at α = 0.05 level

3.1 Flowering Phenology

Stem clipping reduced the initial time to flowering in all cultivars (Figure 2). Herbivory treatment and cultivar had significant effects on flowering phenology (Table 2). Stem clipping reduced the average flowering phenology score to 0.8, compared to an average score of 2.4 in control plants. Among all cultivars, S had the lowest average score (0.6), suggesting a longer time to flowering. The difference in average scores between Sparkler cultivars (0.9) was greater than the difference between Queen cultivars (0.1) suggesting greater variation in time to flowering in Sparklers than in Queens.

Table 2. Significance (*P* values) from the two-way Analysis of Variance for flowering phenology in five *Cleome* cultivars (Queen Rose, Queen White, Sparkler Rose, Sparkler White, and Solo) in response to three simulated herbivory experiments (stem clipping, whole leaf defoliation, partial leaf defoliation)

Simulated herbivory	Factor[a]	df	*P* values[b]
Stem clipping	cv	4	<0.001 ***
	trt	1	<0.001 ***
	cv x trt	4	0.045 *
Whole leaf defoliation	cv	4	<0.001 ***
	trt	2	<0.001 ***
	blk	2	0.111 ns
	cv x trt	8	0.101 ns
Partial leaf defoliation	cv	4	<0.001 ***
	trt	3	0.855 ns
	blk	2	0.698 ns
	cv x trt	12	0.952 ns

[a] cv = cultivar, trt = treatment, blk = block.

[b] *, ***, ns = significant, highly significant, and not significant to α = 0.05.

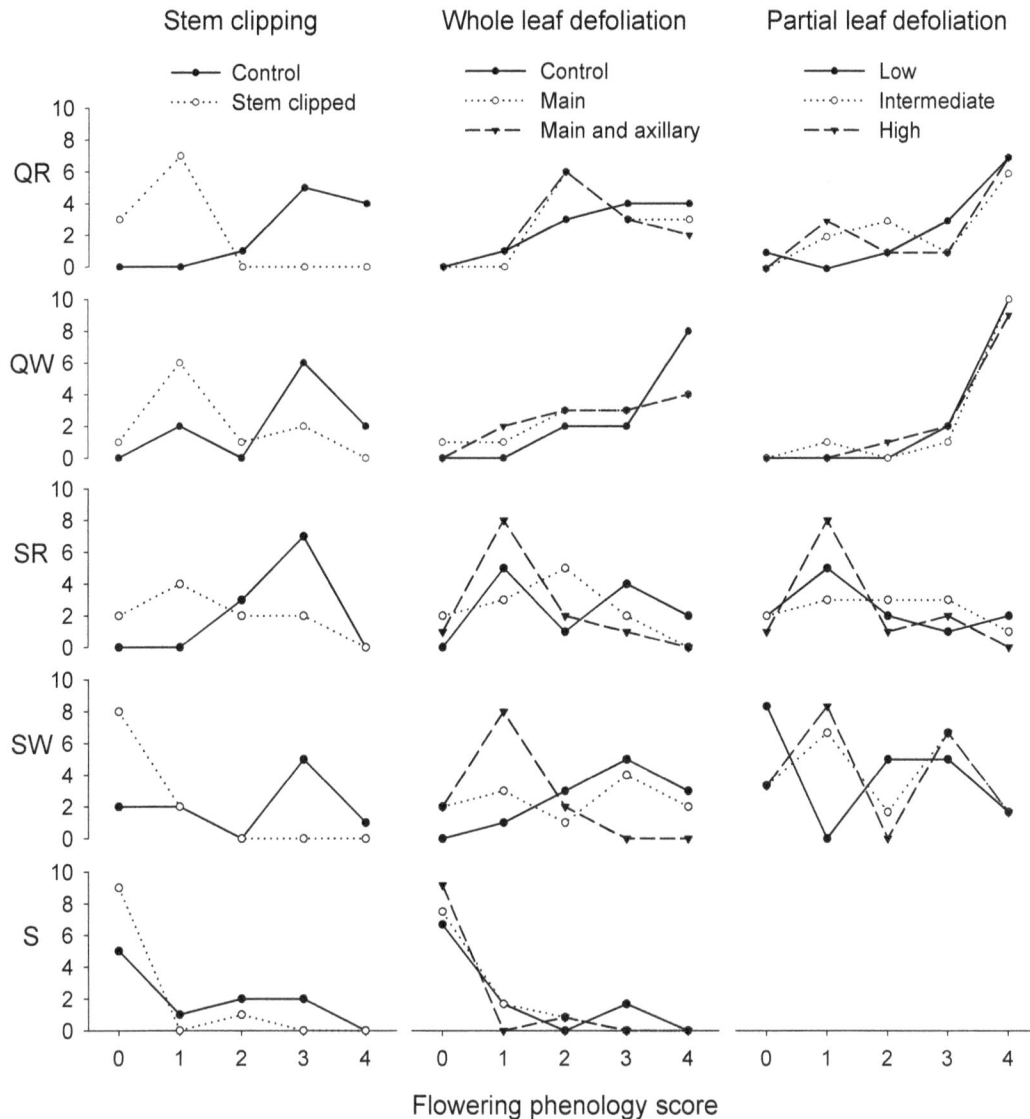

Figure 2. Frequency distribution of *Cleome* cultivars scored by flowering phenology (0-4 scores; see text). Ten plants per cultivar were scored in the stem clipping experiment, twelve in the whole and partial leaf defoliation experiments. Cultivars: QR = 'Queen Rose', QW = 'Queen White', SR = 'Sparkler Rose', SW = 'Sparkler White', S = 'Solo'. S was excluded from partial leaf defoliation experiment (see text)

There was also a significant interaction between cultivar and stem clipping treatment in flowering phenology score (Table 2). QR control plants scored the highest flowering phenology score, but were only the third highest among treated plants (Figure 3). QW and SR control plants scored similarly and responded equally to the stem clipping treatment, with an average reduction in flowering phenology score of 1.4. Moreover, though S was most delayed to initiate flowering (scoring the lowest flowering phenology scores), after treatment SW flowering was delayed equally when compared to S treated plants.

In the whole leaf defoliation experiment, cultivar and treatment had significant effects on time to flowering, but no significant effect of the interaction of cultivar and treatment was detected (Table 2). Queen cultivars had the highest scores for flowering phenology, meaning they initiated flowers the fastest (Figure 3). The cultivar S was the most delayed in flowering with the lowest score for flowering phenology (0.4). In regards to the response to simulated herbivory, time to flowering was delayed as the intensity of damage increased. From control to complete removal of main and axillary leaves, there was a 17.4% drop in the mean flowering phenology scores.

In the partial leaf defoliation experiment, time to flowering was significantly affected by cultivar but, despite the intensity of the simulated herbivory damage, no treatment effects were detected (Table 2). Nonetheless, least mean scores in flowering phenology scores were still greater in Queens than in Sparklers (Figure 3).

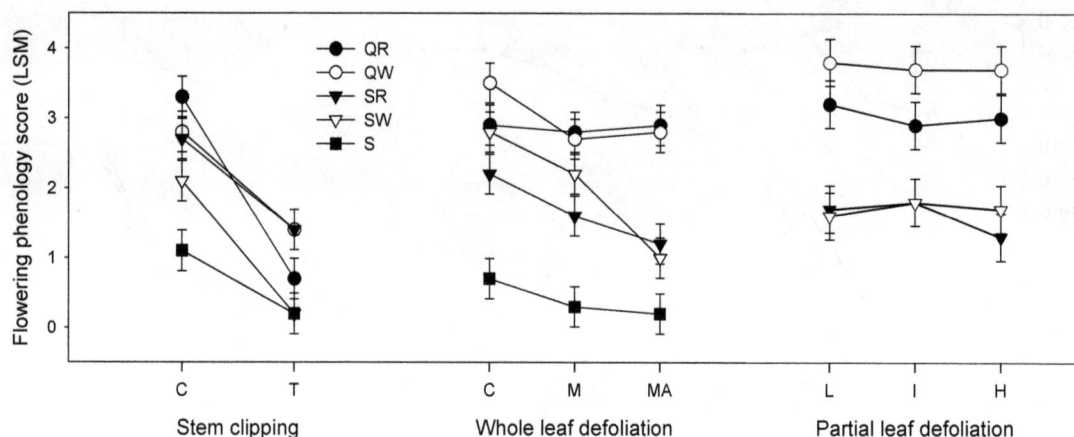

Figure 3. Least square mean plots (± S.E.) of flowering phenology scores of five *Cleome* cultivars (QR = Queen Rose, QW = Queen White, SR = Sparkler Rose, SW = Sparkler White, S = Solo) 30 days after stem clipping (C = control, T = treated; n = 10), whole leaf defoliation (C = control, M = main leaves removed, MA=main and axillary leaves removed) and partial leaf defoliation (L = low, I = intermediate, H = high levels of leaf area removed) simulated herbivory experiments

3.2 Biomass

In the stem clipping experiment, mean total biomass/plant was reduced from 19.7 g (control) to 14.3 g (treated). Since the herbivory simulation consisted of removing half of the plant, it would be expected that total biomass in treated plants would equal 50% of the total biomass of control plants. In this experiment, however, treated plants averaged 72.5%, instead of 50%, total biomass of control plants. Total biomass in treated plants ranged from 1.4 g in S to 21.3 g in QW. In control plants, it ranged from 6.9 in S to 27.2 in QR. All treated cultivars, except S, produced >50% total biomass of control plants. SW (80.9%) and QW (79.0%) produced the largest percent of total biomass relative to control plants, followed by SR (69.0%). On the other hand, S did not compensate after herbivory; the biomass of treated plants was only 20.8% of control plants.

Stem clipping significantly reduced average vegetative and reproductive biomass per plant (Table 3). Vegetative (Figure 4A) and reproductive (Figure 4B) biomass was, on average, less in treated plants than in control plants. Vegetative biomass averaged 15.2 g for control and 13.0 g for treated, whereas reproductive biomass was 4.4 g (control) and 1.3 g (treated). The greatest difference in vegetative biomass was observed in S with 4.8 g more (control) than in treated plants (Figure 4A). Interestingly, this cultivar had one of the smallest differences in biomass of inflorescence between control and treated plants, only 0.65 g more in control plants than in treated plants (Figure 4B). The greatest difference in reproductive biomass in treated and control plants was observed in QR (5.8 g), which had the smallest difference in vegetative biomass (0.8 g). This indicates that the effect of herbivory may be stronger in different traits depending on the cultivar. After simulated herbivory, growth is affected most dramatically in S, while reproduction is most affected in QR.

Table 3. Significance (P values) from the univariate, unrestricted, general linear model analyses for biomass and shoot number of five *Cleome* cultivars (Queen Rose, Queen White, Sparkler Rose, Sparkler White, and Solo) in response to three simulated herbivory experiments (stem clipping, whole leaf defoliation, partial leaf defoliation).

Simulated Herbivory	Factor[a]	df	Biomass[b]			Shoot number[c]		
			Veg[d]	Rep	Rep:Tot	Tot	Flow	Flow:Tot
Stem	cv	4	<0.001***	<0.001***	<0.001***	<0.001***	0.205ns	0.012*
clipping	trt	1	<0.001***	<0.001***	<0.001***	<0.001***	<0.001***	0.010*
	cv x trt	4	<0.001***	0.063ns	0.084ns	<0.001***	0.457ns	0.009***
Whole	cv	4	<0.001***	<0.001***	<0.001***	<0.001***	<0.001***	0.001***
leaf	trt	4	<0.001***	<0.001***	0.120ns	0.900ns	0.002***	0.011*
defoliation	blk	2	0.001***	0.073ns	0.987ns	<0.001***	0.003***	<0.001***
	cv x trt	2	0.354ns	0.248ns	0.248ns	0.040*	0.621ns	0.221ns
	cv x blk	8	0.312ns	0.021**	0.03**	0.226ns	0.581ns	0.784ns
	trt x blk	4	0.297ns	0.339ns	0.666ns	0.046**	0.111ns	0.113ns
	cv x trt x blk	16	0.935ns	0.622ns	0.237ns	0.262ns	0.448ns	0.685ns
Partial	cv	4	<0.001***	<0.001***	<0.001***	<0.001***	0.177ns	0.01*
leaf	trt	3	0.113ns	0.386ns	0.102ns	0.856ns	0.244ns	0.636ns
defoliation	blk	2	0.004***	0.855ns	0.160ns	0.978ns	0.430ns	0.655ns
	cv x trt	12	0.860ns	0.989ns	0.987ns	0.419ns	0.943ns	0.614ns
	cv x blk	8	0.804ns	0.662ns	0.698ns	0.954ns	0.912ns	0.976ns
	trt x blk	6	0.782ns	0.208ns	0.106ns	0.728ns	0.649ns	0.828ns
	cv x trt x blk	24	0.755ns	0.321ns	0.384ns	0.452ns	0.230ns	0.527ns

[a] cv = cultivar, trt = treatment, blk = block.

[b] Veg = vegetative biomass, Rep = reproductive biomass, Rep:Tot = reproductive effort, ratio of reproductive to total biomass.

[c] Tot = total number of axillary branches per plant, Flow = number of flowering axillary branches per plant, Flow:Tot = reproductive allocation, ratio of number of flowering branches to total number of branches per plant.

[d] *, ***, ns = significant, highly significant, and not significant to $\alpha = 0.05$.

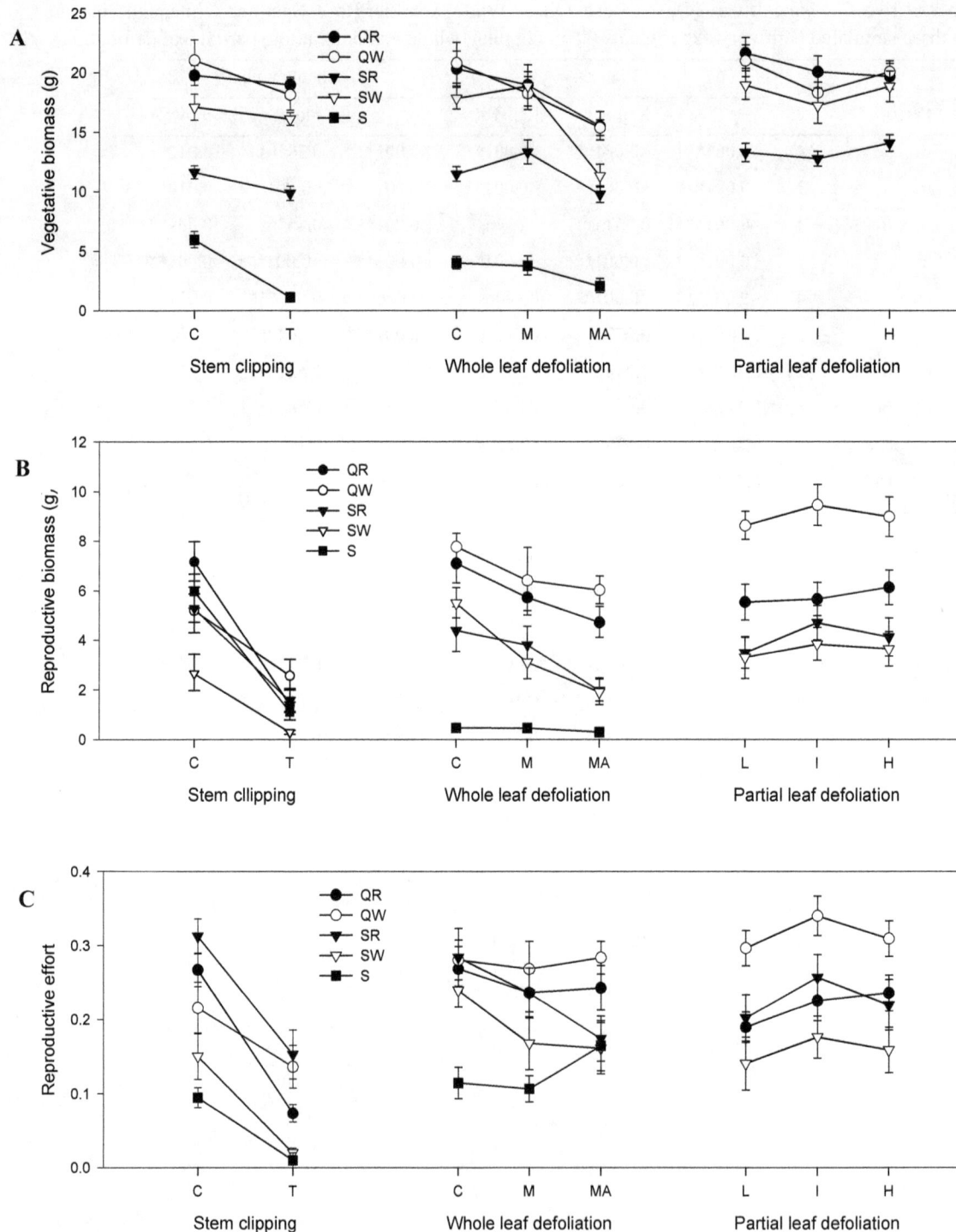

Figure 4. Mean (±S.E. bars) biomass (g) of (A) vegetative and (B) reproductive structures, (C) reproductive effort (ratio of inflorescence biomass to total biomass) in *Cleome* cultivars (QR = Queen Rose, QW = Queen White, SR = Sparkler Rose, SW = Sparkler White, S = Solo) in response to three simulated herbivory experiments (stem clipping, whole leaf defoliation, partial leaf defoliation). C = control, T = terminal half of the main stem clipped, M = main leaves removed, MA = both main and axillary leaves removed, L = low level damage (20% of leaf area removed), I = intermediate level damage (40-50% of leaf area removed), and H = high level damage (80% of leaf area removed). S was excluded from the partial leaf defoliation experiment (See text)

Reproductive effort, defined as the proportion of biomass allocated to inflorescences, averaged 21.0% (control) and 8.6% (treated) in the stem clipping experiment. A significant effect of herbivory x cultivar was not detected ($P = 0.084$; Table 3), suggesting a trend for cultivars to retain their relative ranks between the treatments. The overall tendency was a reduction in reproductive effort across all cultivars after stem clipping. Among control plants, SR allocated the most biomass to reproduction (31.7%), followed by QR (27.0%), QW (20.6%), and SW (15.3%) (Figure 4C). Among treated plants, the proportion allocated to reproductive effort was still high in SR (15.8%) and low in SW (2.1%), but QW (13.7%) allocated more to reproduction than QR (7.4%). S had the lowest reproductive effort with 10.0% biomass in control plants and only 4.2% in treated plants.

Total biomass in the whole defoliation experiment ranged from 2.3 g in S plants treated with the highest herbivory level to 28.5 g of control QW plants. QW and QR had the highest average total biomass across all treatments (24.5 g), followed by SW and SR (17.2 g) and S (3.7 g). Except for SR, the trend among cultivars was towards a decrease in total biomass as the intensity of herbivory increased. The average total biomass for control, intermediate, and high herbivory intensities was 19.9 g, 18.5 g, and 13.8 g, respectively. SR was the exception, producing the greatest biomass at the intermediate level of herbivory (17.1 g) rather than in the control (15.9 g) or highest level of herbivory treatment (11.7 g).

In the whole leaf defoliation experiment, vegetative biomass declined as the level of herbivory increased. The average vegetative biomass in control plants was 14.8 g. At intermediate herbivory levels, it averaged 14.6 g, whereas it significantly decreased to 10.8 g in plants at the highest herbivory level. The two exceptions to this trend were SR and SW, which had the highest vegetative biomass at intermediate levels of herbivory (Figure 4A).

The average reproductive biomass across all cultivars decreased from 5.0 to 3.9 g as whole leaf defoliation intensity increased. Unlike for total biomass and vegetative biomass, this trend was consistent among all cultivars. The relative ranks of cultivars in terms of their reproductive biomass is also similar to that observed for total biomass and vegetative biomass: QW and QR produced the most (6.3 g), followed by SW and SR (3.4 g) and S (0.4 g).

Average reproductive effort across all cultivars was highest for control plants (23.7%) and lowest for the intermediate whole leaf defoliation level (20.2%). Three distinct patterns of tolerance were observed among cultivars in terms of allocation of biomass to reproductive effort (Figure 4C). QR and QW showed no difference in reproductive effort across herbivory treatments, 24.9 % and 27.7 % biomass allocated to reproduction, respectively. SR and SW showed a decrease in reproductive effort as the intensity of herbivory increased. In SR, there was 38.7% reduction in reproductive allocation in plants treated with the maximum herbivory intensity relative to the control group. In SW, the reduction was 32.7%. However, in S there was a slight increase in reproductive allocation as herbivory intensity increased from 11.4% (control) to 16.4% in plants with both main and axillary leaves removed.

Total biomass in the partial leaf defoliation experiment ranged from 16.8 g in SR to 29.6 g in QW for the lowest intensity herbivory treatment. In general, there were no significant differences among treatments in total biomass produced. However, there was a slight increase in total biomass as intensity of treatment increase in SR. At 20% leaf area removal, total biomass was 16.8 g. At 40% and 80% leaf area removal, there was an increase of 0.7 g and 1.4 g, respectively. The other three cultivars showed a slight decrease in total biomass at intermediate levels of herbivory relative to the control group.

Like in the whole leaf defoliation experiment, no differences in vegetative and reproductive biomass were detected among partial leaf defoliation treatments. However, differences among cultivars were highly significant (Table 3). SR averaged the least vegetative biomass (13.4 g), followed by SW (18.3 g), QW (19.8 g) and QR (20.5 g) (Figure 4). In terms of reproductive biomass, SW produced the least (3.6 g), followed by SR (4.1 g), QR (5.8 g), and QW (9.0 g) (Figure 4B).

On average the highest reproductive effort in the partial leaf defoliation experiment was observed for the 40% leaf area removal treatment with an average of 20% biomass allocated to inflorescences across all cultivars (Figure 4), though significant treatment effects were not detected (Table 3). This suggests some degree of compensation at intermediate levels of herbivory. In addition, differences in reproductive effort across cultivars were highly significant (Table 3). SW allocated 15.8% of biomass to reproduction, whereas QW allocated 31.5%.

3.3 Shoots

Stem clipped plants produced significantly fewer axillary shoots than control plants (P < 0.001, Table 3; Figure 5A). Among control plants, SW produced the highest number of axillary shoots per plant (13), while SR produced the least (7). Among treated plants, S produced the least number of axillary shoots per plant (two), while the highest number of shoots was produced by QW (six) instead of SW (four). The shift in relative ranking of cultivars depending on the treatment is denoted by the highly significant effect of the cultivar by treatment interaction ($P < 0.001$, Table 3).

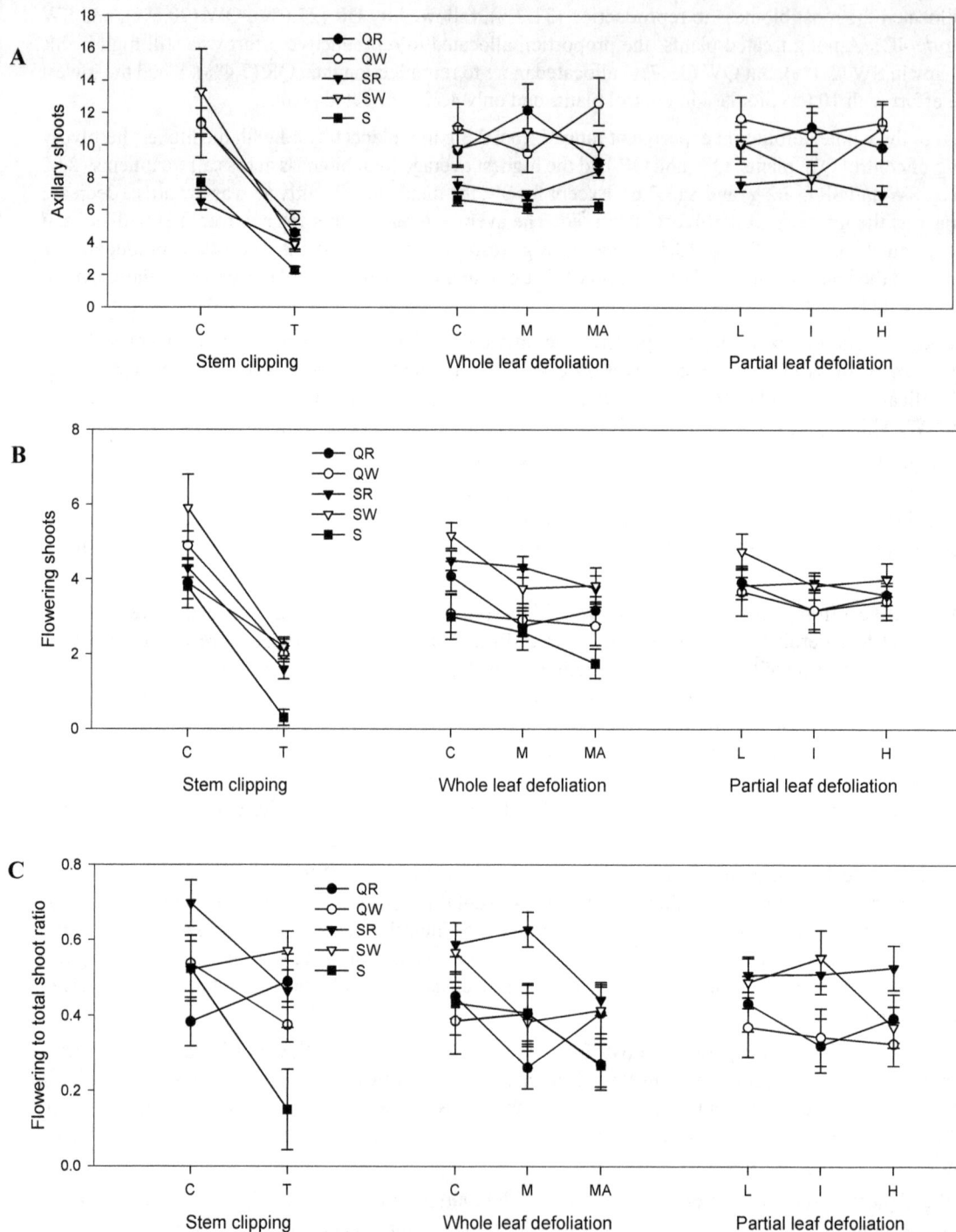

Figure 5. Mean (±S.E. bars) number of (A) axillary shoots, (B) flowering shoots, and (C) reproductive allocation (ratio of flowering branches to total number of branches) in *Cleome* cultivars (QR = Queen Rose, QW = Queen White, SR = Sparkler Rose, SW = Sparkler White, S = Solo) in response to three simulated herbivory experiments (stem clipping, whole leaf defoliation and partial leaf defoliation). C = control, T = terminal half of the main stem clipped, M = main leaves removed, MA = both main and axillary leaves removed, L= low level damage (20% of leaf area removed), I = intermediate level damage (40-50% of leaf area removed), and H = high level damage (80% of leaf area removed). S was excluded from the partial leaf defoliation experiment (See text)

The number of flowering shoots in the control plants was more than twice the number of shoots in stem clipped plants (Figure 5B). No significant cultivar or interaction effects were detected for number of flowering shoots (Table 3). Among control plants, SW produced the most flowering shoots (six), followed by QW (five), and QR and SR (four). Among treated plants, all cultivars but S, produced, on average, two flowering shoots after herbivory.

Since stem clipping produced on average a reduction in total number of shoots without a significant change in the number of flowering shoots, there was a significant effect of treatment on reproductive allocation, the proportion of flowering shoots to total number of shoots per plant ($P = 0.01$, Table 3). A highly significant interaction of cultivar and herbivory in reproductive allocation ($P = 0.009$; Table 3) confirms that reaction norms vary significantly depending on cultivar and that the cultivars' relative ranks are not maintained across treatments (Figure 5C). Furthermore, reproductive allocation was greater in treated plants than control plants of QR and SW (Figure 5C). In the other three cultivars, treated plants had lower reproductive allocation percentages than control plants.

There was no significant difference in the number of total shoots/plant (n = 9) produced across whole leaf defoliation levels. No differences among all three levels of damage where observed in S and SW, producing an average across treatments of 6 and 10 shoots/plant, respectively (Figure 5A). In QR and SW the trend was an increase in total number of shoots in the intermediate levels of herbivory, whereas for QW and SR there was a slight decrease relative to control and the high intensity treatment.

The number of flowering shoots in the intermediate and high intensity whole leaf defoliation levels was less than in the control group (Figure 5B). The average number of flowering shoots in the control group was four flowering shoots/plant, whereas the average number of flowering shoots for the intermediate and high levels of herbivory were three shoots/plant. SR and SW produced the greatest number of flowering shoots among all cultivars (n = 4), and S produced the least (n = 2).

In terms of reproductive allocation, the ratio of flowering:total shoots, cultivars ranged from 48.5% in the control group to 36.0% in the maximum intensity in the whole leaf defoliation level. QW and S responded very similarly with the lowest ratio (27.0%) allocated to reproduction at the highest level of herbivory. Relative to the other cultivars, SR allocated more to reproductive shoots with an average of 55.3% flowering shoots across all treatments and the largest amount (62.8%) allocated to reproduction in the intermediate herbivory level (Figure 5).

In the partial leaf defoliation experiment, no significant differences were detected among treatments in the total number of axillary shoots per plant, the number of flowering shoots and the proportion of flowering to total shoots (Table 3). Moreover, cultivar effects were not detected for number of flowering shoots per plant, but were significant for total number of shoots ($P < 0.001$) and the proportion of flowering shoots to total shoots ($P = 0.01$). Total number of shoots/plant ranged from 7 to 12. The cultivar SR had the least mean number of total shoots (n = 8), followed by SW and QR, each with 10, and QW (n = 11) (Figure 5A). Across all cultivars and treatments, the average number of flowering shoots (n = 4). In terms of the proportion of flowering shoots to total shoots, SR and SW had a higher proportion (49.3%) than QR and QW (36.4%). There was a slight increase in the proportion of flowering shoots to total shoots in SW, and a slight decrease in QR, at the intermediate level of herbivory.

3.4 Ovules

Cultivars and treatment significantly affected the number of ovules produced per flower in the stem clipping experiment (Table 4). Control plants of QR, QW, and SR produced the most ovules (n = 87) per flower (Figure 6). The effect of herbivory was least on SR, which only experienced a 10% reduction in the number of ovules per flower compared to control plants. Treated S plants had a 52% reduction, whereas QR and QW had an average of 41% reduction in ovule number. Missing counts of ovules for treated plants of SW are due to morphological abnormalities in the ovary. Morphological abnormalities in ovaries were significantly different across treatment ($P < 0.001$), with a greater proportion of abnormalities in treated plants (53%) versus control plants (13%). However, no significant effect of cultivar was detected in the proportion of reproductive abnormalities ($P = 0.144$).

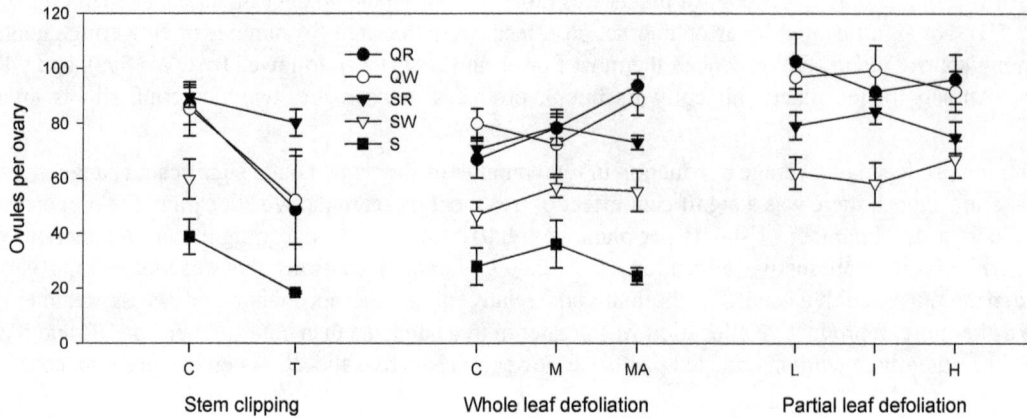

Figure 6. Mean (±S.E. bars) number of ovules per flower of five *Cleome* cultivars (QR = Queen Rose, QW = Queen White, SR = Sparkler Rose, SW = Sparkler White, S = Solo) in response to three different simulated herbivory experiments (stem clipping, whole leaf defoliation, and partial leaf defoliation). C = control, T = terminal half of the main stem clipped, M = main leaves removed, MA = main and axillary leaves removed, L = low level damage (20% of leaf area removed), I = intermediate level damage (40-50% of leaf area removed), and H = high level damage (80% of leaf area removed). 'Solo' was excluded from partial leaf defoliation experiment (See text)

On average, there was a suggestive trend towards number of ovules per flower increasing as the intensity of whole leaf defoliation increased ($P = 0.084$, Table 4). The average number of ovules for the control plants was: 68 ovules/flower, 75 ovules/flower in plants with only the main leaves removed, and 78 ovules/flower in plants with both main and axillary leaves removed. The strongest case of overcompensation is observed in QR with a 40% increase of ovules per flower in the high intensity treatment relative to the control group (Figure 6). Nonetheless, this trend was not the same for all cultivars. QW and QR produced the most ovules per flower (94 and 93 ovules/flower, respectively). SR produced on average 86 ovules per flower, whereas SW produced 62. S produced on average 34 ovules per flower.

Table 4. Significance (*P* values) from the analysis of variance for ovules and ovary lengths of five *Cleome* cultivars (Queen Rose, Queen White, Sparkler Rose, Sparkler White, and Solo) in response to three simulated herbivory experiments (stem clipping, whole leaf defoliation, and partial leaf defoliation)

Simulated herbivory	Factor[a]	Ovules	Ovary length[b]
Stem clipping	ovlen[c]	<0.001***	-
	cv	0.003***	0.031*
	trt	<0.001***	0.804ns
	cv x trt	0.024*	0.364ns
Whole leaf defoliation	ovlen	0.166ns	-
	cv	<0.001***	<0.001***
	trt	0.086ns	0.999ns
	cv x trt	0.192ns	0.511ns
Partial leaf defoliation	ovlen	0.126ns	-
	cv	<0.001***	<0.001***
	trt	0.78ns	0.442ns
	cv x trt	0.861ns	0.282ns

[a] ovlen = ovary length, cv = cultivar, trt = treatment.

[b] Ovary length as response variable, therefore no *P* value available for ovary length.

[c] *, ***, ns = significant, highly significant, and not significant to $\alpha = 0.05$.

As the intensity of partial leaf defoliation increased, there was a reduction in the number of ovules from 85 to 82 ovules/flower, though significance was not detected ($P = 0.78$, data not shown). However, highly significant differences were detected among cultivars for the number of ovules per flower. SW produced the fewest ovules per flower (n = 62), while QR produced the most (n = 97) (Figure 6). There is no information available from SW treated plants because the ovaries collected were severely damaged by fungi, preventing ovule count.

4. Discussion

Cleome cultivars may exhibit varying degrees of tolerance, from undercompensation to overcompensation, in response to simulated herbivory and that there is variation in the response depending on the trait measured and the pattern of simulated herbivory. Compensatory ability in response to herbivory is a topic of substantial debate (Belsky, 1986; Cox, 2004). Some researchers state that most plant species exhibit reduced growth and fertility after herbivore attacks (Crawley, 1983) and evidence supporting compensation is lacking (Belsky, 1986). On the other hand, more recent studies in a wider range of species and environmental conditions suggest that, at least under certain conditions, species can increase their reproductive output after herbivory (Paige, 1999; Juenger, Lennartsson & Tuomi, 2000; Stastny, Scharmer, & Elle, 2005; Sun, Ding, & Ren, 2009; Wise & Abrahamson, 2005). In cleomes, compensatory ability varied among cultivars and pattern of herbivory.

Cleome cultivars exhibited variable degrees of tolerance across the experiments (Figures 3-6). For example, though complete recovery of removed biomass was not attained during the length of the stem clipping experiment, treated cultivars gained > 50% biomass by the end of the experiment suggesting a tendency to tolerate herbivory (Figure 4). Moreover, despite removal of up to 80% leaf area in the partial leaf defoliation, no treatment effects were detected in flowering phenology, biomass, shoot number or ovules/flower indicating compensation after herbivory in all these traits (Tables 2-5, Figures 3-6). These results are consistent with other studies that show compensation in invasive species. Schierenbeck, Mack, & Sharitz (1994) found that, in the presence of herbivores, the alien species *Lonicera japonica* produced greater biomass than its native counterpart *L. sempervirens*. Rogers and Siemann (2004) reported that, unlike natives, invasive ecotypes of Chinese tallow tree (*Sapium sebiferum*) compensate for root damage. However, there are also studies that do not find differences in compensation by invasive and non-invasive species. For example, no differences in tolerance to herbivory were detected between American and European populations of *Solidago canadensis* (Van Kleunen & Schmid, 2003).

Not only were highly significant differences among cultivars detected in most traits across all three herbivory experiments, but also more importantly, we observed variability among cultivars of the same series (Table 4). For example, SW produced greater vegetative biomass than SR after stem clipping and partial leaf defoliation and QR produced less reproductive biomass than QW in stem clipping and partial leaf defoliation experiments (Figure 4). In some instances, the difference between cultivars of the same series was greater than the difference between the series. For example, the proportion of biomass allocated to reproductive structures in clipped plants was greater in QW and SR than in QR and SW, whereas in the partial leaf defoliation experiment, QR allocated as much biomass to reproductive structures as SR and SW (Figure 4). Our results are consistent with previous findings that report variability in compensatory ability and invasive potential among closely related species. In a study of two closely related *Conyza* species, Thébaud et al. (1996) reported higher reproductive effort in *Conyza canadensis* than in *C. sumatrensis*, but stronger inflorescence stems and longer reproductive life-span in *C. sumatrensis* increased seed set in a wider range of environments. Gerlach and Rice (2003) observed that the greater invasive potential of *Centaurea solstitialis* relative to *C. sulphurea* and *C. melitensis* was due to this species' increase in the number of flower heads in response to stem clipping. These small physiological and morphological differences can result in large differences in invasiveness (Thébaud, Finzi, Affre, Debussche, & Escarre, 1996; Gerlach & Rice, 2003; Soti & Volin, 2010).

Variation in tolerance to herbivory, as observed among *Cleome* cultivars, is expected because patterns of plant response to herbivory depend on heritable, morphological, species-specific characters (Li, Shibuya, Yogo, & Hara, 2004). Genetic variation in tolerance to herbivory among natural populations has been documented in several species including *Gentianella campestris* (Juenger et al., 2000) and *Ipomoea purpurea* (Fineblum & Rausher, 1995). It has also been documented for natural populations of invasive species *Spartina alternifolia* (Garcia-Rossi, Rank, & Strong, 2003) and *Alliaria petiolata* (Bossdorf et al., 2004). Nonetheless, genetic variation in herbivore tolerance is often ignored in development of invasive species control. For example, the variation in tolerance and palatability of invasive species can have significant implications for the effectiveness of biological control agents (Bossdorf et al., 2004). This is because highly tolerant species can support large populations of biological control agents without a negative impact in their population size (Müller-Schärer et al., 2004; Garcia-Rossi et al., 2003).

In addition to genetic variation among populations, differing patterns of herbivory in *Cleome* can result in completely different responses (Figures 2-6). Herbivore treatment effects were significant for most of the traits in the stem clipping and whole leaf defoliation experiments, but treatment effects were not detected in the partial leaf defoliation experiment (Table 3). The most dramatic and negative effect was observed on reproductive biomass in the stem clipping experiment resulting in undercompensation with less biomass allocated to reproduction in treated plants than in controls (Figure 4). The decline in biomass is less dramatic in the whole leaf defoliation experiment than in the stem clipping experiment. In the whole leaf defoliation experiment, for example, treated S plants produced as much biomass as controls, whereas those in the stem clipping experiment had significantly less biomass (Figure 4). The least effect on biomass was observed in plants in the partial leaf defoliation experiment. In this experiment, there was no detectable treatment effect across variables measured, though there was a trend towards overcompensation through an increase in biomass allocated to reproduction. Similar trends were observed in the number of flowering shoots/plant where stem clipping had more dramatic and detrimental effects than whole leaf and partial leaf defoliation experiments (Figure 5).

Differences in response to various patterns of herbivory treatment were consistent with findings from other studies. In a simulated herbivory experiment on *Alliaria petiolata*, apical meristem damage and leaf herbivory (folivory) result in different magnitudes of growth and reproduction response (Bossdorf et al., 2004). Moreover, a combination of herbivore treatments may be necessary to describe more accurately the response to herbivory of a species in a new range. For example, in *Solidago canadensis*, a combination of clipping and jasmonic acid treatments better simulates natural response to herbivory damage than either treatment alone (van Kleunen, Ramponi, & Schmid, 2004). In *Lonicera japonica*, maximum compensation occurred only after exposure to herbivory by both insects and mammals (Schierenbeck et al., 1994).

Cultivar x treatment interactions were more striking in the stem clipping experiment than in the whole leaf defoliation experiments and no significant interaction was detected in the partial leaf defoliation experiment (Table 3). In the stem clipping experiment, total biomass allocated to reproduction was less in treated plants (Figure 4), but the effect of treatment varied across cultivars being greatest in S and less severe in QW. In the same experiment, QR and SR overcompensated by producing a greater proportion of flowering shoots compared to controls (Figure 5). This could be the result of short axillary shoots producing flowers sooner in QR and SR than in the other cultivars. If the shoots contributing to the proportion of flowering : total number of shoots are small enough, they may be of significant impact on fitness despite contributing little to total reproductive biomass. In the whole leaf defoliation experiment, there were trends suggesting differences among cultivars in the magnitude of the herbivory effect. For example, there was a greater decline in the biomass allocated to reproduction in SR and SW than in QR and QW, but an increase in S (Figure 4). There was also a slight increase in the vegetative biomass of SR and SW treated with the low intensity relative to the high intensity level of whole leaf defoliation, but this pattern was not observed in QW, QR or S. An interaction between cultivars x herbivory treatments was less detectable in the partial leaf defoliation experiment where vegetative and reproductive biomass varied little across treatments. Interestingly, plants tended to increase biomass allocation to reproduction with herbivory; QW, SR and SW maxed at intermediate levels of herbivory, whereas QR maxed at high levels of herbivory (Figure 4).

Cultivar and herbivory also affected the number of ovules produced/flower (Table 4). Even after including ovary length in the statistical model, strong cultivar effects on the number of ovules/ovary were detected across all experiments. Herbivory effects on ovule counts were highly significant in the stem clipping experiment, marginally significant in the whole leaf defoliation experiment, and non-significant in the partial leaf defoliation experiment suggesting a range of tolerance from undercompensation to compensation. Clipped plants produced significantly fewer ovules/flower than controls, but the magnitude of the effect differed significantly across cultivars; treated SR plants produced as many ovules per ovary as controls, whereas QR, QW and S produced fewer ovules in treated plants (Figure 6). Cultivar by herbivory effects were non-significant in the whole leaf and partial leaf defoliation experiments, but a trend towards overcompensation through an increase in number of ovules/flower was detected in plants in the whole leaf defoliation experiment especially in QR. These results contrast with studies that focus specifically on male and female fitness components. For example, studies in Ipomopsisaggregata have detected overcompensation in total fitness when estimates of the effect of herbivory include male fitness components (Gronemeyer, Dilger, Bouzat, & Page, 1997), but they have not detected overcompensation in female fitness components of total fruit or seed set (Paige, 1999). In a different species, *Ipomopsis arizonica*, reproductive success after herbivory was due solely to male fitness component instead of female fitness components (Paige, Williams, & Hickox, 2001).

5. Conclusion

Reallocation of resources to growth or reproduction after herbivory can be advantageous for species colonizing new areas (Blundell & Peart, 2001; Erneberg, 1999; Rogers & Siemann, 2004; Sun et al., 2009). However, establishing the connection between response to herbivory and invasive potential is extremely difficult. In some cases, herbivory can be a strong barrier limiting plant invasions (Lambrinos, 2002; D'Antonio, 1993), but in other cases it may have minor impact on plant establishment (Thébaud et al., 1996). Moreover, herbivory is highly variable across year and environments, and greatly dependent on community type (Lambrinos, 2002; Thébaud et al., 1996). Many factors including deer paths, weather conditions and the availability of other plants will determine whether *Cleome* is attacked by herbivores (City of Lakeway, TX, 2005). Therefore, protocols to assess invasive potential of ornamental crops in terms of tolerance to herbivory should start with pilot studies that compare simulated and natural herbivory patterns, determine the most important traits associated with tolerance, and correlate greenhouse with field response to herbivory.

Development of adequate protocols to evaluate the invasive potential of ornamental crops before market is advisable because various patterns of herbivory may result in different responses. It is expected that feeding modes of different herbivores will determine which traits are most important in determining tolerance to herbivory (Strauss & Agrawal, 1999). Further experiments with these species could focus on changes in chemical (defense compounds) and morphological barriers to herbivory (plant pubescence, spines). Experiments designed to assess tolerance to herbivory in cultivars developed by ornamental breeding programs may need to include different patterns of simulated and natural herbivory (Tiffin & Inouye, 2000; Lehtilä, 2003; Inouye & Tiffin, 2003) to provide thorough information on how cultivars respond to different types of herbivores encountered in natural conditions. Though simulated experiments are practical (Hjältén, 2004), especially in the context of breeding programs trying to evaluate invasive potential of ornamental crops, natural herbivory experiments may be needed to understand the complex ecological processes and biotic interactions leading to species invasions.

The difference in tolerance observed among cleome cultivars of the same series and species suggest the importance of evaluating each cultivar independently to assess its response to herbivory. In general, cultivars within a series are expected to share similar phenotypes except for flower color. This study shows that this assumption deserves re-examination for invasive potential because cultivars within a series are not genetically homogeneous. It may be necessary to evaluate all cultivars within the series to determine whether an entire series could be marketed as less invasive than another.

In addition to the different effects caused by type of herbivory and cultivars, the ability to detect tolerance to herbivory and compensation is determined by other factors including the traits examined and the conditions in which plants are grown. We focused on biomass allocation, branching and ovule production in our study, but other traits, such as time to flowering, fruit and seed production, and paternal fitness could be studied to determine more thoroughly how cleomes respond to herbivory. We also performed our studies in the greenhouse with sufficient lighting, nutrients, water and minimal competition among plants. All these factors may have an effect on the levels of tolerance observed in our study because resource levels can also affect tolerance to herbivory (Wise & Abrahamson, 2005). Moreover, rapid evolutionary change in tolerance to herbivores in species undergoing range expansion (Müller-Schärer et al., 2004) will surely complicate our ability to predict which cultivars could become invasive.

To the best of our knowledge, this is the first study that focuses on the differences in tolerance to herbivory of cultivars of a commercial ornamental crop prior to naturalization with reference to potential implications for plant breeding programs and the efforts to minimize the distribution of invasive species of ornamental origin. Our study documents variation in tolerance to herbivory among cultivars and shows that compensatory ability varies with type of herbivory and the cultivar affected. This variation could be further explored in plant breeding programs that aim to develop species with reduced invasive potential, but much more research would be needed to directly correlate the impact of compensation on invasive potential. Development and implementation of protocols to evaluate invasive potential of crops, including tolerance to herbivory, before marketing a cultivar as less invasive could be advantageous to controlling invasive species of ornamental value.

Acknowledgments

This manuscript publication is Scientific Journal Series No. 0612190163 of the Department of Horticultural Science, University of Minnesota; research was funded by the Minnesota Agricultural Experiment Station. The authors thank Tami Van Gaal at Wagner's Greenhouse, Minneapolis, Minnesota for technical assistance.

References

Bailey, L. H. (1927). *The standard cyclopedia of horticulture* (Vol. 1: A-E). New York, NY: The Macmillan Company.

Ball, V. (1965). *The Ball red book* (11th ed.). Batavia, Illinois: Ball Publishing.

Belsky, A. J. (1986). Does herbivory benefit plants? A review of the evidence. *Amer. Nat., 127*, 870-892. http://dx.doi.org/10.1086/284531

Bergquist, J., Bergstrom, R., & Zakharenka, A. (2003). Responses of young Norway spruce (*Piceaabies*) to winter browsing by roe deer (*Capreoluscapreolus*): Effects on height growth and stem morphology. *Scand. J. Forest Res., 18*, 368-376. http://dx.doi.org/10.1080/0282758031005431

Bigger, D. S., & Marvier, M. A. (1989). How different would a world without herbivory be? A search for generality in ecology.*Integrative Biol., 1*, 60-67. http://dx.doi.org/10.1002/(SICI)1520-6602(1998)1:2%3C60::AID-INBI4%3E3.0.CO;2-Z

Blossey, B., & Nötzold, R. (1995). Evolution of increased competitive ability in invasive nonindigenous plants: a hypothesis. *J. Ecol., 83*, 887-889. http://dx.doi.org/10.2307/2261425

Blundell, A. G., & Peart, D. R. (2001). Growth strategies of a shade-tolerant tropical tree: the interactive effect of canopy gaps and simulated herbivory. *J. Ecol., 89*, 608-615. http://dx.doi.org/10.1046/j.0022-0477.2001.00581.x

Bossdorf, O., Schröder, S., Prati, D., & Harald, A. (2004). Palatability and tolerance to simulated herbivory in native and introduced populations of *Alliariapetiolata* (Brassicaceae). *Amer. J. Botany, 91*, 856-862. http://dx.doi.org/10.3732/ajb.91.6.856

Chamberlain, C. J. (1932). Methods in plant histology (5th Rev. ed.). Chicago, Illinois: The University of Chicago Press.

City of Lakeway, TX. (2005). *Deer resistant plants; Public communications*. Retrieved from http://cityoflakeway.com/about.asp

Crawley, M. J. (1983). *Herbivory. The dynamics of animal plant interactions*. Oxford: Blackwell. http://dx.doi.org/10.1146/annurev.en.34.010189.002531

Crawley, M. J. (1989). Insect herbivores and plant population dynamics.*Ann. Rev. Entomology, 34*, 531-564.

Cronquist, A. (1988). *The evolution and classification of flowering plants* (2nd ed.). Brooklyn: New York Botanical Garden.

D'Antonio, C. M. (1993). Mechanisms controlling invasion of coastal plant communities by the alien succulent *Carpobrotusedulis*. *Ecol., 74*, 83-95. http://dx.doi.org/10.2307/1939503

de Mazancourt, C., Loreau, M., & Dieckmann, U. (2001). Can the evolution of plant defense lead to plant-herbivore mutualism? *Amer. Naturalist, 158*, 109-123. http://dx.doi.org/10.1086/321306

Erneberg, M. (1999). Effects of herbivory and competition on an introduced plant in decline. *Oecologia, 118*, 203-209. http://dx.doi.org/10.1007/s004420050719

Garcia-Rossi, D., Rank, N., & Strong, D. R. (2003). Potential for self-defeating biological control?Variation in herbivore vulnerability among invasive *Spartina* genotypes. *Ecol. Appl., 13*, 1640-1649. http://dx.doi.org/10.1890/01-5301

Gerlach, J. D., & Rice, K. J. (2003). Testing life history correlates of invasiveness using congeneric plant species. *Ecol. Appl., 13*, 167-179. http://dx.doi.org/10.1890/1051-0761(2003)013%5B0167:TLHCOI%5D2.0.CO;2

Gleason, H. A., & Cronquist, A. (1991). *Manual of vascular plants of northeastern United States and adjacent Canada* (2nd ed.). Brooklyn: New York Botanical Garden.

Gronemeyer, P. A., Dilger, B. J., Bouzat, J. L., & Paige, K. N. (1997). The effects of herbivory on paternal fitness in scarlet gilia: better moms also make better pops. *Amer. Naturalist, 150*, 592-602. http://dx.doi.org/10.1086/286083

Hester, A. J., Millard, P., Baillie, G. J., & Wendler, R. (2004). How does the timing of browsing affect above- and below-ground growth of *Betulapendula*, *Pinussylvestris* and *Sorbusaucuparia*? *Oikos, 105*, 536-550. http://dx.doi.org/10.1111/j.0030-1299.2004.12605.x

Hjältén, J. (2004). Simulating herbivory: problems and possibilities. *Ecol. Studies, 173*, 243-255. http://dx.doi.org/10.1007/978-3-540-74004-9_12

Hochwender, C. G., Marquis, R. J., &Stowe, K. A. (2000). The potential for and constraints on the evolution of compensatory ability in *Asclepias syriaca*. *Oecologia, 122*, 361-370. http://dx.doi.org/10.1007/s004420050042

Inouye, B. D., & Tiffin, P. (2003). Measuring tolerance to herbivory with natural or imposed damage: a reply to Lehtilä. *Evolution, 57*, 681-682.

JMP IN. (1989-2000). Version 4. Cary, North Carolina: SAS Institute, Inc.

Joshi, J., & Vrieling, K. (2005). The enemy release and EICA hypothesis revisited: incorporating the fundamental difference between specialist and generalist herbivores. *Ecol. Letters, 8*, 704-714. http://dx.doi.org/10.1111/j.1461-0248.2005.00769.x

Juenger, T., Lennartsson, T., & Tuomi, J. (2000). The evolution of tolerance to damage in *Gentianellacampestris*: natural selection and the qualitative genetics of tolerance. *Evol. Ecol., 14*, 393-419. http://dx.doi.org/10.1023/A:1010908800609

Keane, R. M., & Crawley, M. J. (2002). Exotic plant invasions and the enemy release hypothesis. *Trends in Ecol. and Evol., 17*, 164-170. http://dx.doi.org/10.1016/S0169-5347(02)02499-0

Kindscher, K. (1987). *Edible wild plants of the prairie: An ethnobotanicalguide.* Lawrence, Kansas: University Press of Kansas.

Lambrinos, J. G. (2002). The variable invasive success of *Cortaderia species* in a complex landscape. *Ecol., 83*, 518-529. http://dx.doi.org/10.1890/0012-9658(2002)083%5B0518:TVISOC%5D2.0.CO;2

Li, B., Shibuya, T., Yogo, Y., & Hara, T. (2004). Effects of ramet clipping and nutrient availability on growth and biomass allocation of yellow nutsedge. *Ecol. Res., 19*, 603-612. http://dx.doi.org/10.1111/j.1440-1703.2004.00685.x

Markkola, A., Kuikka, K., Rautio, P., Härmä, E., Roitto, M., & Tuomi, J. (2004). Defoliation increases carbon limitation in ectomycorrhizal symbiosis of *Betulapubescens*. *Oecologia, 140*, 234-240. http://dx.doi.org/10.1007/s00442-004-1587-2

Müller-Schärer, H., Schaffner, U., & Steinger, T. (2004). Evolution in invasive plants: implications for biological control. *Trends in Ecol. and Evol., 19*, 417-422. http://dx.doi.org/10.1016/j.tree.2004.05.010

Nau, J. (1999). *Ball culture guide.The encyclopedia of seed germination* (3rd ed.). Batavia, Illinois: Ball Publishing.

Paige, K. (1999). Regrowth following ungulate herbivory in *Ipomopsisaggregata*: geographic evidence for overcompensation. *Oecologia, 118*, 316-323. http://dx.doi.org/10.1007/s004420050732

Paige, K., Williams, B., & Hickox, T. (2001). Overcompensation through the paternal component of fitness in *Ipomopsisarizonica*. *Oecologia, 128*, 72-76. http://dx.doi.org/10.1007/s004420100647

Reichard, S. H., & White, P. (2001).Horticulture as a pathway of invasive plant introductions in the United States.*BioScience, 51*, 103-113. http://dx.doi.org/10.1641/0006-3568(2001)051%5B0103:HAAPOI%5D2.0.CO;2

Rogers, W. E., & Siemann, E. (2003). Effects of simulated herbivory and resources on Chinese tallow tree (*Sapiumsebiferum*, Euphorbiaceae) invasive of native coastal prairie.*Amer. J. Bot., 90*, 243-249. http://dx.doi.org/10.3732/ajb.90.2.243

Rogers, W. E., & Siemann, E. (2004). Invasive ecotypes tolerate herbivory more effectively than native ecotypes of the Chinese tallow tree. *Sapiumsebiferum*. *J. Appl. Ecol., 41*, 561-570. http://dx.doi.org/10.1111/j.0021-8901.2004.00914.x

Schierenbeck, K. A., Mack, R., & Sharitz, R. R. (1994). Effects of herbivory on growth and biomass allocation in native and introduced species of *Lonicera*.*Ecology, 75*, 1661-1672. http://dx.doi.org/10.2307/1939626

Soti, P. G., & Volin, J. C. (2010). Does water hyacinth (*Eichhorniacrassipes*) compensate for simulated defoliation? Implications for effective biocontrol. *Biol. Control, 54*, 35-40. http://dx.doi.org/10.1016/j.biocontrol.2010.01.008

SPSS Inc. (2001). *SPSS for Windows.*Version 11. Chicago, Ilinois: SPSS Inc.

Stastny, M., Schaffner, U., & Elle, E. (2005). Do vigour of introduced populations and escape from specialist herbivores contribute to invasiveness? *J. Ecol., 93*, 27-37. http://dx.doi.org/10.1111/j.1365-2745. 2004.00962.x

Steffey, J. (1984). Strange relatives: The Caper family. *Amer. Horticulturalist, 63*, 4-7.

Still, S. (1994). *Manual of herbaceous ornamental plants* (4th ed.). Champaign, Illinois: Stipes Publishing Company.

Strauss, S. Y., & Agrawal, A. A. (1999). The ecology and evolution of plant tolerance to herbivory. *Trends in Ecol. and Evol., 14*, 179-185. http://dx.doi.org/10.1016/S0169-5347(98)01576-6

Sun, Y., Ding, J., & Ren, M. (2009). Effects of simulated herbivory and resource availability on the invasive plant, *Alternantheraphiloxeroides* in different habitats. *Biol. Control, 48*, 287-293. http://dx.doi.org/10.1016/j.biocontrol.2008.12.002

Thébaud, C., Finzi, A. C., Affre, L., Debussche, M., & Escarre, J. (1996). Assessing why two introduced *Conyza* differ in their ability to invade Mediterranean old fields. *Ecol., 77*, 791-804. http://dx.doi.org/10.2307/2265502

Tiffin, P., & Inouye, B. D. (2000). Measuring tolerance to herbivory: accuracy and precision of estimates made using natural versus imposed damage. *Evol., 54*, 1024-1029. http://dx.doi.org/10.1111/j.0014-3820.2000. tb00101.x

Tiffin, P., & Rausher, M. D. (1999). Genetic constraints and selection acting on tolerance to herbivory in the common morning glory, *Ipomoea purpurea. Amer. Naturalist, 154*, 700-716. http://dx.doi.org/10.1086/303271

U.S. Department of Agriculture.Plants Database. (2005). *Cleome hassleriana.* Retrieved from http://plants.usda.gov/

vanKleunen, M., & Schmid, B. (2003). No evidence for an evolutionary increased competitive ability in an invasive plant. *Ecol., 84*, 2816-2823. http://dx.doi.org/10.1890/02-0494

vanKleunen, M., Ramponi, G., & Schmid, B. (2004). Effects of herbivory simulated by clipping and jasmonic acid on *Solidago canadensis. Basic and Appl. Ecol., 5*, 173-181. http://dx.doi.org/10.1078/1439-1791-00225

Whitson, T. D. (Ed). (1991). *Weeds of the west.* Las Cruces, New Mexico: The Western Society of Weed Science in Cooperation with the Western United States Land Grant Universities.

Wise, M. J., & Abrahamson W. G. (2005). Beyond the compensatory continuum: environmental resource levels and plant tolerance of herbivory. *Oikos, 109*, 417-428. http://dx.doi.org/10.1111/j.0030-1299.2005.13878.x

6

Impact of New Natural Biostimulants on Increasing Synthesis in Plant Cells of Small Regulatory si/miRNA With High Anti-Nematodic Activity

Victoria Anatolyivna Tsygankova[1], Galyna Alexandrovna Iutynska[2], Anatoliy Pavlovych Galkin[3] & Yaroslav Borisovych Blume[3]

[1] Department Cell Signal System, Institute of Bioorganic Chemistry and Petrochemistry, National Academy of Sciences of Ukraine, Ukraine

[2] Department of General and Soil Microbiology, Zabolotny Institute of Microbiology and Virology, National Academy of Sciences of Ukraine, Ukraine

[3] Department Genomics and Molecular Biotechnology, Institute of Food Biotechnology and Genomics, National Academy of Sciences of Ukraine, Ukraine

Correspondence: Victoria Anatolyivna Tsygankova, Department Cell Signal System, Institute of Bioorganic Chemistry and Petrochemistry, National Academy of Sciences of Ukraine, Kyiv, 02094, Murmanska str., 1, Ukraine. E-mail: vTsygankova@ukr.net

Abstract

Plant endoparasitic cyst nematode *Heterodera schachtii* Schmidt, gallic nematode *Meloidogyne incognita* and stem nematode *Ditylenchus destructor* damage various agricultural crops. The application of ecologically safe natural biostimulants with bioprotective properties is a newer approach for increasing plant resistance to parasitic nematodes. The molecular-genetic analysis of biostimulants action on plant genome is necessary for creation of new effective bioregulators for plant protection against phytopathogenic organisms. In our field and greenhouse experiments, we investigated the influence of new natural biostimulants Avercom and its derivatives on plant protection against nematodes *Meloidogyne incognita* and *Ditylenchus destructor*. Considerable increase of resistance to nematodes and productivity of cucumber and potato were observed for plants treated by biostimulant Avercom and its derivatives. Impact of biostimulants Radostim-super and Avercom on increase of resistance of sugar beet and cucumber sprouts to nematodes *Heterodera schachtii* and *Meloidogyne incognita* was studied in the laboratory conditions. Comparative analysis of morpho-physiological signs of control and experimental plants showed that plants treated by Radostim-super and Avercom were more viable and resistant to these nematodes as compared to control sprouts. In the molecular-genetic experiments, we studied the impact of these biostimulants on inducing synthesis of small regulatory si/miRNA, which plays key role in plant immune protection. Using method Dot-blot hybridization we studied degree of homology between si/miRNA with mRNA populations, isolated from plants untreated and treated with new natural biostimulants. We found considerable difference in the degree of homology (6-28%) between populations of mRNA and si/miRNA from nematode-infected plants that were either untreated or treated with biostimulants. We have also investigated silencing of translation of mRNA activity of si/miRNA in the wheat embryo cell-free system of protein synthesis. In these experiments, we found high inhibitory activity (38-65%) of si/miRNA from plants treated by biostimulants as compared to low inhibitory activity (15-20%) of si/miRNA from untreated plants. Obtained differences in the degree of homology between populations of mRNA and si/miRNA from untreated and treated with biostimulants plants, which were infected by nematode, and also the high inhibitory activity of si/miRNA from plants treated by biostimulants confirm that these biostimulants induce synthesis of anti-nematodic si/miRNA in plants, resulting in considerable increase of their resistance to these phytopathogens.

Keywords: *Heterodera schachtii*, *Meloidogyne incognita*, *Ditylenchus destructor*, natural biostimulants, anti-nematodic si/miRNA, the degree of homology between mRNA and si/miRNA, silencing activity of si/miRNA in the wheat embryo cell-free system of protein synthesis, plant resistance to nematodes

1. Introduction

Over the last ten years a key role of short interfering RNA (siRNA) and microRNA (miRNA) in the TGS and PTGS - the basic processes of plant development and adaptation to stress-factors of environment, is disclosed (Angaji et al., 2010; Chen, 2009; Filipowicz et al., 2005; Hamilton et al., 2002; Luna et al., 2012; Mirouze et al., 2011; Park et al., 2002; Rasmann et al., 2012; Vaucheret et al., 2001; Zhang et al., 2007). The miRNA is generated from pre-miRNA precursor of ~70 nucleotides (nt) derived from one strand of distinct genomic loci by two rounds of endoribonuclease cleavage by RNase III-like enzymes named Drosha and Dicer (Lee et al., 2003; Mourelatos et al., 2002). The siRNA of ~22-24-nt is generated from longer double-stranded RNA (dsRNA) molecules (derived from repetitive sequences such as transposons and transgenes) through their cleavage by RNase III endoribonuclease named Dicer (Hamilton et al., 2002).

In a process of PTGS also called RNA interference (RNAi) si/miRNA with antisense structure to mRNA functions in a dual role: 1) together with site-specific multi-subunit RNase, referred to as RNA-induced silencing complex (RISC), and with AGO (Argonaute) proteins si/miRNA determines an age period of endogenous mRNA molecule in each eukaryotic cell and 2) together with RISC and AGO proteins si/miRNA participates in enzymatic cleavage or in silencing of translation of homologous mRNA of pathogenic organisms providing protection against pathogens and parasites (Bakhetia et al., 2005; Chen, 2009; Fabian et al., 2010; Filipowicz et al., 2005; Hamilton et al., 2002; Park et al., 2002; Vaucheret et al., 2006; Zhang et al., 2007) Taking part in DNA methylation and histone modification during TGS, and silencing of translation of mRNA of various pathogenic organisms during PTGS, si/miRNA contributes to epigenetic inheritance of plant resistance to diseases (Calarco et al., 2012; Luna et al., 2012; Mirouze et al., 2011; Rasmann et al., 2012; Tsygankova, 2012).

Numerous studies have shown that during infection of plants by pathogenic organisms the changes in small RNA populations (the main components of plant immune system) occur (Hewezi et al., 2008; Katiyar-Agarwal et al., 2006; Padmanabhan et al., 2009; Patel et al., 2010). Targets for si/miRNA are mRNA transcripts of plant genes which expression is induced during infection (therefore damage of plants by phytopathogens raises), or highly homologous mRNA of pathogenic organisms (Baum et al., 2007; Hewezi et al., 2008; Katiyar-Agarwal et al., 2006; Li et al., 2012; Padmanabhan et al., 2009; Patel et al., 2010).

In the case of plant protection against pathogens and parasites, the number of si/miRNA molecules produced in plant cells in response to a mass infection is not sufficient to provide effective protection. There are two approaches to increase synthesis of si/miRNA in response to pathogen or parasite attacks; these are either to insert additional genes of si/miRNA in the cells using genetic transformation or to activate the synthesis of endogenous si/miRNA in plant cells by specific inductors, for example by phytohormones (Bakhetia et al., 2005; Gheysen et al., 2006; Padmanabhan et al., 2009; Spoel et al., 2012; Zhang et al., 2011).

In our previous investigations we have elaborated and proposed the new strategy of nematode disease management: increase of plant resistance to nematodes by the way of inducing of RNA-interference process (RNAi or PTGS) in plant cells, i.e. inducing synthesis of si/miRNA using new ecologically safe polycomponent biostimulants with bioprotective and immune-modulating effects. In our laboratory and field experiments we found that biostimulants significantly increased plant resistance to viral pathogens, nematodes and insect herbivores through stimulation of synthesis in plant cells of immune-protective small regulatory si/miRNA (Tsygankova, Andrusevich et al., 2011; Tsygankova, Galkin et al., 2011; Tsygankova, Andrusevich, Beljavskaja et al., 2012). In these works we used Dot-blot hybridization for the study of changes in the degree of homology between populations of cytoplasmic RNA and small regulatory si/miRNA, isolated from rape, sugar beet, wheat and cucumber plants of the first generation, infected by parasitic nematodes *Heterodera schachtii* and *Meloidogyne incognita*, as well as wheat and nut plants of the second generation, infected by pathogenic micromycetes *Fusarium graminearum* and *Fusarium oxysporum f. ciceris* (Tsygankova, 2012; Tsygankova, Andrusevich et al., 2012; Tsygankova, Ponomarenko et al., 2012; Tsygankova, Stefanovska, Andrusevich et al., 2012; Tsygankova, Stefanovska, Galkin et al., 2012; Tsygankova, Andrusevich, Beljavskaja et al., 2012). Obtained differences in the degree of homology between mRNA and si/miRNA populations we used as genetic markers of increase of plant resistance to phytopathogens. Silencing of translation of mRNA activity of si/miRNA was also verified in the experiments with wheat embryo cell-free system of protein synthesis, which is widely used along with other cell-free systems (the rabbit reticulocyte lysate system and cell-free system from syncytial blastoderm Drosophila embryos) for in vitro study of mRNA translation and for investigation of silencing activity of si/miRNA on in vitro inhibition of mRNA translation (Maniatis et al., 1982; Promega, 1991; Tang et al., 2003; Tuschl et al., 1999).

To induce the synthesis of endogenous si/miRNA we treated plant seeds by new natural polycomponent biostimulants with bioprotective effect - Avercom and its derivatives, created at Zabolotny Institute of Microbiology and Virology, NAS of Ukraine and biostimulants Biogene, Stimpo, and Regoplant, Radostim-super created at the Institute of Bioorganic Chemistry and Petrochemistry, NAS of Ukraine in association with National Enterprise Interdepartmental Science and Technology Center "Agrobiotech" of the NAS and the Ministry of Education, Science and Sport of Ukraine (Iutynska, 2012; Tsygankova, Ponomarenko et al., 2012; Tsygankova, Stefanovska, Galkin et al., 2012; Tsygankova, Andrusevich, Beljavskaja et al., 2012; Tsygankova, Stefanovska, Andrusevich et al., 2012).

Aims of the present work are: a) study of impact of new natural biostimulants on morpho-physiological signs of resistance of sugar beet, cucumber and potato plants to parasitic nematodes: *Heterodera schachtii*, *Meloidogyne incognita* and *Ditylenchus destructor* in the field, greenhouse and laboratory conditions, b) determination of degree of homology between mRNA and si/miRNA populations, isolated from sugar beet, cucumber and potato plants, which were untreated and treated with biostimulants and infected by parasitic phytonematodes *Heterodera schachtii*, *Meloidogyne incognita* and *Ditylenchus destructor*, and c) investigation of silencing activity of si/miRNA populations, isolated from untreated and treated with biostimulants plants, on translation mRNA, isolated from infected plants and from parasitic nematodes, in the wheat embryo cell-free system of protein synthesis.

2. Materials and Methods

2.1 Plant Growing and Treatment

In our greenhouse, field and laboratory molecular-genetic experiments the sugar beet *Beta vulgaris L.*, cucumber *Cucumis sativus* of cultivar Gravina and potato *Solanum tuberosum* of cultivar Bellarosa plants infected by the parasitic nematode *Heterodera schachtii* Shmidt, gallic nematode *Meloidogyne incognita* Chitwood and stem nematode of potato *Ditylenchus destructor* Thome respectively were used.

We investigated bioprotective anti-nematodic effects of new polycomponent biostimulants Avercom and its derivatives (contain metabolites of the soil streptomycete *Streptomyces avermitilis* UCM Ac-2179, i.e. antiparasitic antibiotic avermectine, aminoacids, free fatty acids, vitamins of the B group, and phytohormones: indole-3-acetic acid, isopentenyl adenine, zeatin, zeatin riboside, brassinosteroids) (Iutynska, 2012; Tsygankova, Andrusevich, Beljavskaja et al., 2012); Radostim-super (contains antiparasitic antibiotic aversectine C - metabolites of the soil streptomycete *S. avermitilis* and metabolites, i.e. aminoacids, fatty acids, polysaccharides, phytohormones, and microelements, of cultivated *in vitro* micromycete *Cylindrocarpon obtusiuscuilum* 680, isolated out of Panax ginseng root system) (Tsygankova, Stefanovska, Galkin et al., 2012; Tsygankova, Stefanovska, Andrusevich et al., 2012).

Experimental plants growing at laboratory, greenhouse and field conditions were treated by biostimulants Avercom and Radostim-super. Avercom was obtained by ethanol extraction from of 7-days biomass of *Streptomyces avermitilis* UCM Ac-2179, the concentration of avermectine is 100 µg/ml, its derivates: Avercom nova-1 contains 50 ml of Avercom with antibiotic avermectine at concentration 100 µg/ml with adding 50 ml of supernatant of liquid culture *Streptomyces avermitilis* UCM Ac-2179 and 0.05 mM of salicylic acid; total content of avermectine is 50 µg/ml; and Avercom nova-2 contains 50 ml of Avercom with antibiotic avermectine in concentration 100 µg/ml and 50 ml of supernatant of liquid culture *Streptomyces avermitilis* UCM Ac-2179 and 0.01 mM of water-soluble chitosan of "Sigma" company; the total concentration of avermectine is 50 µg/ml.

Sugar beet and cucumber seeds were sprouted in Petri dishes (9.5 cm in diameter) in nematode-free aqueous medium (control) or with a suspension of nematodes *H. schachtii* and *M. incognita* eggs (at the concentration of 20-50 nematode eggs/ 20 seeds). The seeds were incubated at 23 °C and the nematode larvae hatched in 5-7 days later in average. Each experiment performed in three replicates. In experiments with sugar beet and cucumber seeds we used 4 variants: 1) seeds incubated on aqueous medium (control), 2) seeds incubated on aqueous medium with biostimulant Radostim-super (at the concentration of 25 µl/ml distilled water with final content of aversectine C - 0.025 µg/ml) and with biostimulant Avercom (at the concentration of 25 µl/ml distilled water with final content of avermectine - 0.05 µg/ml), 3) seeds incubated on aqueous medium with a suspension of nematode eggs, 4) seeds incubated on aqueous medium with biostimulant Radostim-super (at the concentration of 25 µl/ml distilled water with final content of aversectine C - 0.025 µg/ml) and with biostimulant Avercom (at the concentration of 25 µl/ml distilled water with final content of avermectine - 0.05 µg/ml) and a suspension of nematode eggs.

Cucumber plants of cultivar Gravina were also growing at greenhouse conditions at a background of artificial contamination with nematode *M. incognita* in a quantity of 700 larvae and eggs in 100 cm³ of soil sample. 7 days after the contamination hollows were made in the substratum, in which 100 ml of 2% solutions of each biostimulants were added; after 2 days the seedlings were planted, roots of seedlings were immersed into solutions of biostimulants for 5 minutes before planting. Damage caused by nematodes was scored in points according to methodical guidelines (Sigareva, 1986).

Investigations with potato of cultivar Bellarosa were carried out in field conditions as little-strip experiments with natural and artificial invasive background, created by planting into hollows 50 g of potato tubers infectected previously by nematode *D. destructor* larvae. Potato plants were treated by biostimulants in the next concentrations: for cultivation of 1 tone of planting material 0.4 l of Avercom or its derivates were dissolved in 20 l of water.

All the experiments were performed in four replicates.

2.2 Identification of Degree of Homology Between si/miRNA and mRNA Populations

Degree of homology between cytoplasmic mRNA and small regulatory si/miRNA populations, isolated from control and experimental plants was determined in the molecular-genetic experiments using Dot-blot hybridization method (Maniatis et al., 1982). Isolation of total RNA from plant cells and separation of high-purity si/miRNA preparations were carried out by our earlier published method (Tsygankova, Andrusevich et al., 2011; Tsygankova, Stefanovska, Andrusevich et al., 2012), the size (of 21-25 nt) of isolated si/miRNA preparations was verified by electrophoresis in a 15% polyacrylamide gel (Maniatis et al., 1982).

Plant si/miRNA labeled *in vivo* with ^{33}P using $Na_2HP^{33}O_4$ before isolation was later used for Dot-blot hybridization with own plant mRNA and with nematode mRNA (Tsygankova, Andrusevich, Beljavskaja et al., 2012; Tsygankova, Stefanovska, Andrusevich et al., 2012). Hybridization was conducted on modified and activated cellulose filters (Whatman 50, 2-aminophenylthioether paper of Company Amersham-Pharmacia Biotech, UK) that form covalent linkages with deposited DNA or RNA unlike the cellulose and nitrocellulose filters that form hydrogen bonds with DNA or RNA enabling to avoid losses of the nucleic acids in the course of the filter washing out (Maniatis et al., 1982).

Radioactivity of hybrid molecules was detected (imp/count per min/20 μg ± SE of mRNA) on glass Millipore AP-15 filter in toluene scintillator using Beckman LS 100C scintillation counter. Degree of homology (%) was determined according to the difference of hybridization between mRNA and si/miRNA from experimental plants and control plants (Tsygankova et al., 2010; Tsygankova, Stefanovska, Andrusevich et al., 2012).

2.3 Determination of Silencing Activity of si/miRNA Populations in the Cell-Free System of Protein Synthesis

Investigation of silencing activity of si/miRNA, isolated from untreated and treated by biostimulants plants, on translation of own plant mRNA or nematode mRNA, was conducted in the wheat embryo cell-free system of protein synthesis, which preparation is described in details elsewhere (Marcus et al., 1974; Tsygankova et al., 2010). Reagents of different companies, namely Amersham-Pharmacia Biotech, UK; New England Biolab, USA; Promega Corporation Inc, USA and Boehringer, Dupont, NEN, USA and Mannheim GmbH, Germany were used for preparation of cell free-system. Determination of inhibition by si/miRNA of protein synthesis on the templates own plant mRNA or nematode mRNA in the cell-free system was carried out according to index of decreasing of incorporation [^{35}S] methionine into proteins and was accounted (in imp./count per min/1mg of protein) on glass filter Millipore AP-15 in toluene scintillator in the scintillation counter LS 100C. Unlabelled si/miRNA were used for testing of inhibitory activity of si/miRNA in the cell-free system of protein synthesis. The silencing activity of si/miRNA (%) was determined as a difference of radioactivity of polypeptides (count per min/1mg of protein) synthesized on template mRNA by experimental plants compared to control plants.

The statistical analysis of the data was carried out by dispersive (Student) method (Bang et al., 2010). The least substantial difference ($LSD_{0.05}$) was also calculated for the field experiments (Dospechov, 1985).

3. Results

3.1 Investigation of Morpho-Physiological Signs of Plant Resistance to Parasitic Nematodes Under Impact of Natural Biostimulants

Biometric researches showed that in the experiments with cucumber plants, which were grown at greenhouse conditions on artificial infectious background created by nematode *M. incognita*, at the end of vegetation period the height of cucumber plants, treated (pre-sowing treatment of seeds and crop spraying) with biostimulant Avercom and its derivatives, exceeds the height of control plants by 10-24% (Table 1). The highest height of the plants was obtained at their treatment by biostimulant Avercom nova-1. The analysis of plant damage by nematodes showed that the plant damage by nematode *M. incognita* reached up to 3.4 points on the control plot. At the same time for cucumber plants treated by biostimulants, considerable decrease of contamination by nematode *M. incognita* (to 29.4% on an artificial background) was observed. In the case of Avercom plant damage by nematodes was not observed, in the case of Avercom derivatives the plant damage was less than 0, 2 - 1, 0 points.

Table 1. Biometric characteristics and degree of damage by nematodes of cucumber plants treated with biostimulants

Experience variant	Height of plants		Damage of plants by nematode *M.incognita*, (in point)
	cm	% in relation to control	
Control (without use of biostimulants)	168±5.9	100	3.4
Avercom	197±6.9	117	0
Avercom nova-1	208±7.3	124	0.2
Avercom nova-2	184±6.4	110	1.0

We have also studied the impact of biostimulants on plant productivity. The yield of cucumbers treated by Avercom and its derivatives was 16-26% higher compared to control. The highest yield was obtained when the plants were treated with Avercom (Table 2).

Table 2. The yield of cucumber plants treated with biostimulants

Experience variant	Yield, kg/m^2	Yield increase	
		kg/m^2	% in relation to control
Control (without use of biostimulants)	6.8	0	100
Avercom	8.6	1.8	126
Avercom nova-1	8.3	1.5	122
Avercom nova-2	7.9	1.1	116
LSD $_{0.05}$	0.2		

In the experiments with potato plants, which were carried out on a natural invasive background and artificial background, created by potato stem nematode *D. destructor*, it was shown that pre-sowing treatment of potato tubers and crop spraying with biostimulants Avercom and its derivatives decreased damage of potato tubers by dytylenchosis and increased crops productivity (Table 3).

Table 3. Damage of potato by ditylenchosis and productivity of potato plants under various conditions of growing

Experiment	Damage of plants by ditylenchosis		Productivity of crops	
	Quantity of sick potato tubers,%	Biological efficiency of biostimulants	Yield, pounds/ha	Yield increase (%) in relation to control
Natural background				
Control (without use of biostimulants)	40.1	-	36375.9	0
Avercom	11.8	70.5	40785.1	12.1
Avercom nova-1	24.7	38.4	37478.2	3.0
Avercom nova-2	34.4	14.2	38580.5	6.1
LSD $_{0.05}$	5.1	-	1763.7	
Artificial invasion background				
Control (without use of biostimulants)	48.0	-	33950.8	100
Avercom	21.5	55.2	37478.2	10.4
Avercom nova-1	34.0	29.2	36596.4	7.8
Avercom nova-2	44.0	8.3	35494.1	4.5
LSD $_{0.05}$	5.1	-	1543.2	

We have obtained the considerable decrease of contamination (to 26.5% on an artificial background) of potato plants by nematode *D. destructor*. Biological efficiency of Avercom reached more than 70% in experiments on a natural nematodic background and 55% on artificial invasion background created by nematodes, the less biological efficacy showed by Avercom nova-1 - 38.4% and 29.2%.

Impact of the biostimulants on potato productivity was also studied. Biostimulants promoted increase in potato productivity up to 12.1% - on natural nematode infection and up to 10.4% - in conditions of artificial infection.

On natural invasive background the most statistically significant increase in productivity (up to 12.1% compared to the control) of potato was achieved at pre-sowing planting material treatment with Avercom. On artificially created invasive background Avercom promoted the increasing of yield of potato by 10.4%.

Similar experiments devoted to impact of biostimulant Radostim-super on morpho-physiological signs and productivity of winter wheat plants have been conducted by us early in field conditions (Figure 1) (Sweere et al., 2011). Results obtained in these experiments testify about positive influence of biostimulant Radostim-super on dynamics of growth and increase of productivity of winter wheat. We conducted similar experiments on other agricultural plants; the considerable increase of photosynthetic activity (i.e. assimilation of CO_2 by plants) was obtained when plants were treated by biostimulant Radostim-super: rye (up to 12.5%), barley (up to 10.3%), oats (up to 13.7%), millet (up to 11.4%) corn (up to 12.9%), rice (up to 13.2%), buckwheat (up to 10.7%) (Sweere et al., 2011). Considerable increase of productivity of these crops from 20% to 65% under the impact of biostimulant Radostim-super was found in these investigations.

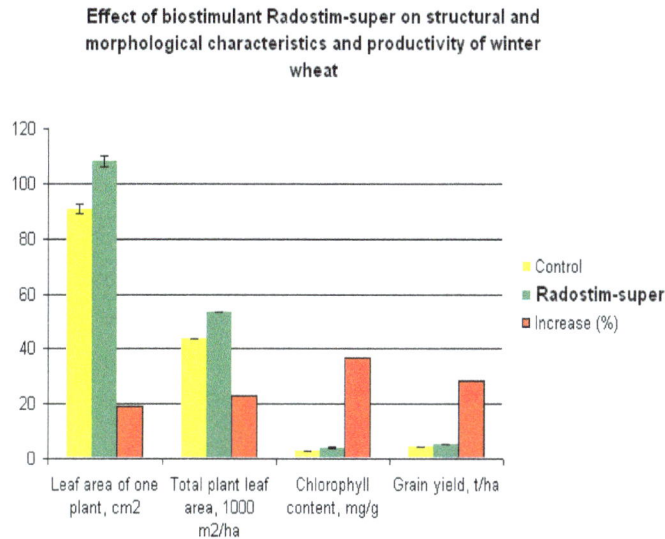

Figure 1. Impact of biostimulant Radostim-super on morpho-physiological signs and productivity of winter wheat plants grown under field conditions (seeds of winter wheat were treated before sowing with biostimulant Radostim-super at the concentration of 25 ml/t of seeds)

We have also tested in the field and greenhouse conditions the bioprotective effects of biostimulant Radostim-super against sugar beet nematode *H. schachtii* (Tsygankova, Stefanovska, Galkin et al., 2012). We have shown that treatment of sugar beet seeds and spraying of crops in vegetation period by Radostim-super considerably decreases the sugar nematode population density in the soil by 74.2%, whereas in control experiments (seeds treated by water) the beet nematode number in soil increased by 22%. In addition to the reduction in nematode numbers, the application of Radostim-super increased the sugar beet yield and sugar yield which were significantly higher than in the control by 40.0 and 6.2 tons/ha, respectively (Tsygankova, Stefanovska, Andrusevich et al., 2012; Tsygankova, Stefanovska, Galkin et al., 2012).

It was found in the laboratory experiments (Figure 2 and Figure 3) that 5-day sugar beet sprouts (obtained from seeds infected by nematode *H. schachtii* and treated with biostimulant Radostim-super) and 5-day cucumber sprouts (obtained from seeds infected by nematode *M. incognita* and treated with biostimulant Avercom) were more viable and resistant to these nematodes as compared to control sprouts (infected by nematodes and untreated with biostimulants).

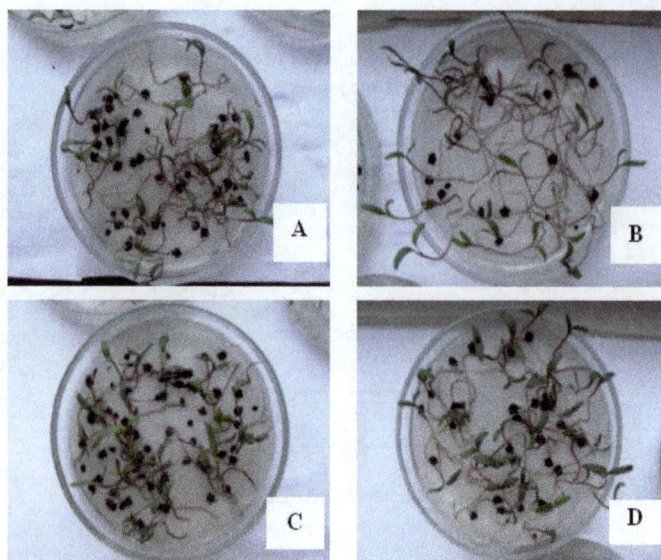

Figure 2. Impact of biostimulant Radostim-super on germination of sugar beet *Beta vulgaris L.* seeds and development of sprouts

A) 5-day sprouts grown on distilled water (control); B) 5-day sprouts grown on infectious background, created by parasitic nematode *H. schachtii*; C) 5-day sprouts treated with Radostim-super and grown on infectious background, created by parasitic nematode *H. schachtii*; D) 5-day sprouts treated with Radostim-super and grown without infectious background.

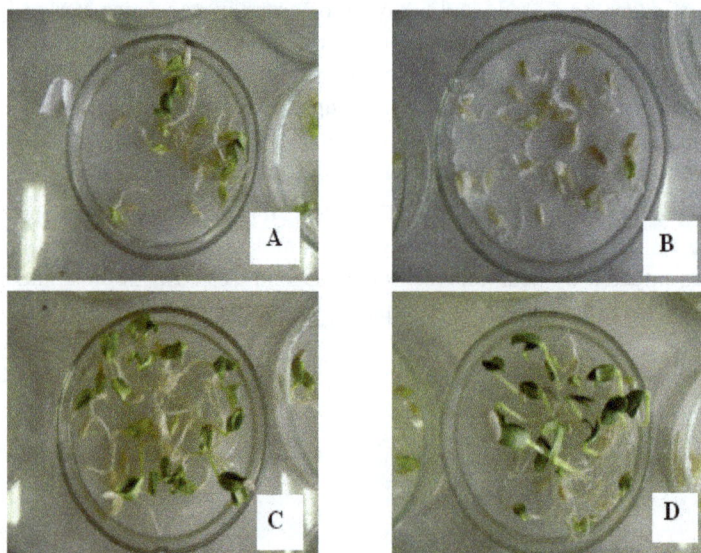

Figure 3. Impact of biostimulant Avercom on germination of cucumber seeds of cultivar Gravina and development of sprouts

A) 5-day sprouts grown on distilled water (control); B) 5-day sprouts grown on infectious background created by parasitic nematode *M. incognita*; C) 5-day sprouts treated with Avercom and grown on infectious background, created by parasitic nematode *M. incognita*; D) 5-day sprouts treated with Avercom and grown without infectious background.

3.2 Impact of Biostimulant Radostim-Super on Changes in the Degree of Homology Between si/miRNA and mRNA, and on Silencing Activity of si/miRNA

Changes in the level of si/miRNA synthesis (according to degree of homology between si/miRNA and mRNA) in the control plants, in the plants treated by Radostim-super, in the plants incubated with nematode *H. schachtii* and in the plants infected by nematode *H. schachtii* and treated with Radostim-super were determined using Dot-blot hybridization method.

Radioautographs on cellulose filters of probes which are hybrid molecules of mRNA isolated from control plants with $[P^{33}]$-si/miRNA isolated from experimental plants treated by biostimulants and grown without infectious background are presented in Figure 4.

Figure 4. Radioautographs on cellulose filters of probes (hybrid molecules of mRNA from control plants with $[P^{33}]$-si/miRNA from experimental plants)

1) mRNA and si/miRNA from control plants; 2) mRNA from control plants and si/miRNA from sugar beet plants, treated with biostimulant Radostim-super; 3) mRNA from control plants and si/miRNA from cucumber plants, treated with biostimulant Avercom; 4) mRNA from control plants and si/miRNA from cucumber plants, treated with biostimulant Avercom nova-1; 5) mRNA from control plants and si/miRNA from cucumber plants, treated with biostimulant Avercom nova-2.

It is shown in Figure 5 that according to degree of homology between populations of mRNA and si/miRNA Radostim-super considerably increased the synthesis of si/miRNA in plants not infected by nematodes, but on the contrary in plants infected by nematodes and untreated with this biostimulant the synthesis of si/miRNA is sharply reduced. Biostimulant Radostim-super increases si/miRNA synthesis in infected plants, but in these plants the level of synthesis is lower compared with level of synthesis in plants not infected by nematodes and not treated with this biostimulant.

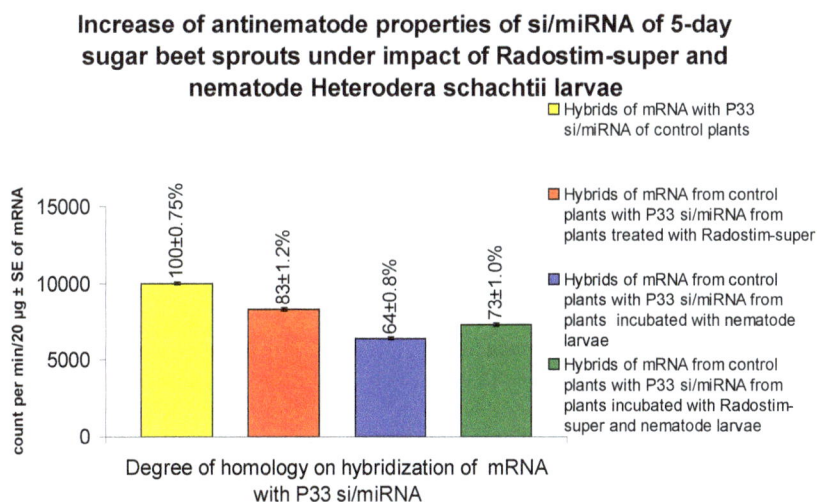

Figure 5. The degree of homology (%) between populations of mRNA and si/miRNA from control and experimental sugar beet plants infected by nematode larvae and treated with biostimulant Radostim-super

Investigation of inhibitory (silencing) activity of si/miRNA on the template of mRNA from plants in the wheat embryo cell-free system of protein synthesis showed (Figure 6) that si/miRNA, isolated from plants not infected by nematodes and treated with Radostim-super, showed high inhibitory activity (82%), close to those of control plants (100%). Obtained results testify that this biostimulant changes si/miRNA population in plant cells.

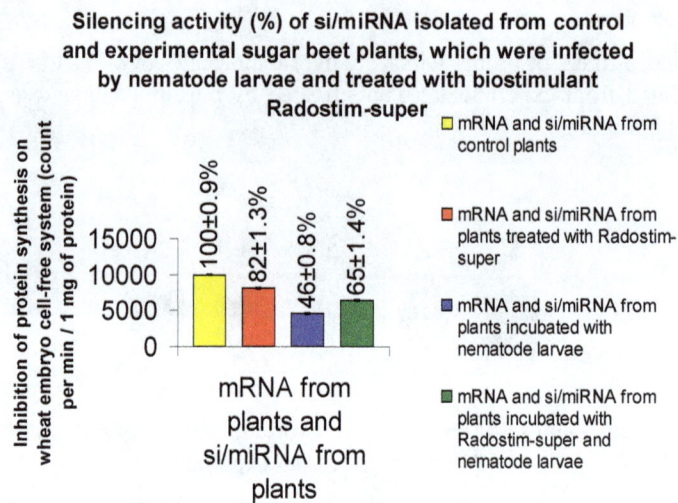

Figure 6. Inhibition of protein synthesis in the wheat embryo cell-free system on the template of mRNA from control and experimental plants by si/miRNA from control and experimental sugar beet plants infected by nematode larvae and treated with biostimulant Radostim-super

The inhibitory activity of si/miRNA, isolated from the same plants, treated by biostimulant, on the template mRNA from nematode larvae (Figure 7) was slightly higher (15%) than that of control si/miRNA, isolated from untreated plants (10%). This shows insignificant homology between plant si/miRNA and nematode mRNA.

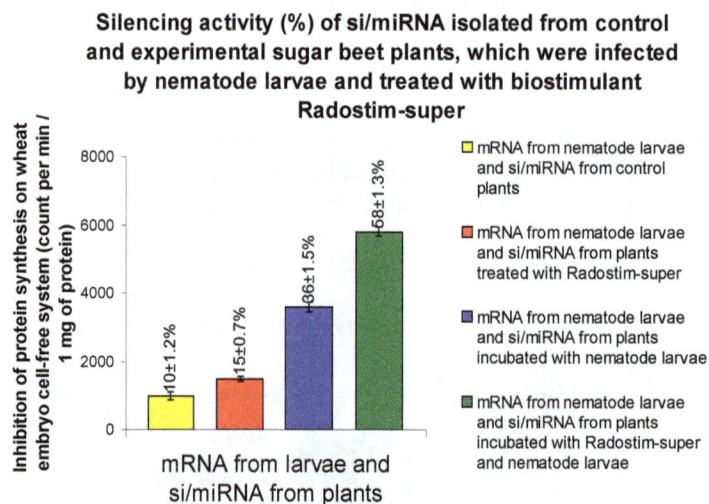

Figure 7. Inhibition of protein synthesis in the wheat embryo cell-free system on the template of mRNA from nematode larvae by si/miRNA from control and experimental sugar beet plants infected by nematode larvae and treated with biostimulant Radostim-super

At the same time (Figure 6 and Figure 7) the inhibitory activity of si/miRNA, isolated from plants, infected by nematode larvae and treated by Radostim-super, considerably increased both on the template of mRNA from the same plants (65%) and on template of mRNA from nematode larvae (58%).

At the same time inhibitory activity of si/miRNA, isolated from plants infected by nematode larvae and untreated by biostimulant, was lower both on the template of mRNA from plants (46%) and on the template of mRNA from nematode larvae (36%).

Obtained results confirm that biostimulant Radostim-super causes reprogramming of plant genome to induce synthesis of si/miRNA specific (antisence) both to own plant mRNA (which expression promotes infection) and to homologous mRNA of nematodes.

3.3 Impact of Biostimulant Avercom and Its Derivatives on Homology Between si/miRNA and mRNA, and on Silencing Activity of si/miRNA

Radioautographs on cellulose filters of probes which are hybrid molecules of mRNA isolated from control plants with [P^{33}]-si/miRNA isolated from experimental plants treated by biostimulants and grown without infectious background are presented above in Figure 4.

Data for the analysis of degree of homology between populations of cytoplasmic mRNA and si/miRNA, isolated from control and experimental potato plants and cucumber, which were grown in greenhouse and field conditions on an artificial infectious background and treated with biostimulant Avercom and its modifications, are presented in Figure 8.

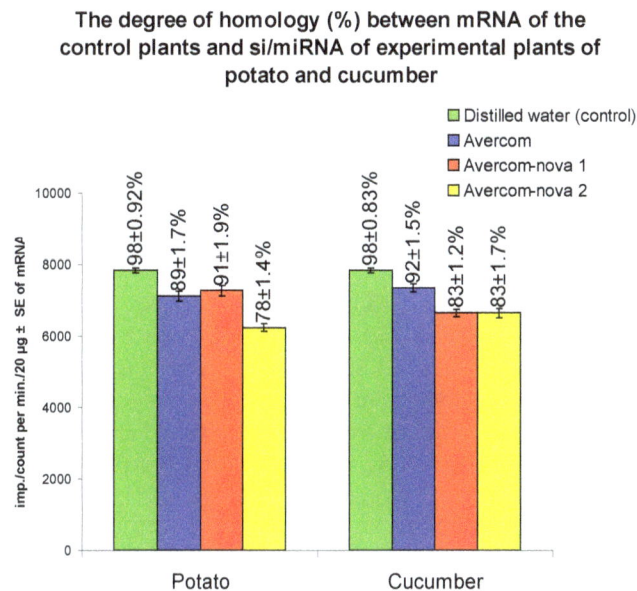

The degree of homology (%) between mRNA of the control plants and si/miRNA of experimental plants of potato and cucumber

Legend:
- Distilled water (control)
- Avercom
- Avercom-nova 1
- Avercom-nova 2

y-axis: imp./count per min./20 µg ± SE of mRNA

Potato bars: 98±0.92%, 89±1.7%, 91±1.9%, 78±1.4%
Cucumber bars: 98±0.83%, 92±1.5%, 83±1.2%, 83±1.7%

x-axis: Potato, Cucumber

Figure 8. Degree of homology (%) between mRNA of the control plants and si/miRNA of the control and experimental potato and cucumber plants treated with biostimulants and infected by parasitic gallic nematode *M. incognita* and stem nematode of potato *D. destructor*

Comparative analysis of degree of homology (%) between si/miRNA and mRNA (Figure 8) obtained in the experimental plants compared to the same values in the control plants, showed that the largest difference in the degree of homology regarding to control plants was observed in experimental plants treated with biostimulants Avercom nova-2 (up to 20% - in potato and up to 15% - in cucumber plants) and Avercom nova-1 (up to 15% - in cucumber and up to 7% - in potato plants), smaller difference in the degree of homology was found in the experimental plants treated with biostimulant Avercom (up to 6% - in cucumber and 9% - in potato plants).

According to experiments in the wheat embryo cell-free system of protein synthesis, results of inhibition of the translation of mRNA from the cucumber and potato plants infected with nematodes *M. incognita* and *D. destructor* and treated with biostimulant Avercom and its derivates show significant increase of silencing activity of si/miRNA (similar to activity of si/miRNA from not infected plants - control N1) isolated from cucumber and potato plants infected by nematodes *M. incognita* and *D. destructor* and treated with these biostimulants (Figure 9).

Silencing activity (%) of si/miRNA isolated from control and experimental potato and cucumber plants treated with biostimulants and infected

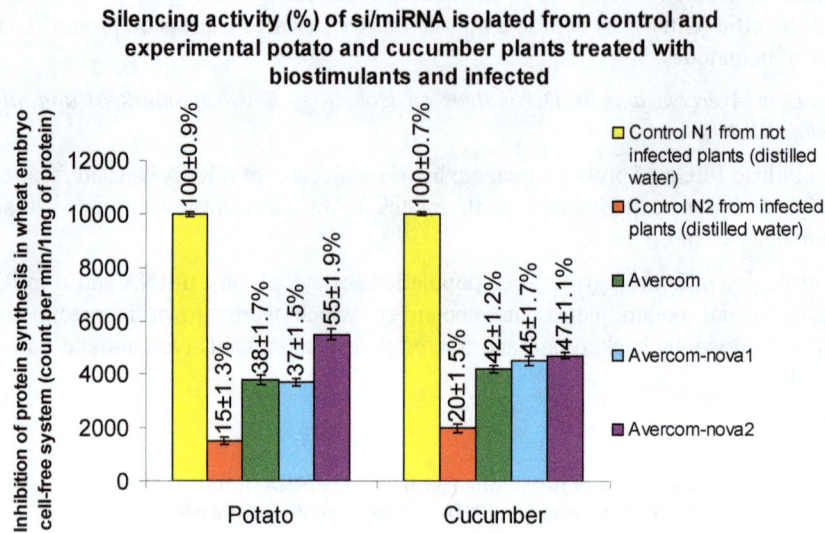

Figure 9. Inhibition of protein synthesis in the wheat embryo cell-free system on the template of mRNA from control and experimental plants by si/miRNA from control and experimental potato and cucumber plants treated with biostimulants and infected by parasitic gallic nematode *M. incognita* and stem nematode of potato *D. destructor*

According to results of inhibition of the translation of mRNA from cucumber and potato plants infected with nematodes *M. incognita* and *D. destructor* the highest silencing activity (compared to control N 1) was shown by si/miRNA isolated from the same plants treated with biostimulants (Figure 9) Avercom nova-2 (up to 55% - in potato and up to 47% - in cucumber plants) and Avercom nova-1 (up to 45% - in cucumber and up to 37% - in potato plants), the lower silencing activity was shown by si/miRNA isolated from experimental plants treated with Avercom (up to 42% - in cucumber and up to 38% - in potato plants). Significantly lower silensing activity (up to 15% - in potato, up to 20% - in cucumber plants) was shown by si/miRNA, isolated from infected plants, which were untreated with biostimulants (control N 2).

4. Discussion

Existing methods for controlling the distribution of nematodes and the reduction in the yield of important crops caused by them are chemically synthesized soil fumigants, nematicides (belonging to the classes of organophosphates and carbamates), and various types of insecticides of natural origin, for example, phytoinsecticide pyrethryn and its synthetic analogs, i.e., pyrethroids (Mitkowski et al., 2003; Oka, 2010; Winter et al., 2006). In most countries worldwide, however, a trend is observed towards practically restricting their use because of their high toxicity to humans and contamination of the environment. Traditional methods to regulate the amount of parasitic nematodes also include various biocontrol technologies, i.e., application of various organic soil fertilisers and industrial waste of vegetable or animal origin, compost, and changes in soil pH (acidification of up to pH 4 or alcalinization of up to pH 8); introduction of antagonistic and competitive microorganisms (bacteria of the strains *Burkholderia cepacia* and *Bacillus chitinosporus* and the fungi micromycetes *Myrothecium verrucaria* and *Paecilomyces lilacinus*) to soil; crop rotation with the development of cultures resistant to nematodes, using biopreparations that contain essential oils of various herbs with an anti-nematodic effect (for example, the oil of sesame, garlic, rosemary, or white pepper); etc. (Oka, 2010). Unfortunately, a combination of the above listed methods can only depress the high viability of this pest class.

Now the success in increasing of plant resistance to nematodes has been reached by genetic engineering and breeding methods (Bleve-Zacheo et al., 2007; Fairbairn et al., 2007; Fuller et al., 2008; Gheysen et al., 2006; Liao et al., 2003; Tsygankova, Andrusevich, Ya, Ponomarenko et al., 2013; Tsygankova, Yemets et al., 2013).

The newer approach for nematode disease management is to increase plant resistance against agricultural pests by new ecologically safe plant growth regulators of natural or synthetic origin, phythohormones, seaweed and plant extracts, and organic compounds (such as sugar or ascorbic acid). In favor of this new approach testify numerous studies (Acquaah, 2007; Aktaruzzaman et al., 2012; Arrigoni et al., 1979; Bleve-Zacheo et al., 2007; Dias-Arieira et al., 2013; Fortum et al., 1983; Khan et al., 2009; Mitkowski et al., 2003; Moghaddam et al., 2012;

Olaiya et al., 2013; Tsygankova, Andrusevich et al., 2012; Tsygankova, Andrusevich et al., 2013; Verhage et al., 2010).

For example, the impact of five commercial products marketed as systemic resistance (SR) and plant growth promotion (PGP) inducers on increase of tomato plant (*Lycopersicon esculentum* Mill.) resistance to pathogenic bacteria or nematodes in greenhouse conditions has been studied by Vavrina et al., 2004.

These SR/PGP inducers included a bacterial suspension [Companion (*Bacillus subtilis* GB03)], two plant defense elicitors with nutrients (Keyplex 35ODP plus Nutri-Phite, and Rezist with Cab'y), natural plant extracts (Liquid Seaweed Concentrate and Stimplex), and synthetic growth regulator (Actigard 50W). Comparative analysis of growth stimulating and bioprotective effects of SR/PGP inducers have shown that highest suppression of bacterial spot [*Xanthomonas campestris* pv. *vesicatoria* (Xcv)] is caused by synthetic regulator Actigard. Other SR/PGP inducers: Companion, Keyplex 35ODP plus Nutri-Phite, Rezist, Cab'y, Liquid Seaweed Concentrate and Stimplex induced only partial suppression of bacterial spot in inoculated tomato plants. The alpha-keto acids plus nutrients (Keyplex 35ODP plus Nutri-Phite) increased plant growth by 14.3% and improved root condition compared to untreated control following exposure to nematodes.

In the present work, we investigated the impact of biostimulants on inducing of RNA-interference process (RNAi or PTGS) in plant cells, i.e. increasing synthesis of si/miRNA with immune-protective anti-nematodic properties. We studied genetic mechanisms of increase of sugar beet, cucumber and potato plant resistance to parasitic nematodes *H. schachtii*, *M. incognita* and *D. destructor* under impact of biostimulants: Radostim-super, Avercom and its derivatives. The experiments we based on the assumption that plants, infected with different types of pathogenic or parasitic organisms, increased the synthesis of si/miRNA specific both to own plant mRNA (which expression rises at the specialized infected plant cells and involves plant developmental processes) and to pathogenic or parasitic highly homologous mRNA (Hewezi et al., 2008; Ithal et al., 2007; Katiyar-Agarwal et al., 2006; Klink et al., 2009; Li et al., 2012; Padmanabhan et al., 2009; Patel et al., 2010). We also assumed that the biostimulants induce synthesis of si/miRNA which improves plant immunity through the specified mechanism of si/miRNA action.

We need to be sure that our assumptions are correct, so that a new generation of biostimulants with the properties of selective activation of synthesis of si/miRNA, which is specific to own plant cell mRNA or to pathogenic or parasitic highly homologous mRNA can be created. Thus we verified the changes in the populations of si/miRNA (according to degree of homology between si/miRNA and mRNA) and compared silencing activity of si/miRNA from biostimulant-treated and untreated plants. The conciderable differences obtained in these experiments in the degree of homology between si/miRNA and mRNA from control and experimental plants and high silencing activity of si/miRNA from plants, which were treated with biostimulants, testify about impact of these biostimulants on reprogramming of plant genome to induce synthesis of immune-protective si/miRNA in plant cells. As a result the resistance of plants to parasitic nematodes considerably rises.

The results of this work correlate and supplement the data of our previous experiments, in which we have conducted numerous investigations on various agricultural plants like rape, wheat, chickpea, corn and soybean, which were grown on invasion background created by pests: ground beetle, *Zabrus tenebrioides*, turnip moth *Scotia segetum* and *Chloropidae spp.*; pathogenic micromycetes: *Mucor spp.*, *Rhizopus spp.*, *Aspergillus spp.*, *Penicillium spp.*, *Trichothecium roseum*, *Fusarium graminearum*, *Fusarium oxysporum* f. *ciceris*, *Alternaria alternata*; phytonematode *Anguina tritici* (Tsygankova, 2012; Tsygankova, Ponomarenko et al., 2012; Tsygankova, Stefanovska, Andrusevich et al., 2012; Tsygankova, Andrusevich et al., 2011; Tsygankova et al., 2013). In these field and greenhouse experiments we have shown that treatment of plant seeds and spraying of crops in vegetation period by biostimulants of natural origin: Regoplant, Stimpo, Radostim and Radostim-super considerably increased (up to 74-98%) plant resistance to above mentioned phytopathogens.

Biological efficiency of biostimulant of microbiological origin Avercom and its composition with elicitors was also tested on early wheat, cucumber and tomato crops in field and greenhouse experiments with natural and artificial invasive backgrounds, created by phytonematodes: *Tylenchorbynchus dubius*, *Pratylenchus pratensis*, *Meloidogyne incognita* and by phytopathogenic micromycete *Fusarium oxysporum* (Iutynska et al., 2011; Tsygankova, Galkin et al., 2011; Tsygankova, Andrusevich, Beljavskaja et al., 2012). High antagonistic activity of biostimulants against all specified phytopathogens was found in the experiments. The plants treated with these biostimulants showed considerably increased resistance (up to 85-100%) to specified phytopathogens.

The molecular-genetic experiments, which were conducted in all these works, showed that increased plant resistance to phytopathogens is caused by inducing effect of these biostimulants on synthesis in plants of si/miRNA immune-protective against nematodes.

5. Conclusion

It was found that in the field, greenhouse and laboratory experiments according to morpho-physiological signs of plants the application of new natural biostimulants: Radostim-super, Avercom and its derivates (containing bioprotective substances - aversectine and avermectine) leads to considerable increase of resistance of sugar beet, cucumber and potato plants to nematodes *H. schachtii, M. incognita* and *D. destructor*.

In the molecular-biological investigations we found conciderable lowering of homology (from 6 to 28%) between si/miRNA and mRNA populations from experimental (infected by these nematodes and treated with biostimulants) plants and control plants.

These differences in degree of homology may be the result of activation of synthesis of small regulatory si/miRNA with high anti-nematodic activity by biostimulants in plants. Increase of silencing activity of si/miRNA (up to 38-65%) of the plants infected by nematodes conforms to this assumption. This effect under the impact of biostimulant Avercom and its derivatives significantly increases plant resistance to parasitic nematodes. Obtained changes in degree of homology between mRNA and si/miRNA populations can be used as genetic markers of increase of plant resistance to phytopathogens.

Acknowledgements

This research has been carried out with support of the STCU project P490 of National Academy of Sciences of Ukraine and of the project "Molecular bases of creation of biologically active and ecologically safe preparations with bioprotective and immunomodulative properties" of the special purpose complex interdisciplinary program of the scientific researches of National Academy of Sciences of Ukraine "Fundamentals of molecular and cellular biotechnology" (confirmed by the decision of Presidium of National Academy of Sciences of Ukraine from 07.07.10, No. 222).

References

Acquaah, G. (2007). *Principles of Plant Genetics and Breeding* (p. 564). Blackwell Publishing Ltd..

Aktaruzzaman, M., Ray, D. B., Hossain, M. F., & Afroz, T. (2012). Effect of Bau-biofungicide and some plant extracts against root-knot (*Meloidogyne javanica*) of papaya. *Int. J. Sustain. Crop Prod., 7*(1), 1-5.

Angaji, S. A., Hedayati, S. S., Hoseinpoor, R., Samadpoor, S., Shiravi, S., & Madani, S. (2010). Application of RNA interference in plants. *Plants Omics Journal, 3*(3), 77-84.

Arrigoni, O., Zacheo, G., Arrigoni-Liso, R., Bleve-Zacheo, T., & Lamberti, F. (1979). Relationship Between Ascorbic Acid and Resistance in Tomato Plants to *Meloidogyne incognita. Phytopathology, 69*(6), 579-581. http://dx.doi.org/10.1094/Phyto-69-579

Bakhetia, M., Charlton, W. L., & Urwin, P. E. (2005). RNA interference and plant parasitic nematodes. *Trends Plant Sci., 10*(8), 362-367. http://dx.doi.org/10.1016/j.tplants.2005.06.007

Bang, H., Zhou, X. K., van Epps, H. L., & Mazumdar, M. (2010). Statistical Methods in Molecular Biology. *Series: Methods in Molecular Biology* (p. 636). New York: Humana Press. http://dx.doi.org/10.1007/978-1-60761-580-4

Baum, T. J., Hussey, R. S., & Davis, E. L. (2007). Root-knot and cyst nematode parasitism genes: the molecular basis of plant parasitism. *Genetic Engineering (N Y), 28,* 17-43. http://dx.doi.org/10.1007/978-0-387-34504-8_2

Bleve-Zacheo, T., Melillo, M. T., & Castagnone-Sereno, P. (2007). The Contribution of Biotechnology to Root-Knot Nematode Control in Tomato Plants. *Pest Technology, Global Science Books, 1*(1), 1-16.

Calarco, J. P., Borges, F., Donoghue, M. T. A., Van Ex, F., Jullien, P. E., Lopes, T., ... Martienssen, R. A. (2012). Reprogramming of DNA Methylation in Pollen Guides Epigenetic Inheritance via Small RNA. *Cell.* http://dx.doi.org/10.1016/j.cell.2012.09.001

Chen, X. (2009). Small RNAs and Their Roles in Plant Development. *Annu. Rev. Cell Dev. Biol., 35,* 21-44. http://dx.doi.org/10.1146/annurev.cellbio.042308.113417

Dias-Arieira, C. R., Santana-Gomes, S., Puerari, H. H., Fontana, L. F., Ribeiro, L. M., & Mattei, D. (2013). Induced resistance in the nematodes control. *African Journal of Agricultural Research, 8*(20), 2312-2318.

Dospechov, B. A. (1985). *Technique of field (with a basis of statistical processing of results of researches).* (5th Ed., p. 51). Moscow: Agropromizdat.

Fabian, M. R., Sonenberg, N., & Filipowicz, W. (2010). Regulation of mRNA Translation and Stability by microRNAs. *Annu. Rev. Biochem., 79*, 351-79. http://dx.doi.org/10.1146/annurev-biochem-060308-103103

Fairbairn, D. J., Cavallaro, A. S., Bernard, M., Mahalinga-Iyer, J., Graham. M. W., & Botella, J. R. (2007). Host delivered RNAi: an effective strategy to silence genes in plant parasitic nematodes. *Planta* (Vol. 226, pp. 1525-1533). http://dx.doi.org/10.1007/s00425-007-0588-x

Filipowicz, W., Jaskiewicz., L., Kolb, F. A., & Pillai, R. S. (2005). Post-transcriptional gene silencing by siRNAs and miRNAs. *Curr. Opin. Struct. Biol., 15*, 331-341. http://dx.doi.org/10.1016/j.sbi.2005.05.006

Fortum, B. A., & Lewis, S. A. (1983). Effects of growth regulators and nematodes on Cylindrocladium blach root rot of soybean. *Plant Desease, 67*, 282-284. http://dx.doi.org/10.1094/PD-67-282

Fuller, V. L., Lilley, C. J., & Urwin, P. E. (2008). Nematode resistance. *New Phytol., 180*, 27-44. http://dx.doi.org/10.1111/j.1469-8137.2008.02508.x

Gheysen, G., & Vanholme, B. (2006). RNAi from plants to nematodes. *Trends in Biotechnol., 25*(3), 89-92. http://dx.doi.org/10.1016/j.tibtech.2007.01.007

Hammilton, A., Voinnet, O., & Chapell, L. (2002). Two classes of short interfering RNA in RNA silencing. *EMBO J., 21*, 4671-4679. http://dx.doi.org/10.1093/emboj/cdf464

Hewezi, T., Howe, P., Mairer, T. R., & Baum, T. J. (2008). Arabidopsis small RNAs and their targets during cyst nematode parasitism. *Mol. Plant-Microbe Interact., 21*, 1622-1634. http://dx.doi.org/10.1094/MPMI-21-12-1622

Ithal, N., Recknor, J., Nettleton, D., Hearne, L., Maier, T., Baum, T. J., & Mitchum, M. G. (2007). Parallel genome wide expression profiling of host and pathogen during soybean cyst nematode infection of soybean. *Mol. Plant–Microbe Interact., 20* (3), 293-305. http://dx.doi.org/10.1094/MPMI-20-3-0293

Iutynska, G. A. (2012). Elaboration of natural polyfunctional preparations with antiparasitic and biostimulating properties for plant growing. *Mikrobiol. J., 74*(4), 3-12.

Iutynska, H. O., Tytova, L. V., Leonova, N. O., Antypchuk, A. F., Brovko, I. S., Eakin, D., ... Pindrus, A. A. (2011). Complex preparations based on microorganisms and plant growth regulators. In S. P. Ponomarenko & H. O. Iutynska (Eds.), *New plant growth regulators: basic research and technologies of application* (pp. 161-207). Kyiv: Nichlava.

Katiyar-Agarwal, S., Morgan, R., Dahlbeck, D., Borsani, O., Villegas, A., Zhu, J. K., ... Jin, H. L. (2006). A pathogen-inducible endogenous siRNA in plant immunity, *Proc Natl Acad Sci USA, 103*(47), 18002-18007. http://dx.doi.org/10.1073/pnas.0608258103

Khan, W., Rayirath, U. P., Subramanian, S., Jithesh, M. N., Rayorath, P., Hodges, D. M, ... Prithviraj, B. (2009). Seaweed Extracts as Biostimulants of Plant Growth and Development. *J. Plant Growth Regul., 28*, 386-399. http://dx.doi.org/10.1007/s00344-009-9103-x

Klink, V. P., & Matthews, B. F. (2009). Emerging Approaches to Broaden Resistance of Soybean to Soybean Cyst Nematode as Supported by Gene Expression Studies. *Plant Physiology, 151*, 1017-1022. http://dx.doi.org/10.1104/pp.109.144006

Lee, Y., Ahn, C., & Han J. (2003). The nuclear RNase III Drosha initiates microRNA processing. *Nature, 425*, 415-419. http://dx.doi.org/10.1038/nature01957

Li, X., Wang, X., Zhang, S., Liu, D., Duan, Y., & Dong, W. (2012) Identification of Soybean MicroRNAs Involved in Soybean Cyst Nematode Infection by Deep Sequencing. *PloS ONE, 7*(6), e39650. http://dx.doi.org/10.1371/journal.pone.0039650

Liao, Y. C., Li, H. P., Zhao, C. S., Yao, M. J., Zhang, J. B., & Liu, J. L. (2006) Plantibodies: a novel strategy to create pathogen-resistant Plants. *Biotechnology and Genetic Engineering Reviews, 23*, 253-271. http://dx.doi.org/10.1080/02648725.2006.10648087

Luna, E., Bruce, T. J. A., Roberts, M. R., Flors, V., & Ton, J. (2012). Next-Generation Systemic Acquired Resistance. *Plant Physiol., 158*, 844-853. http://dx.doi.org/10.1104/pp.111.187468

Maniatis, T., Fritsch, E. F., & Sambrook, J. (1982). *Molecular cloning: A laboratory manual* (p. 545). New York: Cold Spring Harbor Lab.

Marcus, A., Efron, D., & Week, D. P. (1974). The wheat embryo cell free system. *Method Enzymol., 30*(2), 749-754. http://dx.doi.org/10.1016/0076-6879(74)30073-0

Mirouze, M., & Paszkowski, J. (2011). Epigenetic contribution to stress adaptation in plants. *Curr. Opin. in*

Plant Biol., 14, 1-8. http://dx.doi.org/10.1016/j.pbi.2011.03.004

Mitkowski, N. A., & Abawi, G. S. (2003). Root-knot nematodes. *The Plant Health Instructor*.

Moghaddam, M. R. B., & Van den Ende, W. (2012). Sugars and plant innate immunity. *J. Experim. Bot.*, 1-10.

Mourelatos, Z., Dostie, J., & Paushkin, S. (2001). miRNPs: a novel class of ribonucleoproteins containing numerous microRNAs. *Genes Dev., 16*, 720-728. http://dx.doi.org/10.1101/gad.974702

Oka, Y. (2010). Mechanisms of nematode suppression by organic soil amendments. *Appl. Soil. Ecol., 44*, 101-115. http://dx.doi.org/10.1016/j.apsoil.2009.11.003

Olaiya, C. O., Gbadegesin, M. A., & Nwauzoma, A. B. (2013). Bioregulators as tools for plant growth, development, defence and improvement. *African Journal of Biotechnology, 12*(32), 4987-4999.

Padmanabhan., Ch., Zhang, X., & Jin, H. (2009). Host small RNAs are big contributors to plant innate immunity. *Curr. Opin. in Plant Biol., 12*, 465-472. http://dx.doi.org/10.1016/j.pbi.2009.06.005

Park.,W., Song, L. J., Messing, R. J., & Chen, X. (2002). CARPEL FACTORY, a dicer homolog, and HEN1, a novel protein, act in microRNA metabolism in Arabidopsis thaliana induced silencing complex (RISC), which targets homologous RNAs for degradation. *Curr. Biol., 12*, 484-1495. http://dx.doi.org/10.1016/S0960-9822(02)01017-5

Patel, N., Hamamaouch, N., Li, C., Hewezi, T., & Hussey, R. S. (2010). A nematode effector protein similar to annexins in host plants. *J. Exp. Bot., 61*, 235-248. http://dx.doi.org/10.1093/jxb/erp293

Rasmann, S., De Vos, M., Casteel, C. L., Tian, D., Halitschke, R., Sun, J. Y., … Jander, G. (2012). Herbivory in the Previous Generation Primes Plants for Enhanced Insect Resistance. *Plant Physiol., 158*, 854-863. http://dx.doi.org/10.1104/pp.111.187831

Sigareva, D. D. (1986). Methodical guidelines on revealing and the account of parasitic nematodes of field crops (p. 41). Kiev: Urojay.

Spoel, S. H., & Dong, X. (2012). How do plants achieve immunity? Defence without specialized immune cells. *Nature Reviews, Immunology, 12*, 89-100. http://dx.doi.org/10.1038/nri3141

Sweere, D., Ponomarenko, S. P., Anishin, L. A., Babayants, O. V., & Hrytsayenko, Z. M. (2011). Research of PGR efficiency in the farm Kyiv-Atlantic-Ukraine and the center of scientific providing of agroindustrial production of the Cherkasy region, Uman State Agrarian University and Odessa Selection and Genetic Institute. In S. P. Ponomarenko & H. O. Iutynska (Eds.), *New plant growth regulators: basic research and technologies of application* (pp. 69-93). Kyiv: Nichlava.

Tang, G., Reinhart, B. J., Bartel, D. P., & Zamore, P. D. (2003). A biochemical framework for RNA silencing in plants. *Genes & Development, 17*, 49-63. http://dx.doi.org/10.1101/gad.1048103

Titus, D. E. (1991). Promega protocols and applications guide (2nd ed., p. 422). USA: Promega Corporation.

Tsygankova, V. A. Andrusevich, Ya. V., Beljavskaja, L. A., Kozyritskaja, V. E., Iutinskaja, H. A., Galkin, A. P., … Boltovskaya, E. V. (2012). Growth stimulating, fungicidal and nematicidal properties of new microbial substances and their impact on si/miRNA synthesis in plant cells. *Mikrobiol. J., 74*(6), 3-12.

Tsygankova, V. A., Andrusevich, Ya. V., & Blume, Ya. B. (2011). Isolation of small regulatory si/miRNA with antinematode activity from plant cells. *Dopovidi Akademii nauk Ukrainy (Rep. Nat. Acad. Sci. Ukr.), 9*, 159-164.

Tsygankova, V. A., Andrusevich, Ya. V., Babayants, O. V., Ponomarenko, S. P., Medkov, A. I., & Galkin, A. P. (2013). Stimulation of plant immune protection against pathogenic fungi, pests and nematodes with growth regulators. *Physiol. and biochem. cultivated plants, 45*(2), 138-147.

Tsygankova, V. A., Andrusevich, Ya. V., Ponomarenko, S. P., Galkin, A. P., & Blume, Ya. B. (2012). Isolation and Amplification of cDNA from the Conserved Region of the Nematode *Heterodera schachtii 8H07* Gene with a Close Similarity to Its Homolog in Rape Plants. *Cytology and Genetics, 46*(6), 335-341. http://dx.doi.org/10.3103/S0095452712060114

Tsygankova, V. A., Galkin, A. P., Galkina, L. A., Musatenko, L. I., Ponomarenko, S. P., &Iutynska, H. O. (2011). Gene expression under regulators' stimulation of plant growth and development. In S. P. Ponomarenko, & H. O. Iutynska (Eds.), *New plant growth regulators: basic research and technologies of application* (pp. 94-152). Kyiv: Nichlava.

Tsygankova, V. A., Musatenko, L. I., Ponomarenko, S. P., Galkina, L. A., Andrusevich, Ya. V. & Galkin, A. P.

(2010). Change of functionally active cytoplasmical mRNA populations in plant cells under growth regulators effect and biological perspectives of cell-free systems of protein synthesis. *Biotechnol., 3*(2), 19-32.

Tsygankova, V. A., Stefanovska, T. R., Andrusevich, Ya. V., Ponomarenko, S. P., Galkin, A. P., Blume, & Ya. B. (2012). Induction of biosynthesis of si/miRNA with antipathogenic and antiparasitic properties in plant cells by growth regulators. *Biotechnol., 3*, 62-74.

Tsygankova, V. A., Stefanovska, T. R., Galkin, A. P., Ponomarenko, S. P., & Blume, Ya. B. (2012). Inducing effect of PGRs on small regulatory si/miRNA in resistance to sugar beet cyst nematode. *Comm. Appl. Biol. Sci., Ghent University (Belgium), 77*(4), 779-788.

Tsygankova, V. A., Yemets, A. I., Iutinska, H. O., Beljavska, L. O., Galkin, A. P., & Blume, Ya. B. (2013). Increasing the Resistance of Rape Plants to the Parasitic Nematode *Heterodera schachtii* Using RNAi Technology. *Cytology and Genetics, 47*(4), 222-230.

Tsygankova, V. A. (2012). Genetic mechanisms of wheat and chickpea (*Cicer arietinum* L.) inheritance of resistance to pathogenic micromycete of *Fusarium L.* genus. *Dopovidi Akademii nauk Ukrainy (Rep. Nat. Acad. Sci. Ukr.), 11*, 185-190.

Tsygankova, V. A., Ponomarenko, S. P., &Blume, Ya. B. (2012). The molecular genetic mechanisms of plant growth regulators' action with bioprotective properties. *Bull. Vavilov Soc. Genet. Breed.Ukr., 10*(1), 86-94. http://dx.doi.org/10.3103/S0095452713040105

Tuschl, T., Zamore, P. D., Lehmann, R., Bartel, D. P., & Sharp, P. A. (1999). Targeted mRNA degradation by double-stranded RNA in vitro. *Genes and development, 13*, 3191-3197. http://dx.doi.org/10.1101/gad.13.24.3191

Vaucheret, H. (2006). Post-transcriptional small RNA pathways in plants: mechanisms and regulations. *Genes & Development, 20*, 759-771. http://dx.doi.org/10.1101/gad.1410506

Vaucheret, H., Béclin, C., & Fagard, M. (2001). Post-transcriptional gene silencing in plants. *J. Cell Sci., 114*, 3083-3091.

Vavrina, C. S., Roberts, P. D., Kokalis-Burelle N., & Ontermaa E. O. (2004). Greenhouse Screening of Commercial Products Marketed as Systemic Resistance and Plant Growth Promotion Inducers. *Hort Science, 39*(2), 433-437.

Verhage, A., van Wees, S. C. M., & Pieterse, C. M. J. (2010). Plant Immunity: It's the Hormones Talking, But What Do They Say? *Plant Physiology, 154*, 536-540. http://dx.doi.org/10.1104/pp.110.161570

Winter, S. M. J., Rajcan, I., & Shelp, B. J. (2006). Soybean cyst nematode: Challenges and opportunities. *Canadian Journal of Plant Science, 86*(1), 25-32. http://dx.doi.org/10.4141/P05-072

Zhang, B., Wang, Q., & Pan, X. (2007). MicroRNAs and their regulatory roles in animals and plant. *J. Cell. Physiol., 210*, 279-289. http://dx.doi.org/10.1002/jcp.20869

Zhang, W., Gao, S., Zhou, X., Chellappan., P, Chen, Z., Zhou, X., & Zhang, X. (2011). Bacteria-responsive microRNAs regulate plant innate immunity by modulating plant hormone networks. *Plant Mol. Biol., 75*, 93-105. http://dx.doi.org/10.1007/s11103-010-9710-8

7

Phytosociological Study and Phytoecologique of Psammophytes of the Coastline of The Region of Tlemcen (Oranie-Algeria)

Stambouli-Meziane H[1], Merzouk A[1] & Bouazza M[1]

[1] Laboratory of Ecology and Management of Natural Ecosystems, Tlemcen, Algeria

Correspondence: Stambouli-Meziane H, Laboratory of Ecology and Management of Natural Ecosystems, Tlemcen, Algeria. E-mail: meziane_hassiba@yahoo.fr

Absract

This study is devoted to the analysis of psammophile of coastal dunes of the region of Tlemcen. The interpretation by the factor analysis of matches (A. F. C.) has allowed us to individualize classes' phytosociologique different. The colonize psammophile, par excellence, the embryonic dunes. Some species colonize the dunes vivid. Finally, other occupies the dunes the most advanced and laid down. Using the data and phytosociologique phytodynamiques, we were able to understand the evolution of this vegetation, and its diversity.

Keywords: Phytosociologie, psammophile, coastline, tlemcen, Algeria, diversity

1. Introduction

The vegetation, the region of Tlemcen, presents a good example of study of plant diversity; and especially an interesting synthesis on the natural dynamics of ecosystems from the shoreline up to the steppe. This study has been launched by several authors. These include mainly: Zeraïa (1981), Dahmani-Megrouche (1997), Quezel (2000), and Bouazza and Benabadji (1998).

The ecosystems Mediterranean coastlines are characterized by climatic constraints and strong soil, salinity, wind, drought and shallow soils or mobile.

The work that we are presenting here concerns the evolution of psammophiles the coastline of the region of Tlemcen. The latter is linked to a high percentage of sand, always higher than 60 %. Although they are located in the northern part; in the South, these formations are well represented and are essentially related to the importance of deposits of sand and the presence of gypsum and salts.

This study has been carried out on the basis of the readings phytosociologiques to determine the narrow affinities of different plant groups. In the second place, the knowledge of this floristic richness allows you to make proposals leading to the preservation and improvement of these fragile environments, to limit the degradation and to promote their development in a rational way.

2. Materials and Methods

The study covers the analysis of the distribution of species in the the coastal region of Tlemcen: study sites were chosen. From the beach Beni Saf up Marsat Ben M'hidi.

For this we chose two areas repartees as follows:

• Representative areas them live dunes and dune embryonic (from the beach Beni-Saf up to Marsat Ben M'hidi).

• Zones representing the semi-fixed dunes (Ghazaouet cement factory station (Beni-Saf).

The study area is characterized by a high floristic diversity which is related to the combination of ecological factors that are also very varied (variation bioclimatic, Action anthropozoogéne).

This study has been carried out on the basis of the readings phytosociologique to determine the narrow affinities of different plant groups. In the second place, the knowledge of this floristic richness allows you to make proposals leading to the preservation and improvement of these fragile environments, to limit the degradation and to promote their development in a rational way.

For this study it was selected 10 stations to study locating in the western part of the North West Algeria Figure.1. These are located between 1°27' and 1°51' west longitude and 34Â°27' and 35°18' north latitude. They are geographically limited:

- to the North by the Mediterranean sea

- to the south by the wilaya of Na'imah

- to the west by the moroccan-algerian border

- to the east by the wilaya of Temouchent

- to the south-east by the wilaya of Sidi Bel Abbes

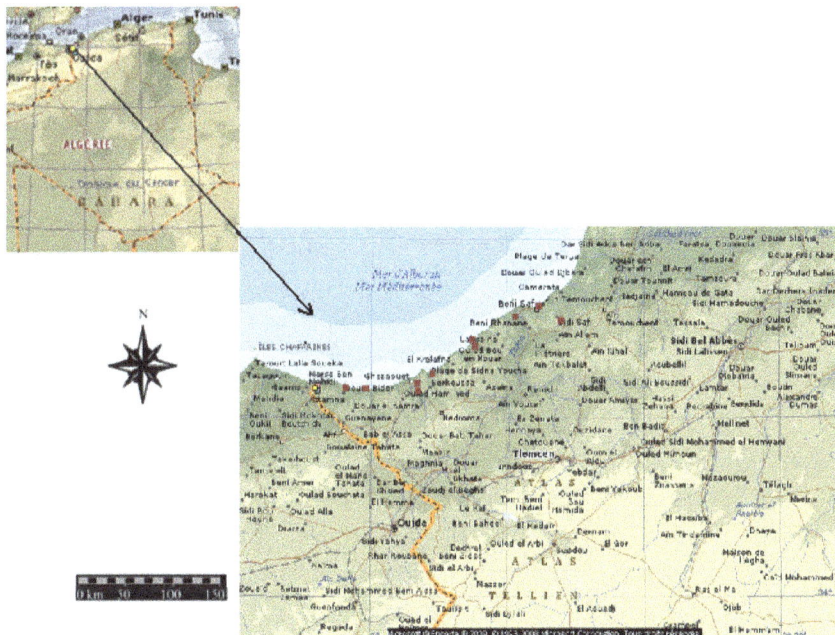

Figure 1. Location of Studies Stations

2.1 Béni Saf

Those lands are limestone lithothamniées rich in fossil shells lumachellique of type post- tablecloths Miocene. rest on these limestones intercalations clays to sandstone Tortonian age (Miocene). The limestones constitute a plateau called "Sidi Safi plateau" from which is calcium carbonate noted for cement plant Beni Saf These limestones are covered with places by volcanic formations of type basaltic, Guardia (1975).

2.2 Rachgoune

The station is located at the mouth of Tafna. These are the dune deposits at "El Guedim" and, on the right bank of the Oued, in these dunes appear basalt flows black color inter stratified with the volcanic tuffs, Guardia (1975).

2.3 Genesis of Sea Dunes

Under the effect of erosion, sand particles are going to move grace to winds to feed the dune ridge of coastline.

2.4 Dunes

The wind pushes the sand which will hang on waste brought by the sea. This forms a hump get bigger and bigger.This is the birth of a dune where embryonic going to develop a ephemeral vegetation based on: *Medicago marina; Cakile maritima; Euphorbia paralias.* According to Favennec (2002), dune is a deposition of sand edified by the wind into coming up against various obstacles such as vegetation and asperities terrain encountered between the beach and the mainland.

The dynamics of dunes depends on the one hand of the Wind speed and the dimension of sand particles and, on the other hand, obstacles which are the vegetation or the reliefs. As a function of the latter we distinguish 04 kinds of dunes.

- **The high dunes**: encountered the vicinity of the sea (beach Rachgoune, Beidar, Egla M'Khaled).

- **Dunes on slopes**: are on slopes exposed to the sea (the valleys Rachgoune).

- **Suspended dunes**: are formed on the cliffs parallel to neighborhood of the sea (Ouled Ayad).

- **Dunes clad**: depots constitutes tackles against of the scree of slope. It is characterized by the vegetation based : *Crucianella maritima, Thymelaea hirsuta* and *Elichrysum stoechas.(* Marsat Ben M'hidi)

The bioclimatic study for two periods (1913-1938) and (1970-2002). Figure **2** showed a vertical indent of each station in direct relation with the Q2 Emberger. Station Ghazaouet, despite falling on of the value of Q2 always under floor lower semi-arid to hot winter.

This climate favors the extension of vegetation therophytic xerophytes.

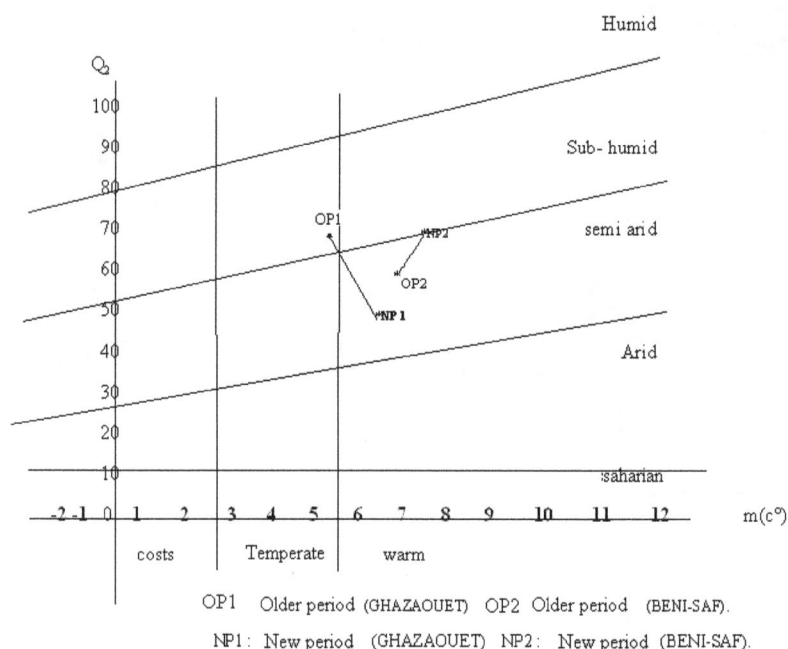

OP1   Older period (GHAZAOUET)   OP2 Older period (BENI-SAF).
NP1 : New period (GHAZAOUET)   NP2 : New period (BENI-SAF).

Figure 2. Temperature and humidity within them different zones

3. Results and Discussion (of the species from the beach)

This analysis focused on 98 surveys in the beach and shoreline (inside). It subdivided the processing of data in two part that corresponds:

- 41 records in the beach

-57 records indoors (coastal)

Plan	1	2	3
Rate of inertia	5,5330	3,8675	3,2664
Eigen values	0,138	0,097	0,082

Examination of factorial maps showing the plans of 2/1 and 3/1 projections shows the existence of 04 contrasting sets we will thus attempt to specify what will be the major ecological factors in the diversification of the sward.

3.1 Plan 2/1:

The negative side: *Lobularia maritima , Chenopodium album, Eryngium maritimum, Asteriscus maritimus, Calycotome spinosa, Chamaerops humilis, Chenopodium album, Hedysarum sp, Juniperus phoenicea*

The positif side: *Centaurium unbellatum, Cladanthus arabicus, Juniperus oxycedrus, Paronychia argentea, Rhamnus lycioides, Scabiosa stellata, Spartium junceum, Trifolium stellatum*

This axis against indifferent to substrates species and independence vis à vis the water factor.

3.2 Plan 3/1:

The negative side: *Gnaphalium luteo-album, Lagurus ovatus, Plantago marina, Quercus coccifera, Silene maritima, Spartium junceum, Ulex parviflorus*

The positif side: *Centaurium unbellatum, Cistus monspeliensis, Cistus salvifolius, Erica multiflora Phagnalon saxatile, Trifolium stellatum*

The positive side of this axis lie in particular species characterizing locations silica, and the negative side revealing a less silica than the first pole.

It seems that this group of species in their vast majority is plants that are most commonly seen in the matorral on siliceous substrate to Cisto-Lavanduletea.

In the center of the axis lie in particular a lot of species of *Cakile maritima, Ammophila arenaria, Calystegia soldanella, Echinophora spinosa Medicago minima, Medicago marina*

These species of psammophytes quintessential (purely psammophilous) that grow on the dunes a strong accumulation Sandy periods of respite from erosion marine; This is a group that is found on the beach of Rechgoune and Marsat Ben M'hidi at an elevation of 0 m on a low slope of zero and with a very low rate of recovery.

As it moves away from the beach, ecological conditions (Climate, soil) take up the top, and allow the installation of annuals and same perennial basis from: *Cynodon dactylon, Silene pseudo-atocion, Lobularia maritima, Teucrium pollium, Matthiola sinuate, Silene conica, Lagurus ovatus, Elichrysum stoechas.*

The behavior of species diversity and vegetation stresses the importance of the stability of the substrate for vegetation and already show that the coast is highly structured depending on the distance to the sea. The before dune and dune pioneer are influenced by a strong dynamic of sand and have a low plant collection, as well as a monotonous flora. Greater stability of the sand allows the species to cover larger areas and encourages, Furthermore, the coexistence of most abundant species.

It appears therefore that A.F.C was able to reveal the classic data of the littoral vegetation as described in 1923 by Kuhnholtz-Lordat. In phytosociological terms, maritime dunes belonging to the Ammophilion alliance, characterized by *Cakile maritima, Eryngium maritimum* et *Calystegia soldanella.*

Vegetation there forms a complex of plant associations arranged parallel to the shore (Figure 3) and richly described by Molinier and Tallon (1965).

The **Agropyretum-mediterraneum** is the typical association forming the belt at the base of the dunes (zone 1). Its characteristic species are all observed in (zone 1), we have: Agropyrum junceum, Polygonum maritimum et Cyperus aegyptiacus.

Ammophiletum arundinaceae is the typical of the growing dune (zone 2 and 3). It is characterized by: *Ammophila arenaria, Medicago marina* et *Echinophora spinosa.* Enfin, le **Crucianelletum** settled on the dunes consolidated (area 4 and 5) and is typically: *Crucianella maritima, Medicago littoralis et Pancratium maritimum.*

The indications of the analysis of vegetation carried out in this study support this vision, but should not consider these associations and their characteristic species too strictly. Indeed, the analyses presented here also highlight the ongoing nature of the changes in vegetation.

Nude beach (zone 0) there is no vegetation. This area is in fact continually swept by waves that do allow no plant to set. The sand is so naked up to the larger amplitude wave zone. Furthermore, as the present low amplitude tides Mediterranean, this portion of the beach is never covered by water and does not allow a particular fauna to settle. Apart from his recent tourist interest and the fact that it represents the pulling of the constitutive sand dunes area, this portion of the Mediterranean beach is low biological interest.

Table 1. Them floristic surveys of the beach and Valleys of Rachgoune

Station: valleys of Rachgoune. Beach and Siga

exposition : north-South

Recovery : 60-70%

altitude (m)		172	180	205	204	214	206	190	209	212	160	185	180	210	200	160
GENRES SPECIES	Survey	1	2	3	4	5	6	7	8	9	10	11	12	13	14	15
Ammophila arenaria (L.) Link.	Poacées										+					
Anagalis arvensis L.	Primulacées	0	0	+	+					+		+				
Asperula hirsuta L.	Rutacées	0	0	+	+	1	1						+	+		
Asphodelus microcarpus Salzm et Viv.	Liliacées	0	0	0	+											
Avena sterilis L.	Poacées	0	0	0								+				
Bromus rubens L.	Poacées											+				
Cakile maritima Scop.	Brassicacées										1					
Calendula arvensis L.	Astéracées	3	0													+
Calycotome spinosa (L.) Link.	Fabacées	2	2	1	+	1	1									
Calystegia soldanella L.	Convolvulacées										1					
Centaurium umbellatum (Gibb). Beck.	Gentianacées	+	1	2	+	+		+								
Chamaerops humilis L.	Palmacées	0	0	0	1											
Chrysanthemum grandiflorum (L.) Batt.	Astéracées									+	+	+				
Chrysanthemum coronarium L.	Astéracées											+				
Cistus monspeliensis L.	Cistacées	2	2	3	4	3	4	+	3	3	2		1	2		1
Cistus salvifolius L.	Cistacées	2	2	2	2		+	1								
Cladanthus arabicus (L). Cass.	Astéracées	2	2	3	4		2	+	+			3	1	2		3
Cuscuta sp (Tourn). L.	Cuscutacées															+
Dactylis glomerata L.	Poacées	1	0		+	+										
Daucus carota L.	Apiacées	2	0	2	+	1										
Daucus carota subsp gummifer Lamk.	Apiacées									1	+					
Echinops spinosus L.	Apiacées									1	+				+	
Echinophora spinosa L.	Apiacées									1			+			
Echium vulgare Tourn.	Borraginacées	3	0		+			+						+	+	
Ephedra fragilis Desf.	Ephedracées									1	+					
Erica multiflora L.	Ericacées	1	1	1	2	2				1	+					
Euphorbia paralias L.	Euphorbiacées									1	1	1				
Euphorbia peplis L.	Euphorbiacées	0	0	0	+											
Fagonia cretica L.	Zygophyllacées	3	0													
Gla diolus segetum Ker-Gawl.	Iridacées										+					
Globularia alypum L.	Globulariacées	2	0			+										
Gnaphalium luteo-album L.	Astéracées	0	0	0	0	+		1				+				1
Hedysarum sp L.	Fabacées	2	1		2			+	+							3
Inula crithmoides L.	Astéracées			+					+							
Juncus maritimus Lamk.	Juncacées									1	+					2
Juniperus oxycedrus L.	Cupressacées	4	3	4	4	4	4	4	4	3	4					4
Juniperus phoenicea L.	Cupressacées	0	0	+	1		+							2		2
Lagurus ovatus L.	Poacées	2	0									1				1
Limonium sinuatum (L.) Mill.	Linacées	0	0	0	0	0	0	+								

Espèce	Famille													
Linum strictum L.	Linacées	3	0					+		+	1	1		
Lygeum spartum L.	Poacées											1		
Marrubium vulgare L.	Lamiacées	0	0	0	0	0	+	+			+			4
Medicago marina L.	Fabacées	0	0	0	0	0	0	0	0	1	2	2	+	3
Medicago minima Grufb.	Fabacées	0	0	0	0	0	0	0	+	1				
Medicago littoralis Rhode.	Fabacées									1				
Mesembryanthemum nodiflorum L.	Aizoacées	0	0	0	+									
Muscari comosum (L.) Mill.	Liliacées	0	0	0	0	0	0	0	+					
Myrtus communis M.	Myrtacées	0	+				+	+		+				
Olea europaea L.	Oléacées	0	0	0	0	0	0	+	+					
Ononis spinosa L.	Fabacées	+	0											3
Ononis natrix L.	Fabacées	0	0	0	0	+			+	1		+		1
Paronychia argentea (Pourr.) Lamk.	Caryophyllacées	2	0	1	+									+
Phagnalon saxatile (L.) Cass.	Astéracées	2	2	3	3	2	3	+		+	2			
Phragmites communis	Poacées									+		+		+
Pinus halepensis L.	Pinacées	1	0										1	
Pinus maritima L.	Pinacées	2	0	3	1		1	4	2	2	1			1
Pistacia lentiscus L.	Oléacées	0	0	+	+	+								
Plantago argentea Desf.	Plantaginacées	3	0									1		
Plantago lagopus L.	Plantaginacées									+	1		+	1
Plantago marina L.	Plantaginacées	0	2	2			+				1			
Plantago psyllium L.	Fabacées	0	1	2			+							
Quercus coccifera L.	Fagacées	3	2	2							3			
Raphanus raphanistrum L.	Brassicacées	0	0	0	+									
Reichardia tingitana (L.) Roth.	Astéracées	0	0	0	+									
Rhamnus alaternus L.	Rhamnacées	0	0	0	0	0	1		1	+				+
Rhamnus lycioides L.	Rhamnacées	1	1			1	+		1	1				
Rosmarinus officinalis L.	Lamiacées								1	+				
Rubia peregrina L.	Rubiacées													1
Rubia sp L.	Rubiacées	0	0	4	4		+							
Salicornia ramosissima L.	Chénopodiacées								+	+				
Scabiosa slellata L.	Dipsacacées	2	1	3	2	1	+		2	+	2			+
Scorpiurus vermlculatus L.	Fabacées								1	+				
Senecio leucanthemifolius Poiret.	Astéracées				+						+	+		
Silene coeli-rosa (L.) A. Br.	Caryophyllacées				+						2		2	1
Silene maritima L.	Caryophyllacées	2	2			2								
Spartium junceum L.	Fabacées	+	1	3	4	3	3			1	2			
Teucrium fruticans L.	Lamiacées	0	1							+	1			
Teucrium polium L.	Lamiacées									+	1	+		
Thymus ciliatus Desf.	Lamiacées									1	+	+		3
Trifolium stellatum L.	Fabacées	0	1	0	+	1		+		+	1		+	
Ulex parviflorus Pourret.	Fabacées	0	0	0	1		+						+	

Table 2. Them floristic surveys of beach of Ouled Ben Ayad (Ghazaouet)

Station : Ouled Ben Ayed					
exposition : north east					
Covering rates : 40-50%					
Substrate : silicious					
Altitude (m)		185	16	40	20
GENRES SPECIES	Survey	1	2	3	4
Alopecurus pratensis L.	Poacées	+		+	
Arenaria emarginata Brot.	Caryophyllacées	+		+	+
Asteriscus maritimus (L.) Less.	Astéracées	1	+	+	+
Atriplex halimus L.	Chénopodiacées	+			+
Avena sterilis L.	Poacées	+		+	+
Bromus madritensis L.	Poacées	1		+	1
Bromus rubens L.	Poacées	+			1
Cakile maritima Scop.	Brassicacées	+		+	+
Centaurea pullata L.	Astéracées	+	+	+	+
Chrysanthemum grandiflorum (L.) Batt.	Astéracées	1	1	+	+
Dactylis glomerata L.	Poacées	1		+	1
Erodium moschatum L.	Géraniacées	+	+	+	+
Frankenia laevigata L.	Frankeniacées	+	+	+	+
Gnaphalium luteo-album L.	Astéracées	+		+	+
Hedysarum sp L.	Fabacées	1	1	1	+
Hippocrepis multisiliquosa L.	Brassicacées	+		+	+
Hordeum murinum Witth.	Poacées	1	1	+	+
Inula crithmoides L.	Astéracées	+		+	+
Lagurus ovatus L.	Poacées	1	1	1	+
Lavatera maritima Gouan.	Malvacées	1	1	+	+
Lobularia maritima (L.) Desv.	Brassicacées	+		+	+
Lolium rigidum Gaud.	Poacées	+		+	1
Lotus ornithopoides L.	Fabacées	+		+	+
Malva sylvestris L.	Malvacées	+		+	+
Matthiola sinuata (L.) R. Br.	Fabacées	1	+		+
Medicago marina L.	Fabacées	+	1	1	+
Medicago littoralis Rhode.	Fabacées	+	+	+	+
Medicago minima Grufb.	Fabacées	+		+	
Oxalis pes-caprae L.	Oxalidacées	+		+	+
Orchis purpurea L.	Orchidacées	+	+		+
Ononis natrix L.	Fabacées		1	+	
Paronychia argentea (Pourr.) Lamk.	Caryophyllacées		+		+
Phagnalon saxatile (L.) Cass.	Astéracées	+		+	+
Plantago lagopus L.	Plantaginacées	+	+		+
Plantago marina L.	Plantaginacées	+		+	+
Plantago psyllium L.	Plantaginacées	+	1		+
Raphanus raphanistrum L.	Brassicacées	1	1	1	+
Reichardia tingitana (L.) Roth.	Astéracées	+		+	+
Senecio leucanthemifolius Poiret.	Astéracées	+		+	+
Silene maritima L.	Caryophyllacées	+		+	+
Suaeda maritima (L.) Dumort.	Chénopodiacées	+	+		+
Trifolium angustifolium L.	Fabacées	+			+
Trifolium stellatum L.	Fabacées	+		+	

Table 3. Them floristic surveys of beach of Beni Saf; Sidi Boucif and Sid Safi

		Station : beach of Beni Saf				Station : beach of Sidi Boucif								Station : beach of Sidi Safi						
		Exposition : North				exposition : North								Exposition : North						
		Covering rates: 05-10%				Covering rates : 20-25%								Covering rates : 10-20%						
Altitude (m)		190	209	212	160	191	191	200	200	210	210	210	200	200	271	273	270	260	215	202	180
GENRES SPECIES	Survey	1	2	3	4	5	6	7	8	9	10	11	12	13	14	15	16	17	18	19	20
Andropogan hirtus L.	Poacées	+		+	1																
Anagallis arvensis L.	Primulacées															+		+		+	
Arenaria emarginata Brot.	Caryophyllacées	1	+		1	1		+	+	1	1	+	+	+							
Asparagus stipularis Forsk .	Liliacées	1		+	1	1	1	+		+	+		+								
Asteriscus maritimus (L.) Less.	Astéracées						+	+	+	+	+		+	+	1	+	1	1		1	+
Atractylis concellata L.	Astéracées	+														+					
Atractylis pycnocephalus L.	Astéracées	+														+					
Atriplex halimus L.	Chénopodiacées				+													+			
Avena sterilis L.	Poacées					+					+		+								
Bellis annua L.	Astéracées					+	+	+	+	+	+										
Bupleurum protractum Hoffm . et Link.	Apiacées	1		1	+																
Bromus rubens L.	Poacées					+					+							1			
Calycotome spinosa (L.) Link.	Fabacées	+		+												+					
Calystegia soldanella L.	Astéracées	1		+	+																
Catananche coerula L.	Astéracées	1	+		+	+	+	+	+	+	+	+		+							
Centaurea pullata L.	Gentianacées						+													+	
Chamaerops humilis L.	Chénopodiacées	+		+	1										3	+				+	1
Chenopodium album L.	Astéracées														+						
Chrysanthemum grandiflorum (L.) Batt.	Astéracées														+						
Chrysanthemum coronarium L.	Cistacées	+		1																	+
Cuscuta sp (Tourn). L.	Cuscutacées						+														+
Dactylis glomerata L.	Poacées					+	+	+													
Daucus carota subsp gummifer Lamk.	Apiacées	1		1	+										+	1		2	1		1
Delphinium peregrinum L.	Renonculacées														+		+				
Echinophora spinosa L.	Apiacées	1		+	+																
Echium vulgare Tourn.	Borraginacées																1				
Ephedra fragilis Desf.	Ephedracées														+	1					
Erica arborea L.	Ericacées	+																			
Erica multiflora L.	Ericacées	1	+		+		1	+		+	+		+		+						
Erodium moschatum (Burm) L'Her.	Géraniacées																				
Eryngium tricuspidatum L.	Apiacées					+		+	+	+		+	+	+							
Eryngium maritimum L.	Apiacées																				1
Fagonia cretica L.	Zygophyllacées														+						
Globularia alypum L.	Globulariacées														2	+		3	+	1	1
Gnaphalium luteo-album L.	Astéracées					+		+	+	+	+		+	+							
Halimium halimifolium (L.) Willk.	Cistacées														1				1	1	+
Hedysarum sp L.	Fabacées	2	3	1	+															+	
Hordeum murinum Witth.	Poacées					+			+												
Inula crithmoides L.	Astéracées		1										+								

Espèce	Famille	1	2	3	4	5	6	7	8	9	10	11	12	13	14	15	16	17	18	19
Juncus maritimus Lamk.	Juncacées														2	1				
Juniperus phoenicea L.	Cupressacées	1	2	2	1	+		+		1					2	2			1	
Lagurus ovatus L.	Poacées														1	+	+	2		1
Lavandula stoechas L.	Lamiacées					+				+				+						
Lavandula dentata L.	Lamiacées	1	+	+	1															
Limonium sinuatum (L.) Mill.	Plumbaginacées														+		+		1	+
Linum strictum L.	Linacées																+			
Lobularia maritima (L.) Desv.	Brassicacées	+	1		+			1						+						
Lotus ornithopoides L.	Fabacées	1	+		1															
Malva sylvestris L.	Malvacées															1				
Marrubium vulgare L.	Lamiacées														3	3		1		
Medicago marina L.	Fabacées							+	+	+	+	+		+		1	1		1	
Medicago littoralis Rhode.	Fabacées							+	+	+	+	+		+						
Mesembryanthemum nodiflorum L.	Aizoacées	1	+		+															
Ononis spinosa L.	Fabacées														1			+		
Periploca laevigata Auct.	Asclépiadacées														+	1	1			
Paronychia argentea (Pourr.) Lamk.	Caryophyllacées					+		+	+	+	+		+	+	+					
Phagnalon saxatile (L.) Cass.	Astéracées	1	+	+	2															
Pinus maritima L.	Pinacées														1	+				+
Pistacia lentiscus L.	Oléacées	1		2	1										+	+	+			
Plantago argentea Desf.	Plantaginacées														+		1			
Plantago marina L.	Plantaginacées	1		+											1	1		+	1	2
Plantago psyllium L.	Plantaginacées							+	+	+	+									
Raphanus raphanistrum L.	Brassicacées					1	+	+	+	+	+	+	+	+	+	1				+
Reseda alba L.	Résédacées																			+
Reichardia tingitana (L.) Roth.	Astéracées					+		+	+	+	+	+	+							
Reseda lutea L.	Résédacées	+		+	+										1			+		
Rhamnus alaternus L.	Rhamnacées															1	2			+
Rhamnus lycioides L.	Rhamnacées					+					+		*							
Rosmarinus officinalis L.	Lamiacées	+	1		1															
Rubia peregrina L.	Rubiacées	+		+																
Rumex bucephalophorus L.	Polygonacées	1		1		+	+	+			+	+								
Ruta chalepensis L.	Rutacées						1	1	1	+		+	+							
Salicornia ramosissima L.	Chénopodiacées																+			1
Satureja graeca L.	Lamiacées	+	1		1															
Scorpiurus vermIculatus L.	Fabacées														+		3	1		
Sedum acre L.	Crassulacées	+		1		+		+	+	+	+	+	+							
Senecio leucanthemifolius Poiret.	Astéracées														1		1			
Silene maritima L.	Caryophyllacées														3	2	1	2	1	1
Smilax aspera L.	Liliacées														4	3	+	+	3	4
Stipa tortilis Desf.	Poacées										+		+							
Taraxacum officinalis L.	Astéracées	+		+		+		+	+	+	+	+			+			1	+	+
Tamarix gallica L.	Tamaricacées																+		1	
Teucrium polium L.	Lamiacées	+	1	+	1											3	+	1	1	1
Trifolium stellatum L.	Fabacées														2	1			+	1
Ulex parviflorus Pourret.	Fabacées										+	+	+							

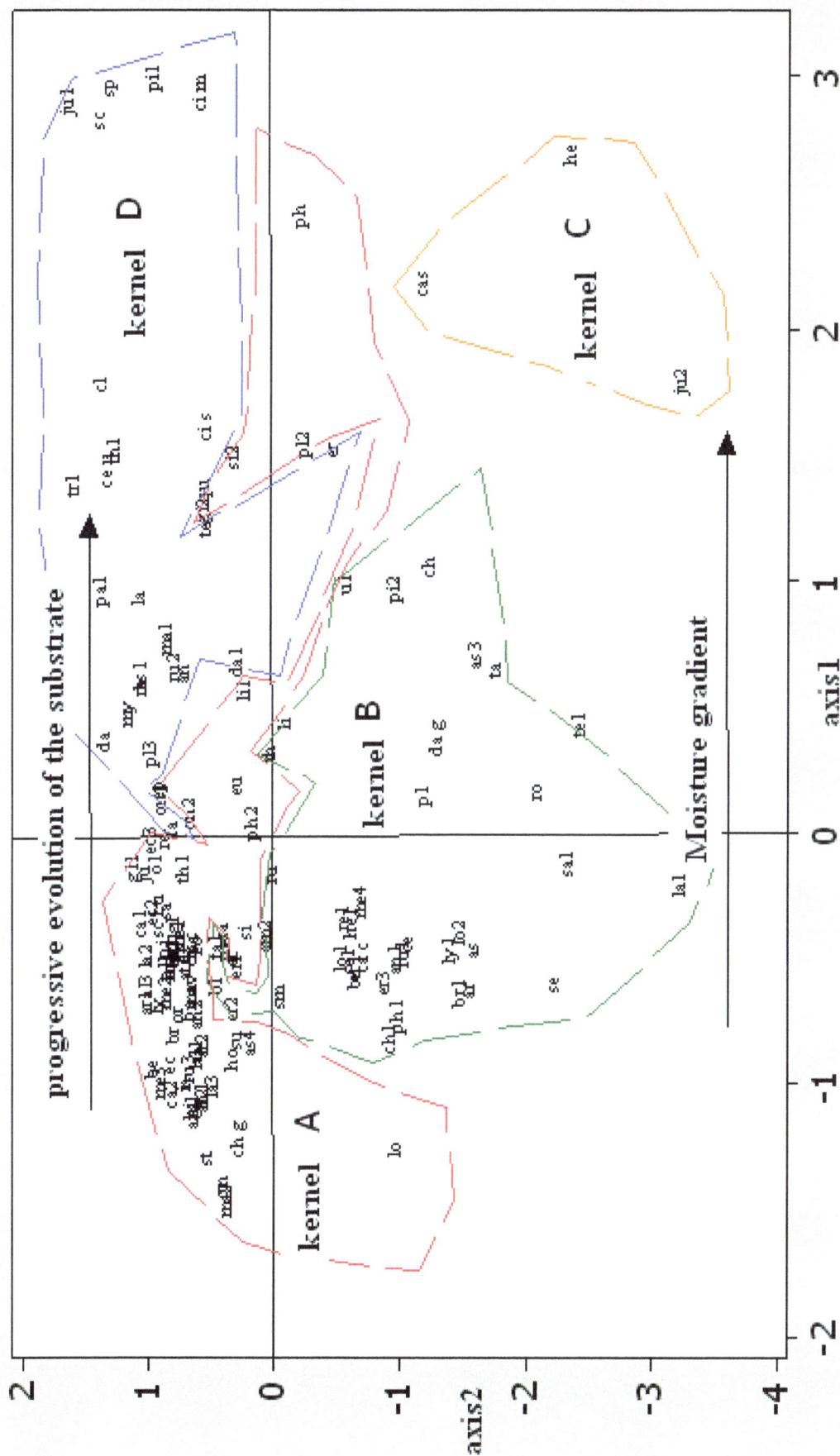

Figure 3. Factorial of species from the beach (axis 2 to axis 1)

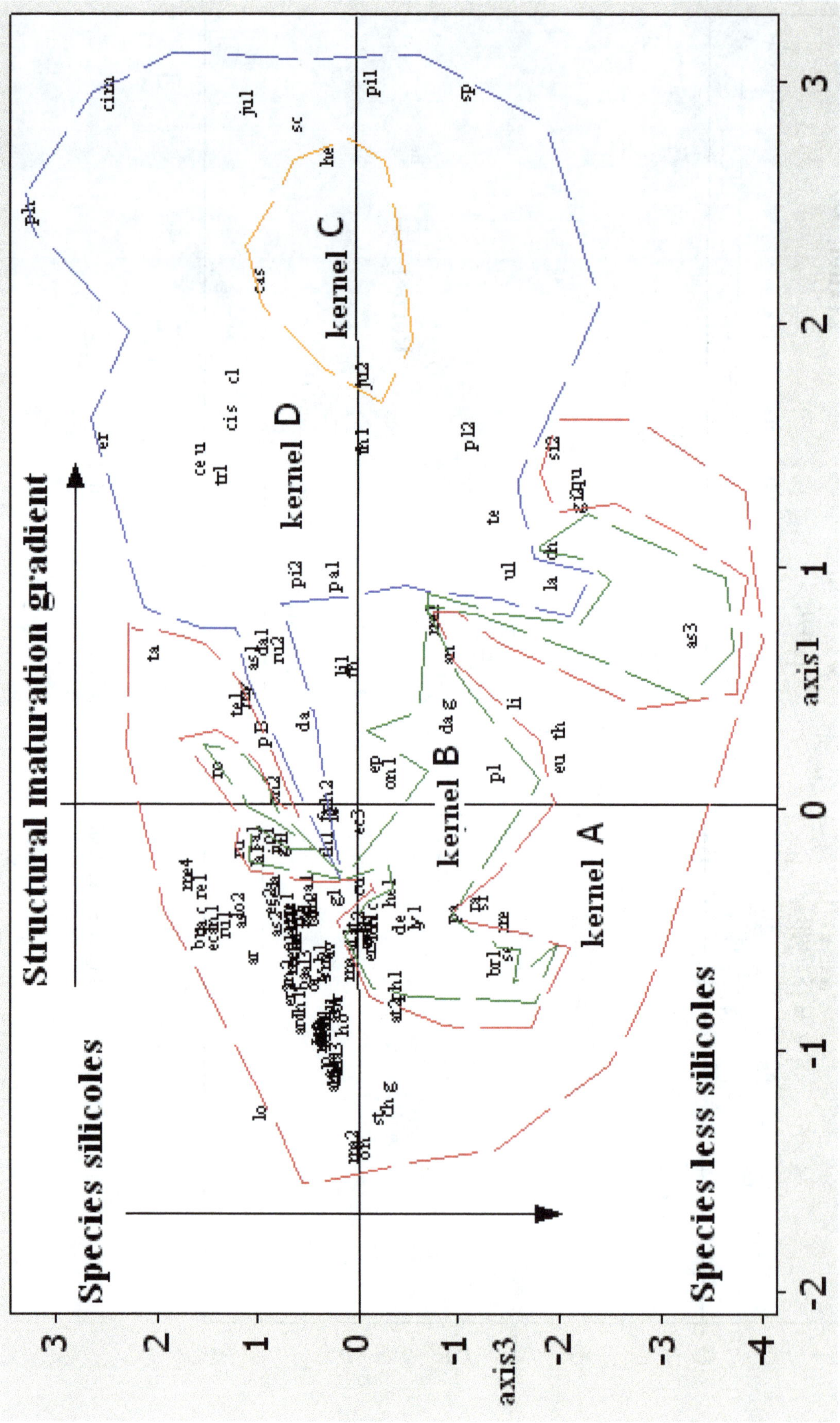

Figure 4. Factorial of species from the beach (axis 3 to axis 1)

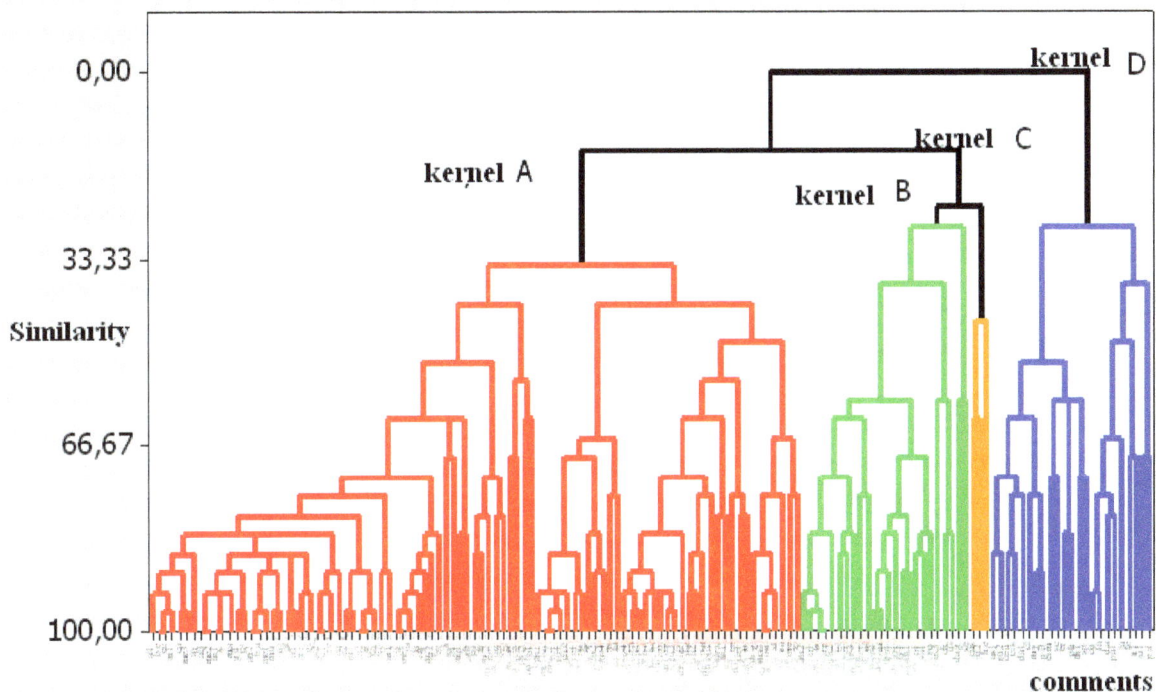

Figure 5. Dendrogram of the species from the beach

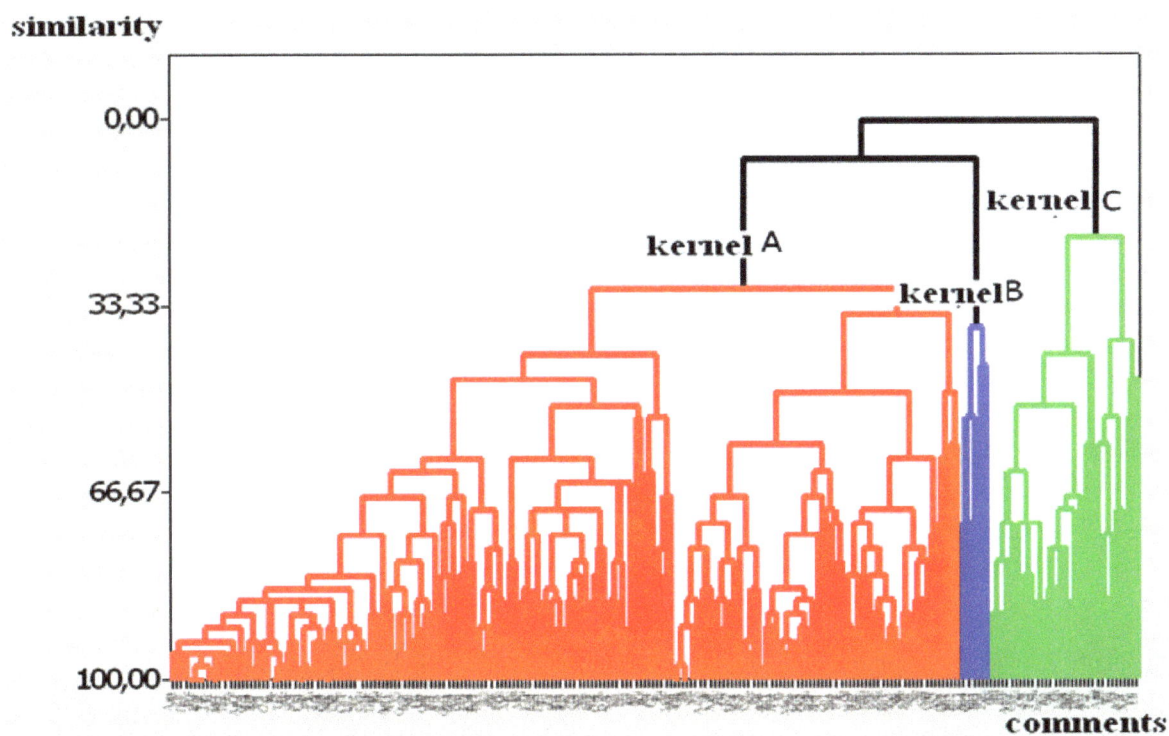

Figure 6. Dendrogram of coastal species

A

stability of the
sand
age of the
dune
Plant cover
Species
diversity

Species of the garrigue

B   ZONE 1   ZONE 2-3   ZONE 4- 5

Agropyretum mediterraneum   **L'Ammophiletum**
Agropyrum junceum   - *Ammophila arenaria*   - *Crucianella maritima*
Cyperus aegyptiacus   -*Medicago marina*   - *Medicago littoralis*
        - *Echinophora spinosa*   - *Pancratium*

maritimum.

C embryo Dune   Dune pioneer   Growing dune   Growing dune

        (Agropyretum)   (Ammophiletum)   Crucianelletum

                        Forest

                Storms   Destruction

                        anthropozoique

        Invasion by   Sand bare

        Ammophila

                        = erosion

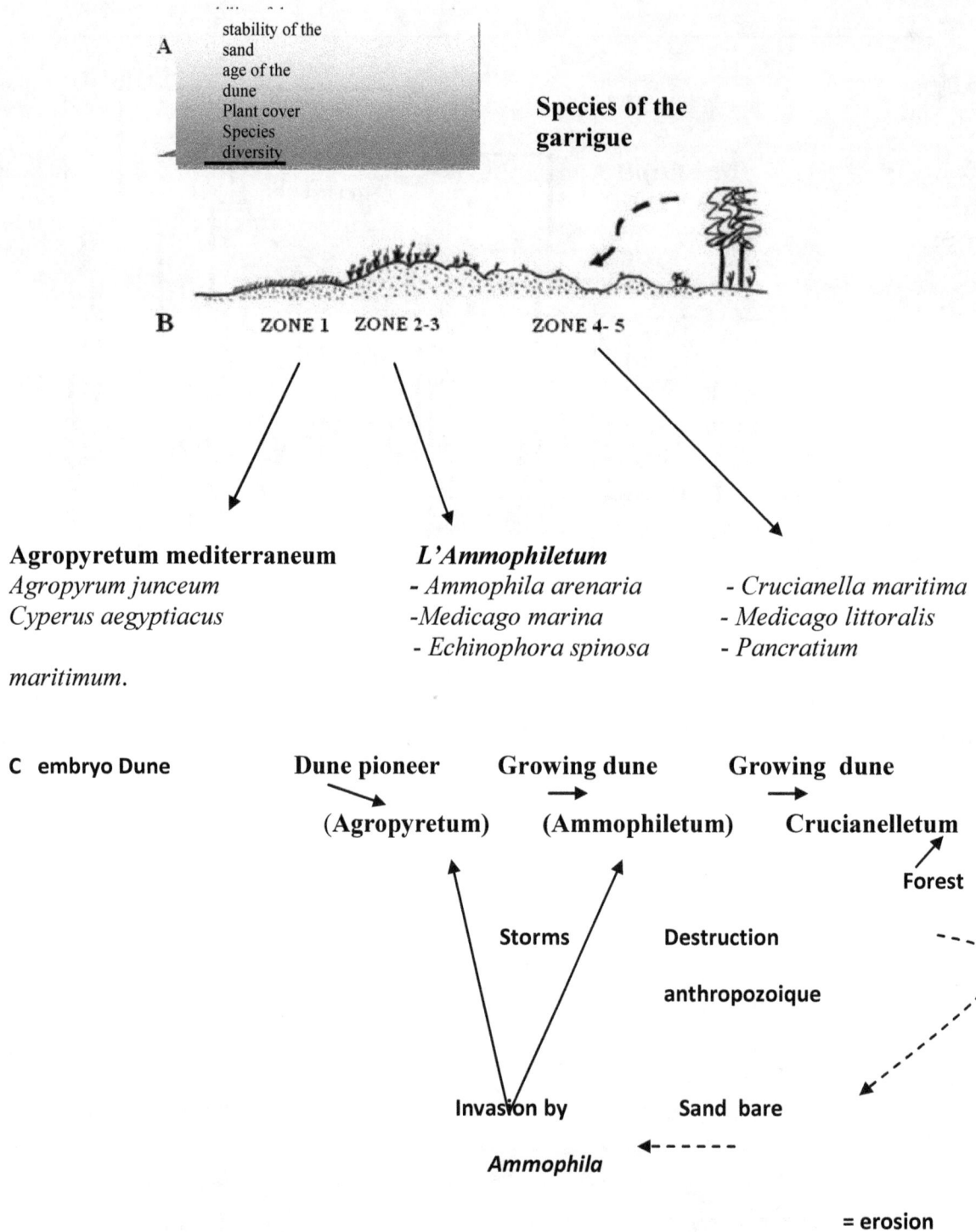

Figure 7. Schematic summary of the ecology of sandy coastline. (KUNHOLTZ – LORDAT 1923)

A: Appearance of the beach and summary of the reciprocal action of physical factors and vegetation. B: Vegetation Associations typical of different areas of the beach, with their characteristic species. These species are typically arranged in their order of appearance. C: Succession of the main plant associations of the sandy coastline. Disturbance (antropozoogenes or natural, tillé line arrows) cause regression of vegetation. The natural dynamics of the system leads to the establishment of a cyclic process.

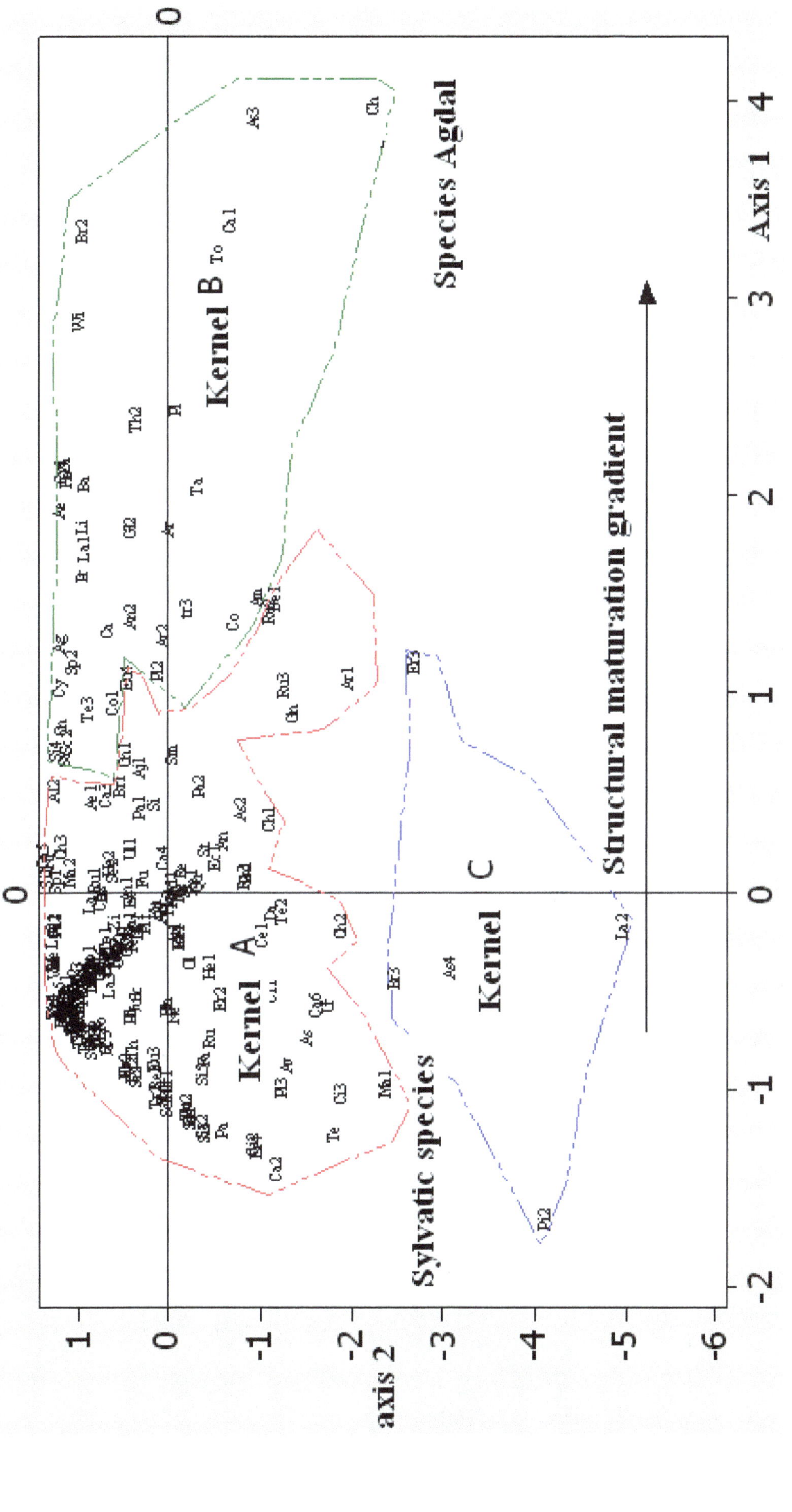

Figure 8. Plant factorial des especes du Littoral (axis 2 to axis 1)

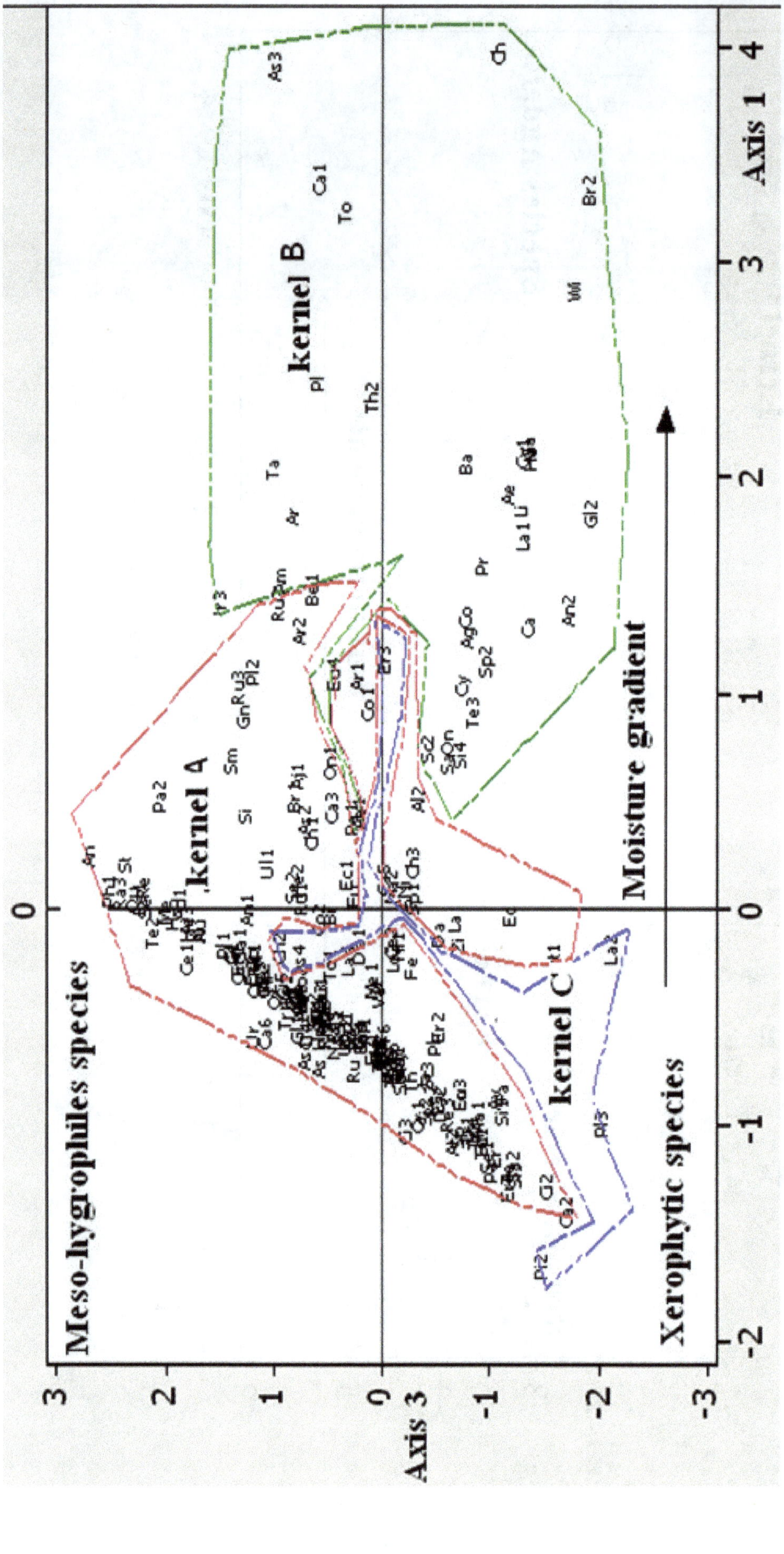

Figure 9. Factorial of coastal species (axis 3 to axis 1)

4. Results and Discussion (of the Species of the Coast)

It is a set of 57 phytosociological surveys and 223 species on the sandy coast of Sid Safi station; Rechgoune and Ghazaouet.

On the coast, the majority of the species belonging to the class of the **Therobrachypodietea**

4.1 Plan 2/1:

Negative side: *Ammoides verticillata, Arisarum vulgare, Asparagus acutifolius, Asteriscus maritimus, Avena sterilis, Cistus ladaniferus, Cistus monspeliensis, Eryngium maritimum, Eryngium tricuspidatum, Marrubium vulgare, Lobularia maritima, Genista numidica, Pistacia lentiscus, Tetraclinis articulata*

Positive side: *Althaea hirsuta, Atractylis carduus, Carthamus coerulus, Centaurea pungens, Retama monosperma*

In this area we find species very opposed by their procession and their vocation. The first batch consists of pre-stock species where a semblance of woodsy atmosphere. The second batch of so-called species mate without any taxonomic convergence

4.2 Plan 3/1 :

Negative side: *Aegilops ventricosa, Agropyron repens, Anthyllis tetraphylla, Avena sterilis, Atractylis concellata, Prasium majus, Plantago lagopus, Pistacia lentiscus, Tetraclinis articulata, Teucrium pseudo-chamaepitys, Echinops spinosus, Eryngium tricuspidatum*

Positive side: *Trifolium compestre, Teucrium polium, Stipa tenacissima, Smilax aspera, Raphanus raphanistrum, Reichardia picriodes, Reichardia tingitana, Quercus coccifera, Phylleria angustifolia, Ophrys apifera, Orchis coriophora, Medicago littoralis*

On the positive side there are relatively meso hygrophilous such as species *Ophrys apifera; Orchis coriophora .* and the negative side are relatively more tolerant species: *Echinops spinosus ; Eryngium tricuspidatum; Atractylis concellata*

This very clear axis to identify by a moisture gradient. Values are so low, they are difficult to interpret. The structuring of the cloud is no worse; Add to this the human conversations at this level leading to homogenization of the flora.

This region corresponds to a fixed dune characterized by a grouping evolved; plastic and weakly psammophilous, differentiated by *Juniperus phoenicea, Pinus maritime, Erica multiflora, Asparagus acutifolius, Asparagus stipularis.*

This advanced considered grouping formed the coastal juniperais. On the site map, it is aimed at altitudes of 100 m to 400 m, on slopes ranging from 5 to 25% and forming a collection of 70-90%; It is noted from the wet to the semi-arid.

The presence of this group of species is explained by its spatial heterogeneity and has his adaptation (R strategies). This plasticity confirms it a wider environmental spectrum, and colonization of the dunes by producing many seed species (*Pinus halepensis*; *Juniperus phoenicea*).

5. Conclusion

The analysis of the A.F.C highlighted 03 vegetal groups that organise themselves on the map 2/1 and 3/1 according to a schema corresponding to the analysis of adaptive strategies (MAC-ARTHUR 1957) in Chaabane (1993). This segregation is a variation of soil moisture and textural and structural elements. Furthermore, Therophytiques nitrophilous species endowed with a strong potential biotic and reciprocal growth settled more easily, there will be designated the R selection, the form's own selection.

The three groupings sets are represented by:

Groups psammophilous by excellence:

Ammophila arenaria, Cakile maritima, Calystegia soldanella, Eryngium maritimum, Medicago minima, Medicago marina

These species pertaining to **Ammophiletea** and **Cakiletea maritimae** class. They occupy the vertices of the beaches in maritime borders and also it means vegetation psammo-halo-nitrophilous therophytes (Chaabane, 1993) that characterize the embryonic dunes.

In moving away from the beach, a very diverse vegetation moved to attach these dunes giving birth to more or less fixed bright dunes. This vegetation is related to the **Therobrachypodietea** class

The interaction of different natural factors has the mosaics of biotopes and vegetation structure. But this arrangement, linked to the variability of the physicochemical characteristics of the substrates, is particularly disturbed by the actions of origins anthropozoogenic.

This pressure results in depletion of the most advanced sets, a loss of 'natural' biodiversity of specific groups of the coast; it caused great difficulty in the individualization says beaches **(Cakiletea maritimae)** or even of dunes fixed sets.

Finally, we say that the future is worrying about maintaining this national heritage. It is important that effective, even drastic measures be taken rapidly to alleviate this pressure of anthropozoic origin. Admittedly, this is not unique to the Algerian coast since it occurs in all countries of the Maghreb and on almost all of the territories, but the reduction of vested coastal areas to natural ecosystems is an aggravating factor.

It is with this concern to reduce the pressure, or even to improve the forest of the spit, that we brought the attention on *Ammophila arenaria*.

This species by their leaf system, its flexibility and its flexibility slows the speed of the wind and allow sand accumulation. It has good resistance to loosening and stabilizes the dunes by its root systems. Lay down the dunes and Ammophila arenaria in the coast plantations offer a picturesque great heritage value and original topography.

References

Bouazza, M., & Benabadji, N. (1998). *Composition floristique et pression anthropozoïque au Sud–Ouest de Tlemce* (pp. 93-97). Rev. Sci. Tech. Univ. Constantine, Algérie.

Chaabane, A. (1993). *Etude de la végétation du littoral septentrional de Tunisie: typologie, syntaxonomie et éléments d'aménagement* (Doctoral dissertation). Es-sciences en Ecologie. Uni. Aix-Marseille III

Dahmani-Megrouche, M. (1997). *Le chêne vert en Algérie. Syntaxonomie phytosociologie et dynamique des peuplements* (Doctoral dissertation, Thèse doct. Es-sciences. UnivHouariBoumediene. Alger).

Favennec, J. (2002). *Guide de la flore des dunes littorales de la Bretagne au Sud des Landes*. Edition sud ouest/ONF.

Guardia, P. (1975). *Géodynamique de la marge alpine du continent africain d'après l'étude de l'Oranie nord-occidentale: relations structurales et paléogéographiques entre le Rif externe, le Tell et l'avant-pays atlasique* (Doctoral dissertation). Thèse 3ème cycle, Univ. Nice.

Kuhnholtz-Lordat, G. (1923). *Les dunes du Golfe du Lion: (essai de géographie botanique)*. Paris: Les Presses Universitaire de France.

Quézel, P. (2000). *Réflexions sur l'évolution de la flore et de la végétation au Maghreb méditerranéen* (Ibis Press Edit). Paris: Ibis Press.

Zeraia, L. (1981). *Essai d'interprétation comparative des données écologiques, phénologques et de production subero-ligneuse dans les forêts de chêne-liège de Provence cristalline (France méridionale) et d'Algérie* (Doctoral dissertation).

8

Woody Vegetation Utilisation in Tembe Elephant Park, Kwazulu-Natal, South Africa

Jerome Y. Gaugris[1,2], Caroline A. Vasicek[2] & Margaretha W. van Rooyen[3]

[1] Centre for Wildlife Management, University of Pretoria, Pretoria, South Africa

[2] Flora Fauna & Man, Ecological Services Ltd., Road Town/Tortola, British Virgin Island

[3] Ekotrust cc. 272 Thatcher's Field, Lynwood, Pretoria, South Africa

Correspondence: Jerome Y. Gaugris, Centre for Wildlife Management, University of Pretoria, 0002 Pretoria, South Africa. E-mail: jeromegaugris@florafaunaman.com

Abstract

A survey of woody plant species utilisation by large (excluding elephants), medium and small browsers,man and "natural damage", was conducted in nine vegetation units of Tembe Elephant Park, KwaZulu-Natal, South Africa. Woody species use and canopy removal were evaluated within two age ranges, (a) recent, ≤ 12 months prior to study and (b) old, > 12 months prior to the study. The results show that recent canopy removal by medium and small browsers was intensive and generally represented one third of height classes available to the agents which were consistently used withinall vegetation types. The overall utilisation pattern indicated that medium and small browsers may be removing the regeneration classof the woody plants layer. Natural damage was found to be considerable and it was hypothesized that it may be linked and possibly amplified by prior elephant utilisation. In conclusion, it is possible to suggest that the regular use of the sapling level by small and medium browsers could promote woodland to grassland retrogression, as was found in east Africa under high densities of animals.

Keywords: browsers, Maputaland, sand forest, tree utilisation, woodland

1. Introduction

Numerous ecological problems have been documented regarding over-concentration of animals; in small or large, fenced or even open systems (Walpole et al., 2004; Western, 2007; Eckhardt et al., 2000; Mosugelo et al., 2002). Among them is the transformation of a woodland landscape into shrubby grassland, or the suppression of woody vegetation growth (Western & Maitumo, 2004). Growing animal populations confined in reserves in Africa has come to the forefront of conservation issues because public, scientific, and conservation opinions are divided on how to manage this problem (Lombard et al., 2001; van Aarde & Jackson, 2007).

When the vegetation type supporting growing animal populations is endangered, poorly known or sensitive to utilisation, understanding the impact of animal populations is essential (Lombard et al., 2001; O'Connor et al., 2007). In Tembe Elephant Park (Tembe), a South African reserve, Sand Forest is considered one of the most valuable vegetation types of the Maputaland – Pondoland – Albany hotspot of biodiversity (Smith et al., 2006). In Tembe, conservation authorities indicated that Sand Forest conservation prevails over animal conservation targets (Matthews, 2006). However, subsequent to the park's fencing in 1989 and successful conservation measures, without limiting larger carnivores, animal populations have grown and appeared unlimited by density dependence (Guldemond & van Aarde, 2007). This is a concern as Sand Forest has been shown to react to large herbivore utilisation (Gaugris & van Rooyen, 2008), and requires that utilisation levels be established to understand the pressure and potential impact.

As Sand Forest origin remains in debate and because arguments exists that it may develop from a woodland succession sequence (Gaugris & van Rooyen, 2008), the present study investigated woody vegetation utilisation by small to large mammalian browsing herbivores in woodlands and Sand Forest of Tembe (excluding elephants) , to quantify utilisation. Human use although anecdotal is also reported. Finally the occurrence of what can be considered as "natural damage" is considered. These aspects are examined by evaluating general woody species utilisation and canopy removal within two different periods (≤ 12 months prior to the study and > 12 months prior to the study) and considering the implications of results observed.

2. Methodology

2.1 Objectives of the Study

The present study was undertaken to measure the utilisation of woody species in a fenced off conservation area. The study was considered urgent as growing numbers of browsing mammalian herbivores were considered to have a significant impact on the regeneration of woody plant species. The present study is therefore a description of the measured utilisation levels of woody plants by small to large herbivores (excluding elephants).

2.2 Study Area

The study area lies in Maputaland, KwaZulu-Natal, South Africa. Tembe's 30,000 ha (proclaimed in 1983) were completely fenced in 1989. Tembe is characterised by sandy plains interspersed with ancient littoral dunes and a north flowing swamp (Muzi Swamp) along the park's eastern boundary. Tembeis covered by a mixture of habitats types such as grasslands, woodlands and Sand Forest patches (Matthews et al., 2001). Summers are hot, wet, and humid, while winters are cool to warm and dry. The mean annual rainfall is 700 mm (Gaugris, 2008).

Tembe's larger mammalian herbivore populations are composed of the following species (re-established species are marked with an asterisk) with the numbers in brackets representing 2000 census – 2005census (Matthews, 2006):

- African elephant *Loxodonta africana* (130 - 179 (195 (Note 1))
- White rhinoceros* *Ceratotheriumsimum* (35 - 43)
- Black rhinoceros* *Dicerosbicornis* (22 - 20)
- Giraffe* *Giraffacamelopardalis* (100 - 131)
- Hippopotamus *Hippopotamusamphibius* (14 - 20)
- Plain's zebra* *Equusquagga* (200 - 176)
- Eland* *Tragelaphusoryx* (40 - 0)
- Buffalo *Syncerus caffer* (60 -100)
- Kudu* *Tragelaphusstrepsiceros* (290 - 532)
- Blue wildebeest* *Connochaetestaurinus* (130 - 434)
- Waterbuck* *Kobus ellipsiprymnus* (360 - 419)
- Impala* *Aepycerosmelampus* (600 - 694)
- Nyala *Tragelaphusangasii* (300 - 1800)
- Bushbuck *Tragelaphus scriptus* (unknown - 40)
- Reedbuck *Reduncaarundinum* (880 - 268)
- Grey duiker *Sylvicapragrimmia* (unknown - 200)
- Red duiker *Cephalophus natalensis* (unknown - 400)
- Suni *Neotragusmoschatus* (estimated > 500)
- Warthog *Phacochoerus africanus* (260 - 300)
- Bush pig *Potamochoerus porcus* (unknown)

3. Methods

3.1 Fieldwork

A total of 135 rectangular plotsof density-dependent size (15 m by 2 m to a maximum of 45 m by 19 m) placed throughout Tembe by stratified random sampling were surveyed between May and October 2004. Plots were located at least 50 m away from management tracks and 100 m away from tourist tracks.

Plots were divided into two halves lengthwise. All woody individuals (defined as plants with an erect to scrambling growth form and ligneous trunk) were identified to the species in the first half, but only individuals' ≥ 0.4 m tall were recorded on the return leg. Live and dead stems were counted and diameters (D) measured at 30 cm or above the basal swelling. Plant height (H) and height to base of the canopy (HBC – defined as the height where the larger lowest branches were found) were measured, followed by the largest canopy diameter (D1) and the canopy diameter perpendicular (D2) to D1. Standing dead trees were measured, while fallen dead trees were reconstructed using best fit relationships based on live trees.

Utilisation was evaluated for each plant and each utilisation episode was scored separately. The following parameters were recorded for each plant (Figure 1):

- Plant state;
- Utilisation type;
- Presumed agent for the observed utilisation event;
- Utilisation event age;
- Estimated percentage of material removed (canopy/bark/roots …) by utilisation event;
- Growth response to the damage/utilisation.

State of the woody plant as encountered

1	Normal growth
2	Normal with branch regrowth from breakage
3	Pollarded (main stem snapped off, height reduced) – tree living, resprouting
4	Pollarded (main stem snapped off, height reduced) – tree living, coppicing
5	Pollarded (main stem snapped off, height reduced) – tree living, no growth response
6	Pushed over, stem intact, still partially rooted - living
7	Pushed over, stem partially broken - living
8	Mostly normal growth with some hedge growth
9	Hedge growth from continuous, regular browsing
10	Coppice growth from larger (older) dead stem
11	Coppice growth from accumulated browsing of young plant
12	Coppice growth from repeated fire
13	Coppice growth from repeated moisture stress
20	Senescent
30	Tree dead - main stem partially broken
31	Tree dead - main stem completely broken (pollarded)
32	Tree dead - main stem pushed over (partially uprooted)
33	Tree dead - main stem debarked
34	Tree dead - main stem intact, accumulated branch removal
35	Tree dead - debarking and branches / stems removed
50	Tree dead - intact - cause of death unknown
51	Tree dead - intact - killed by moisture stress
52	Tree dead - intact - dead from shading
53	Tree dead - intact - dead from high light
54	Tree dead - killed by combination of moisture stress and branch removal
55	Tree dead - killed from combination of shading and branch removal
56	Tree dead - killed by fire
60	Tree dead - totally uprooted
70	Top kill - drought dieback
71	Top kill - frost dieback
72	Top kill - dieback from debarking
80	Windfall
90	Live – deciduous leaf loss
91	Dying some branches still alive
92	Hedge growth from human utilisation
93	Tree dead, pushed over and broken, not uprooted

Agent (Agt.) of utilisation

1	Elephant
2	Giraffe
3	Kudu
4	Eland
5	Black rhinoceros
6	Nyala
7	Impala
8	Bushbuck
9	Grey duiker
10	Red duiker
11	Suni
12	Unidentifiable mega browsers (elephant, giraffe)
13	Unidentifiable large/medium size browsers (kudu, nyala, eland, etc)
14	Unidentifiable medium/small size browsers (impala, bushbuck, duiker etc)
15	Moisture stress
16	Flooding
17	Shading
18	High light intensity
19	Fire
21	Wind
22	Accidental
23	Unknown
24	Human
25	Insects
26	Cane rat
27	Lightning
28	Cattle
29	Porcupine
30	Goats

Type of utilisation observed

1	Whole plant (canopy and roots) utilized
2	Whole canopy utilized (roots still intact in ground)
3	Leaves and small twigs removed
4	Leaves, twigs, small branches, and large branches removed
5	Branch ends bitten off
6	Leaves plucked off
7	Leaves stripped
8	Parts of leaves removed
9	Only young leaves and leaf buds removed
10	Only mature leaves removed
11	Only senescent leaves removed
12	Bark removed
13	Roots removed
14	Flowers removed
15	Fruit / seeds removed
16	Dieback of main vertical branches/stems from top down
17	Dieback of horizontal branches/branch ends
18	Main stem/s cut
20	Accidental damage
21	No use / not damaged
22	Fire
23	Lightning
24	Pushed over and main stem broken
25	Pushed over and main stem intact

Growth responses (G.R.) to branch removal, stem breaking and debarking

1	Coppice growth
2	Wound regrowth
3	Main stem resprouting
4	No coppice or regrowth - vigour appears unaffected
5	No coppice or regrowth - vigour appears reduced (tree dying)
6	Hedge growth
7	Mostly hedge growth with some normal growth
8	Mostly normal growth with some hedge growth
9	Tree dead

Age of utilization (Age)

1	< 1 month
2	> 1 – 2 months
3	> 2 – 4 months
4	> 4 – 6 months
5	> 6 – 12 months
6	> 12 – 24 months
7	> 24 months
8	Continuous Regular Use

Debarking – circumference (Brk.)

1	1 % - 10 %	
2	11 % - 25 %	
3	26 % - 50 %	
4	51 % - 75 %	of the circumference
5	76 % - 90 %	of the stem removed
6	91 % - 99 %	
7	100 %	

Canopy volume removal

1	1 % - 10 %
2	11 % - 25 %
3	26 % - 50 %
4	51 % - 75 %
5	76 % - 90 %
6	91 % - 99 %
7	100 %

Debarking - stem height (Brk.)
Percentage of Stem Height

0.1	1 % - 10 %	
0.2	11 % - 25 %	
0.3	26 % - 50 %	
0.4	51 % - 75 %	of the height of stem
0.5	76 % - 90 %	removed
0.6	91 % - 100 %	
0.7	Whole stem plus branches	

The coding is derived from a code database used for other studies in KwaZulu-Natal, Northern Maputaland, South Africa, and only the codes relevant to the present study are displayed

Figure 1. Copy of the data capture coding sheet used for the study

3.2 Data Analysis

Plants were assigned to one of eight height classes (< 0.1 m, 0.1 to < 0.5 m, 0.5 to < 1.5 m, 1.5 to < 3.0 to < 5.0 m, 5.0 to < 8.0 m, 8.0 to < 12 m, ≥ 12 m) representative of vegetation structure (Gaugris & van Rooyen, 2011). Utilisation events affecting only canopy volume were distinguished from those representing overall utilisation events (including canopy removal, bark damage, stem or branch breakages, uprooting and other types of damage). Canopy volume removal events were separated between recent events (12 months prior to fieldwork), and old events, for all older events.All analyses were conducted for each agent category for each vegetation unit:

- Large browsers (giraffe)

- Medium browsers (kudu, eland, black rhinoceros)

- Small browsers (suni, red duiker, common duiker, nyala, impala, bushbuck)

- Man

- Natural damage: considered as all instances where elephant or otheragent'sutilisation couldbe excluded and where a natural cause (wind, drought, fire, lightning or light conditions …) could be considered as most likely reason for observed damage.

The range of height classes available to an agent was all height classes where any utilisation event was documented for that agent during our study. It was calculated as the sum of height classes for all woody species with potential for use per vegetation unit for the agent. For example if six woody species were sampled and known to be used by the agent, and all eight height classes known to have been used by the agent for these species during our study, 48 size classes are considered as available in the vegetation type. However, if the agent can only access the lower four size classes, only 24 size classes were available to the agent.

Available canopy volume and canopy volume removal per height class per woody species per vegetation unit were estimated using Walker's method of 1976. The number of height classes where canopy removal events by an agent was observed was counted at vegetation unit level and expressed as a percentage of total number of height classes available to that agent. The number of height classes where at least 50% of the canopy volume was removed was calculated similarly per agent at vegetation unit level.

4. Results

The Sand Forest classification in Gaugris and van Rooyen (2008) was followed. A total of 168 species were encountered (84% of Tembe's documented woody species) through measurement of 12,915 woody plants.

4.1 Sand Forest Association

In the Sand Forest association and *Afzeliaquanzensis* clumps, medium and small browsers affected 20 to 76% of woody species during the recent period (Table 1). The older canopy removal values showed that natural damage affected the greatest number of woody species (40 to 78% of species). Utilisation marks appeared less noticeable after 12 months for medium (40 to 60% less species appearing as used) and small browsers (88 to 100% less species appearing as used).

Table 1. The number of woody species utilised by various agents in the vegetation associations of Tembe Elephant Park

Vegetation Unit			*Afzeliaquanzensis* clumps (VT 01.1.1)		Short Sand Forest (VT 01.2.1)		Tall Sand Forest (VT 01.2.2)		Mature Sand Forest (VT 01.2.3)	
No. of Species sampled			25		60		71		53	
No. of Species used by	Age	Type	No	(%)	No	(%)	No	(%)	No	(%)
Large Browsers	Recent	CV	0	0.0	0	0.0	0	0.0	0	0.0
	Old	CV	0	0.0	1	1.7	0	0.0	1	1.9
	All ages	O U	0	0.0	1	1.7	0	0.0	1	1.9
Medium Browsers	Recent	CV	5	20.0	25	41.7	33	46.5	20	37.7
	Old	CV	3	12.0	11	18.3	20	28.2	8	15.1
	All ages	O U	5	20.0	28	46.7	37	52.1	27	50.9
Small Browsers	Recent	CV	19	76.0	25	41.7	35	49.3	20	37.7
	Old	CV	0	0.0	2	3.3	4	5.6	0	0.0
	All ages	O U	19	76.0	25	41.7	38	53.5	22	41.5
Man	Recent	CV	0	0.0	0	0.0	0	0.0	0	0.0
	Old	CV	0	0.0	2	3.3	4	5.6	0	0.0
	All ages	O U	0	0.0	2	3.3	5	7.0	0	0.0
Natural Damage	Recent	CV	0	0.0	12	20.0	17	23.9	5	9.4
	Old	CV	10	40.0	47	78.3	51	71.8	28	52.8
	All ages	O U	10	40.0	51	85.0	52	73.2	29	54.7

Values are given for canopy removal (number of species where a percentage of canopy volume (CV) was removed) for the two periods evaluated (Recent: within 12 months prior to the study and Old: > 12 months prior to the study) and for the overall utilisation (O U), including all utilisation events, but time was undetermined.

The overall utilisation analysis showed that small browsers are the greatest users in the *Afzeliaquanzensis* clumps. However, in the Sand Forest association, marks from natural damage appeared to affect more woody species than any other agents (Table 1). Human activity signs were old.

During the recent period, medium browsers utilised up to 25.91% of height classes available to them (Table 2). Small browsers used the greatest part of the canopy available to them in Sand Forest (up to 60.47% of height classes available). Canopy removal by natural damage was low. In the old period, canopy removal marks from small and medium browsers appeared to disappear with time (Table 2). Human-linked canopy removal took place in the past (Table 2). Signs of canopy removal by natural damage accumulated over time, and up to 55% of available height classes were "used" in Short Sand Forest.

Table 2. The number and percentage of height classes (HC) utilised by the various agents in the Sand Forest association of Tembe Elephan t Park

Agent	Range of height classes used	Age	Type	Vegetation Units									
				Afzeliaquanzensis clumps (VT 01.1.1)					Short Sand Forest (VT 01.2.1)				
				NHCS*	Total HC utilisation		HC use where> 50 % of HC used		NHCS*	Total HC utilisation		HC use where> 50 % of HC used	
					(No)	(%)	(No)	(%)		(No)	(%)	(No)	(%)
Large Browsers	03 - 06	Recent	CV	36	0	0.00	0	0.00	125	0	0.00	0	0.00
		Old	CV	36	0	0.00	0	0.00	125	1	0.80	0	0.00
		All ages	O U	36	0	0.00	0	0.00	125	1	0.80	0	0.00
Medium Browsers	02 - 06	Recent	CV	45	5	11.11	1	2.22	146	32	21.92	1	0.68
		Old	CV	45	3	6.67	1	2.22	146	12	8.22	1	0.68
		All ages	O U	45	5	11.11	4	8.89	146	35	23.97	12	8.22
Small Browsers	01 - 05	Recent	CV	43	26	60.47	0	0.00	130	43	33.08	1	0.77
		Old	CV	43	1	2.33	0	0.00	130	0	0.00	0	0.00
		All ages	O U	43	27	62.79	26	60.47	130	43	33.08	21	16.15
Man	02 - 07	Recent	CV	48	0	0.00	0	0.00	154	0	0.00	0	0.00
		Old	CV	48	0	0.00	0	0.00	154	1	0.65	1	0.65
		All ages	O U	48	0	0.00	0	0.00	154	2	1.30	0	0.00
Natural Damage	01 - 08	Recent	CV	49	0	0.00	0	0.00	163	15	9.20	1	0.61
		Old	CV	49	14	28.57	1	2.04	163	90	55.21	2	1.23
		All ages	O U	49	14	28.57	13	26.53	163	96	58.90	78	47.85

Agent	Range of height classes used	Age	Type	Tall Sand Forest (VT 01.2.2)					Mature Sand Forest (VT 01.2.3)				
				NHCS*	Total HC utilisation		HC use where> 50 % of HC used		NHCS*	Total HC utilisation		HC use where> 50 % of HC used	
					(No)	(%)	(No)	(%)		(No)	(%)	(No)	(%)
Large Browsers	03 - 06	Recent	CV	164	0	0.00	0	0.00	90	0	0.00	0	0.00
		Old	CV	164	0	0.00	0	0.00	90	1	1.11	0	0.00
		All ages	O U	164	0	0.00	0	0.00	90	1	1.11	0	0.00
Medium Browsers	02 - 06	Recent	CV	193	50	25.91	2	1.04	109	26	23.85	2	1.83
		Old	CV	193	28	14.51	2	1.04	109	8	7.34	2	1.83
		All ages	O U	193	64	33.16	17	8.81	109	40	36.70	21	19.27
Small Browsers	01 - 05	Recent	CV	173	68	39.31	1	0.58	96	30	31.25	0	0.00
		Old	CV	173	4	2.31	0	0.00	96	0	0.00	0	0.00
		All ages	O U	173	72	41.62	32	18.50	96	32	33.33	23	23.96
Man	02 - 07	Recent	CV	206	0	0.00	0	0.00	114	0	0.00	0	0.00
		Old	CV	206	7	3.40	1	0.49	114	0	0.00	0	0.00
		All ages	O U	206	8	3.88	1	0.49	114	0	0.00	0	0.00
Natural Damage	01 - 08	Recent	CV	224	27	12.05	0	0.00	121	5	4.13	0	0.00
		Old	CV	224	123	54.91	3	1.34	121	46	38.02	0	0.00
		All ages	O U	224	128	57.14	90	40.18	121	49	40.50	33	27.27

NHCS* = Number of height classes sampled in the range used by the agent;

HC = Height Classes;

No. = Number.

The number of height classes utilised is represented range utilised by the agent, the number of height classes in that range where utilisation of at least 50% of the height class was observed for at least one species. Values are given for canopy removal utilisation events (CV) for the two periods evaluated (Recent: within 12 months prior to the study and Old: > 12 months prior to the study) and for the overall utilisation (O U), including all utilisation events, but time was undetermined.

In terms of overall utilisation, values mirrored those of old canopy removal, although the percentage of height classes used was usually higher for all agents (Table 2), especially when considering the number of height classes where utilisation affected ≥ 50% of a height class canopy volume. More than 25% of height classes lost > 50% of their volume through natural damage.

4.2 Woodland Vegetation Group

Recent woody species use was greatest for small and medium browsers (Table 3). Small browsers used more species in Closed Woodland Thicket and Sparse Woodland on Sand than any other agent, and used ≥ 50% of woody species available in all woodland units. Natural damage affected less woody species than other agents in most instances. When considering old canopy removal, utilisation signs by medium and small browsers were disappearing. However, > 50% of woody species showed old natural damage signs in three woodland vegetation types.

Table 3. The number of woody species utilised by the various agents in the vegetation associations of Tembe Elephant Park

Vegetation Unit			Closed Woodland Thicket (VT 02.1.0)		Closed Woodland on Clay (VT 02.2.0)		Closed Woodland on Sand (VT 02.3.0)		Open Woodland on Sand (VT 03.1.0)		Sparse Woodland on Sand (VT 04.1.0)	
No of Species sampled			29		116		115		92		40	
No of Species used by	Age	Type	No	(%)	No	(%)	No	(%)	No	(%)	No	(%)
Large Browsers	Recent	CV	0	0.0	1	0.9	5	4.3	11	12.0	0	0.0
	Old	CV	0	0.0	1	0.9	0	0.0	4	4.3	0	0.0
	All ages	O U	0	0.0	3	2.6	5	4.3	14	15.2	1	2.5
Medium Browsers	Recent	CV	8	27.6	66	56.9	60	52.2	43	46.7	11	27.5
	Old	CV	5	17.2	35	30.2	37	32.2	23	25.0	4	10.0
	All ages	O U	8	27.6	89	76.7	78	67.8	60	65.2	15	37.5
Small Browsers	Recent	CV	16	55.2	79	68.1	76	66.1	72	78.3	21	52.5
	Old	CV	1	3.4	2	1.7	4	3.5	2	2.2	2	5.0
	All ages	O U	16	55.2	82	70.7	80	69.6	73	79.3	22	55.0
Man	Recent	CV	1	3.4	1	0.9	3	2.6	0	0.0	0	0.0
	Old	CV	0	0.0	0	0.0	1	0.9	0	0.0	1	2.5
	All ages	O U	1	3.4	2	1.7	4	3.5	0	0.0	1	2.5
Natural Damage	Recent	CV	1	3.4	28	24.1	20	17.4	38	41.3	7	17.5
	Old	CV	8	27.6	56	48.3	63	54.8	43	46.7	14	35.0
	All ages	O U	8	27.6	59	50.9	67	58.3	52	56.5	16	40.0

Values are given for canopy removal (number of species where a percentage of canopy volume (CV) was removed) for the two periods evaluated (Recent: within 12 months prior to the study and Old: > 12 months prior to the study) and for the overall utilisation (O U), including all utilisation events, but time was undetermined.

The overall woody species utilisation showed that the medium browsers used mainly the Closed Woodlands on Clay and Sand and the Open Woodland on Sand (more than 50% of available woody species). Overall utilisation by small browsers was consistently > 50% of sampled woody species in all woodland units. Natural damage affected > 50% of woody species in three woodland units. Recent signs of human utilisation appeared in Closed

Woodlands.

Small browsers affected the greatest number of height classes (Table 4) followed by natural damage and medium browsers. Utilisation by large browsers was seldom encountered. Natural damage affected a substantialnumber of height classes where canopy removal was ≥ 50% in the Closed Woodland on Sand and Clay, the Open Woodland on Sand and the Sparse Woodland on Sand. Old canopy removal events followed similar trends as Sand Forest, whereby signs of utilisation by medium and small browsers tended to disappear. Natural damage canopy removal affected approximately a third of height classes throughout the Woodlands.

Table 4. The number and percentage of height classes (HC) utilised by the various agents in the Woodland association of Temb e Elephant Park

				Vegetation Units								
				Closed Woodland Thicket (VT 02.1.0)					Closed Woodland on Clay (VT 02.2.0)			
Agent	Range of height classes used	Age	Type	NHCS*	Total HC utilisation		HC use where > 50 % of HC used		NHCS*	Total HC utilisation		HC use where > 50 % of HC used	
					(No)	(%)	(No)	(%)		(No)	(%)	(No)	(%)
Large Browsers	03 - 06	Recent	CV	41	0	0.00	0	0.00	261	1	0.38	0	0.00
		Old	CV	41	0	0.00	0	0.00	261	1	0.38	0	0.00
		All ages	O U	41	0	0.00	0	0.00	261	3	1.15	0	0.00
Medium Browsers	02 - 06	Recent	CV	50	16	32.00	0	0.00	333	131	39.34	5	1.50
		Old	CV	50	11	22.00	0	0.00	333	55	16.52	5	1.50
		All ages	O U	50	14	28.00	9	18.00	333	189	56.76	105	31.53
Small Browsers	01 - 05	Recent	CV	43	24	55.81	0	0.00	321	150	46.73	0	0.00
		Old	CV	43	1	2.33	0	0.00	321	2	0.62	0	0.00
		All ages	O U	43	26	60.47	18	41.86	321	156	48.60	74	23.05
Man	02 - 07	Recent	CV	50	1	2.00	0	0.00	342	1	0.29	0	0.00
		Old	CV	50	0	0.00	0	0.00	342	0	0.00	0	0.00
		All ages	O U	50	1	2.00	0	0.00	342	2	0.58	1	0.29
Natural Damage	01 - 08	Recent	CV	51	2	3.92	0	0.00	364	48	13.19	6	1.65
		Old	CV	51	16	31.37	0	0.00	364	107	29.40	8	2.20
		All ages	O U	51	17	33.33	9	17.65	364	119	32.69	50	13.74

				Closed Woodland on Sand (VT 02.3.0)				
Agent	Range of height classes used	Age	Type	NHCS*	Total HC utilisation		HC use where > 50 % of HC used	
					(No)	(%)	(No)	(%)
Large Browsers	03 - 06	Recent	CV	262	5	1.91	0	0.00
		Old	CV	262	0	0.00	0	0.00
		All ages	O U	262	5	1.91	2	0.76
Medium Browsers	02 - 06	Recent	CV	317	115	36.28	2	0.63
		Old	CV	317	56	17.67	2	0.63
		All ages	O U	317	145	45.74	60	18.93
Small Browsers	01 - 05	Recent	CV	317	145	45.74	1	0.32

Agent	Range	Age	Type	NHCS*	(No)	(%)	(No)	(%)
		Old	CV	317	4	1.26	1	0.32
		All ages	O U	317	153	48.26	78	24.61
Man	02 - 07	Recent	CV	332	3	0.90	0	0.00
		Old	CV	332	1	0.30	0	0.00
		All ages	O U	332	4	1.20	0	0.00
Natural Damage	01 - 08	Recent	CV	373	36	9.65	8	2.14
		Old	CV	373	132	35.39	12	3.22
		All ages	O U	373	134	35.92	74	19.84

Agent	Range of height classes used	Age	Type	Open Woodland on Sand (VT 03.1.0)					Sparse Woodland on Sand (VT 04.1.0)				
				NHCS*	Total HC utilisation		HC use where > 50 % of HC used		NHCS*	Total HC utilisation		HC use where > 50 % of HC used	
					(No)	(%)	(No)	(%)		(No)	(%)	(No)	(%)
Large Browsers	03 - 06	Recent	CV	188	12	6.38	0	0.00	48	1	2.08	0	0.00
		Old	CV	188	4	2.13	0	0.00	48	0	0.00	0	0.00
		All ages	O U	188	15	7.98	1	0.53	48	1	2.08	0	0.00
Medium Browsers	02 - 06	Recent	CV	246	91	36.99	0	0.00	72	18	25.00	4	5.56
		Old	CV	246	40	16.26	0	0.00	72	4	5.56	1	1.39
		All ages	O U	246	122	49.59	44	17.89	72	20	27.78	8	11.11
Small Browsers	01 - 05	Recent	CV	243	145	59.67	3	1.23	73	34	46.58	0	0.00
		Old	CV	243	14	5.76	0	0.00	73	5	6.85	0	0.00
		All ages	O U	243	148	60.91	71	29.22	73	34	46.58	26	35.62
Man	02 - 07	Recent	CV	254	0	0.00	0	0.00	74	0	0.00	0	0.00
		Old	CV	254	0	0.00	0	0.00	74	1	1.35	0	0.00
		All ages	O U	254	0	0.00	0	0.00	74	1	1.35	0	0.00
Natural Damage	01 - 08	Recent	CV	273	85	31.14	6	2.20	77	12	15.58	8	10.39
		Old	CV	273	109	39.93	10	3.66	77	23	29.87	4	5.19
		All ages	O U	273	132	48.35	63	23.08	77	32	41.56	22	28.57

NHCS* = Number of height classes sampled in the range used by the agent;

HC = Height Classes;

No = Number.

The number of height classes utilised is represented in three ways, a total number of height classes utilised within the range utilised by the agent, the number of height classes in that range where utilisation of at least 50% of the height class was observed for at least one species. Values are given for canopy removal utilisation events (CV) for the two periods evaluated (Recent: within 12 months prior to the study and Old: > 12 months prior to the study) and for the overall utilisation (O U), including all utilisation events, but time was undetermined.

Overall utilisation values were generally higher than those for canopy removal and the percentage of height classes where at least 50% of the individuals' available canopy was used was noticeably higher, especially from natural damage (19.48%) in Sparse Woodland on Sand (Table 4).

5. Discussion

While utilisation of trees by elephant is easily observed and has previously been informedfor Tembe (Guldemond & van Aarde, 2007), medium and small mammalsbrowsing, less easily observed, had not yet been considered. Results indicate that utilisationproved quite intense, although this should not be surprising. A long term study in Zululand in Ithala Game reserve (Ithala) observed that browsers and "natural damage" generate three times more impact than elephants, and that the combined impact, over a period of eight years changed both woody species composition and population structures (Wiseman et al., 2004). In all Tembe's Woodlands and to a lesser extent in Sand Forest, small and medium browsers utilised a sizeable portion of available height classes (generally 1/3 to 2/3 of height classes available). Considering these animals' size, the height classes utilised are the smaller ones, which are important for recruitment, such as seedlings and saplings. In Ithala, the combination of elephants and browsers pressure was such that plant species composition changed from species with climax state lifecycle traits (long lived, low recruitment) to more pioneer like species (high recruitment, shorter lifespans) and also promoted the recruitment of less desirable browse species (Wiseman et al., 2004). Other studies have shown that small mammals herbivory is usually not a limiting factor on its own, although it slows regeneration in cases of high densities (Barnes, 2001; Walpole et al., 2004; Western & Maitumo, 2004; Western, 2006). In Kenya's Masaai Mara National Reserve, 73% of woody species were utilised by small browsers, which changed species composition and abundance andfacilitated some invasive species (Walpole et al., 2004). In Botswana, small to medium browsers were considered responsible for changing vegetation morphology (Styles & Skinner, 2001). The overall utilisation levels by small and medium browsers in Tembe are approaching Kenyan values or even exceeding them in some vegetation types. Therefore, the risk that further small and medium browser population increases would lead to homogenisation of some vegetation units cannot be discarded. Utilisation of woodlands in Tembe already showed that elephants homogenised Open Woodlands, with a risk of forcing succession towards Sparse Woodlands (Guldemond & Van Aarde, 2007). We consider that elephant utilisation assisted by the heavy smaller browser utilisation in Tembe's woodland conditions seenhere is likely to push succession from dense woodlands to sparse ones. This is likely to progressin much the same ways as happenedin East Africa (Western & Maitumo, 2004; Birkett & Stevens-Wood, 2005) unless a management action is taken to limit herbivore populations.

The question of mark accumulation appears quite straightforward. In an unfenced situation, natural migratory movements of animal populations, following rainfall and food availability afford plants time to recover after utilisation events (van Aarde & Jackson, 2007; Wiseman et al., 2004). Likewise, low animal population numbers make repeated use a rare event, and not the norm. However, in Tembe, fences restrict migratory movements, and the size of the park does not allow simulated migratory movements (van Aarde & Jackson, 2007), therefore, repeated utilisation events appear unsurprising unless it exceeds what can be considered as normal considering woody decay rates (Sheil & Salim, 2004). Medium and small browsers utilise many height classes but their actions appear to leave little durable signs of utilisation, which is logical as they tend to defoliate rather than break. Comparatively, elephants, which usually defoliate by breaking branches, stems, or even uprooting whole trees (Wiseman et al., 2004; O'Connor et al., 2007) leave a long-lasting imprint on vegetation that would be observed for long periods. Natural damage marks accumulation is, however, a puzzle.

"Natural damage" intensity was generally a concern. This includes all natural phenomenon that can potentially "damage" or even kill trees (wind, fire, lightning, moisture conditions, drought, light conditions, disease, etc.), which are part of a natural system under normal conditions or subsequent to catastrophic events (Lindenmayer et al., 2006). In Tembe's forests and woodlands, natural damage affects a considerable number of species (27 to 58% of species in woodlands and 40 to 85% of species in Sand Forest) in an obvious manner (32 to 41% of height classes available in woodlands and 28 to 59% of height classes available in Sand Forest). This aspect was also established in Ithalawhere natural damage was considered to represent 37% of all utilisation events (Wiseman et al., 2004). Of concern here is that marks appear to also accumulate over time (Tables 1, 2, 3, 4) at a rate that must exceed what it should be for a natural system in the absence of a natural disturbance in both Sand Forest and woodlands. While it is acknowledged that marks observation is influenced by wood decay rate (Sheil & Salim, 2004) and that in slow decay areas, marks can be observed for much longer periods and therefore falsely give impression that marks accumulate, in Short Sand Forest, 85% of woody species are affected by "natural damage" and 55.00% of available height classes displayed evidence of canopy removal (Table 2). This appears extremely high for a vegetation type where fire hardly occurs and where wind should have not much effect because of large tall trees absence (Izidine et al., 2003; Matthews, 2006). As the plants abundance and Short Sand Forest vegetation height make it particularly suited to elephant utilisation, akin to thickets in Addo Elephant Park (Lombard et al., 2001),in the present study, considering utilisation levels, we hypothesise that "natural damage" marks observed may be by default promoted by elephant browsing in the vegetation type.

6. Conclusion

In conclusion, as demonstrated for other areas, small and medium mammalian browsers appear to have a considerable influence on all vegetation types of Tembe. Although this impact appears to rarely last for more than a year in terms of canopy volume removal (defoliation), observed levels of use are close to levels measured elsewhere that were followed by a retrogression sequence with significant associated vegetation changes. As elephant impact is also considered significant in Tembe (Gaugris & van Rooyen, 2011; Gaugris et al., 2012), a clear danger exists of forcing Tembe's vegetation into a phase of rapid changes. Subsequent to the present study field work in 2004, larger carnivores (Lions and wild dogs) were reintroduced in Tembe, which should be a key step to restore natural limits to herbivore populations' growth. A follow up study to evaluate whether utilisation decreased as a result would be a worthwhile effort to consider the effectiveness of higher order predators' reintroduction to control herbivore impact on vegetation state and vegetation dynamics.

References

Barnes, M. E. (2001). Effects of large herbivores and fire on the regeneration of *Acacia erioloba* woodlands in Chobe National Park, Botswana. *African Journal of Ecology, 39*, 340-350. http://dx.doi.org/10.1046/j.1365-2028.2001.00325.x

Birkett, A., & Stevens-Wood, B. (2005). Effect of low rainfall and browsing by large herbivores on an enclosed savannah habitat in Kenya. *African Journal of Ecology, 43*, 123-130. http://dx.doi.org/10.1111/j.1365-2028.2005.00555.x

Eckhardt, H. C., Wilgen, B. W., & Biggs, H. C. (2000). Trends in woody vegetation cover in the Kruger National Park, South Africa, between 1940 and 1998. *African Journal of Ecology, 38*, 108-115. http://dx.doi.org/10.1046/j.1365-2028.2000.00217.x

Gaugris, J. Y. (2008). *The impacts of herbivores and humans on the utilisation of woody resources in conserved versus non-conserved land in Maputaland, northern KwaZulu-Natal, South Africa*. University of Pretoria, Pretoria, South Africa. Retrieved from http://upetd.up.ac.za/thesis/available/etd-06052008-162658

Gaugris, J. Y., & van Rooyen, M. W. (2008). A spatial and temporal analysis of Sand Forest tree assemblages in Maputaland, South Africa. *South African Journal of Wildlife Research, 38*, 171-184. http://dx.doi.org/10.3957/0379-4369-38.2.171

Gaugris, J. Y., & van Rooyen, M. W. (2011). The effect of herbivores and humans on the Sand Forest species of Maputaland, northern KwaZulu-Natal South Africa. *Ecological Research, 26*, 365-376. http://dx.doi.org/10.1007/s11284-010-0791-2

Gaugris, J. Y., Vasicek, C. A., & van Rooyen, M. W. (2012). Herbivore and human impacts on woody species dynamics in Maputaland, South Africa. *Forestry An International Journal of Forest Research, 85*, 497-512. http://dx.doi.org/10.1093/forestry/cps046

Guldemond, R. A. R., & van Aarde, R. J. (2007). The impact of elephants on plants and their community variables in South Africa's Maputaland. *African Journal of Ecology, 45*, 327-335. http://dx.doi.org/10.1111/j.1365-2028.2007.00714.x

Izidine, S., Siebert, S., & van Wyk, A. E. (2003). Maputaland'slicuati forest and thicket, botanical exploration of the coastal plain south of Maputo Bay, with an emphasis on the Licuati Forest Reserve. *Veld & Flora, 89*, 56-61.

Lindenmayer, D. B., Franklin, J. F., & Fischer, J. (2006). General management principles and a checklist of strategies to guide forest biodiversity conservation. *Biological Conservation, 131*, 433-443. http://dx.doi.org/10.1016/j.biocon.2006.02.019

Lombard, A. T., Johnson, C. F., Cowling, R. M., & Pressey, R. L. (2001). Protecting plants from elephants: botanical reserve scenarios within the Addo Elephant National Park, South Africa. *Biological Conservation, 102*, 191-201. http://dx.doi.org/10.1016/S0006-3207(01)00056-8

Matthews, W. S. (2006). *Contributions to the ecology of Maputaland, southern Africa, with emphasis on Sand Forest*. PhD thesis. University of Pretoria, Pretoria, South Africa.

Matthews, W. S., van Wyk, A. E., van Rooyen, N., & Botha, G. A. (2001). Vegetation of the Tembe Elephant Park, Maputaland, South Africa. *South African Journal of Botany, 67*, 573-594.

Mosugelo, D. K., Moe, S. R., Ringrose, S., & Nellemann, C. (2002). Vegetation changes during a 36-year period in northern Chobe National Park, Botswana. *African Journal of Ecology, 40*, 232-240. http://dx.doi.org/10.1046/j.1365-2028.2002.00361.x

O'Connor, T. G., Goodman, P. S., & Clegg, B. (2007). A functional hypothesis of the threat of local extirpation of woody plant species by elephant in Africa. *Biological Conservation, 136*, 329-345.

http://dx.doi.org/10.1016/j.biocon.2006.12.014

Shaw, M. T., Keesing, F., & Ostfeld, R. S. (2002). Herbivory on Acacia seedlings in an East African savanna. *Oikos, 98*, 385-392. http://www.jstor.org/stable/3547179

Sheil, D., & Salim, A. (2004). Forest Tree Persistence, Elephants, and Stem Scars. *Biotropica, 36*, 505-521. http://dx.doi.org/10.1646/1599

Smith, R. J., Goodman, P. S., & Matthews, W. (2006). Systematic conservation planning: a review of perceived limitations and an illustration of the benefits, using a case study from Maputaland, South Africa. *Oryx, 40*, 400-410. http://dx.doi.org/10.1017/S0030605306001232

Styles, C. V., & Skinner, J. D. (2000). The influence of large mammalian herbivores on growth form and utilization of mopane trees, *Colophospermummopane*, in Botswana's Northern Tuli Game Reserve. *African Journal of Ecology, 38*, 95-101. http://dx.doi.org/10.1046/j.1365-2028.2000.00216.x

van Aarde, R. J., & Jackson, T. P. (2007). Megaparks for metapopulations: Addressing the causes of locally high elephant numbers in southern Africa. *Biological Conservation, 134*, 289-299. http://dx.doi.org/10.1016/j.biocon.2006.08.027

Walker, B. H. (1976). An approach to the monitoring of changes in the composition and utilisation of woodland and savanna vegetation. *South African Journal Wildlife Research, 6*, 1-32.

Walpole, M. J., Nabaala, M., & Matankory, C. (2004). Status of the Mara Woodlands in Kenya. *African Journal of Ecology, 42*, 180-188. http://dx.doi.org/10.1111/j.1365-2028.2004.00510.x

Western, D. (2006). A half a century of habitat change in Amboseli National Park, Kenya. *African Journal of Ecology, 45*, 302-310. http://dx.doi.org/10.1111/j.1365-2028.2006.00710.x

Western, D., & Maitumo, D. (2004). Woodland loss and restoration in a savanna park: a 20-year experiment. *African Journal of Ecology, 42*, 111-121. http://dx.doi.org/10.1111/j.1365-2028.2004.00506.x

Wiseman, R., Page, B. R., & O'Connor, T. G. (2004). Woody vegetation change in response to browsing in Ithala Game Reserve, South Africa. *South African Journal of Wildlife Research, 34*, 25-37. http://hdl.handle.net/10520/EJC117186

Notes

Note 1. Hypothesis based on combination of known group count, total area count and informed guess.

9

Acclimatized Apparatus Enhanced Seed Germination in *Stevia rebaudiana* Bertoni

Raji A. Abdullateef[1,3], Mohamad bin Osman[2] & Zarina bint Zainuddin[1]

[1] Department of Biotechnology, Kulliyyah of Science, International Islamic University, Malaysia

[2] Faculty of Plantation and Agrotechnology, UniversitiTeknologi MARA (UiTM), Shah Alam, Selangor, Malaysia

[3] Sinwan Agricultural Research and Development Institute, Kwara State, Nigeria

Correspondence: Raji A. Abdullateef, Sinwan Agricultural Research and Development Institute, Ahli Sunnah close, Gbagba, Airport road, Ilorin, Kwara State, Nigeria. Email: abdullateef_raji@yahoo.com

Abstract

Stevia rebaudiana bertoni produces sweet glycosides with zero calorie and has strong health and dietary implications. With these properties, it has the potential to substitute sugar. However, poor seed germination in this plant constitute obstacle towards large scale propagation, thereby causing plant materials to be scarce and costly. High percent seed germination could be induced via simulation of favorable climatic environment. Thus, new protocols and prototype tagged 'seed germination apparatus', inclusive of conducive factors, were developed. Additionally, Seed viability level was also tested using 1 % tetrazolium chloride. ANOVA revealed significant differences between treatments at $p < 0.05$. The apparatus influenced high yield of about 67 % seed germination, while the viability test showed 69 % viable seeds. Acclimatized condition, owing to innovated seed germination apparatus, showed high impact on seed germination in stevia.

Keywords: apparatus, climatic simulation, poor seed germination, prototype, sweet glycosides, stevia, zero calorie

1. Introduction

Stevia rebaudiana Bertoni, also known as sweet leaf, sugar leaf or "stevia", belongs to the family Asteracea and it is one of the 300 species of the genus. The species exist as herbs and shrubs (Soejarto, Compadre, Medon, Kamath, & Kinghorn, 1983).

Stevia rebaudiana Bertoni is considered as a good substitute for sugar. It is a sweetener with zero calorie, indigestible in the human digestive tract because the compounds cannot be chemically breaking down. Thus, *Stevia rebaudiana* Bertoni constitutes a safe sweetener for diabetic patients (Strauss, 1995). The poor seed germination- 10% (Sakaguchi & Kan, 1992) and 36.3% (Goettemoeller & Ching, 1999) - problem with this plant caused obstacles towards establishment of *Stevia rebaudiana* Bertoni on a large-scale basis. This resulted in scarcity of the plant materials in terms of availability and cost effectiveness. Poor germination rates do not mean the seeds are dead. Non-germinability in seeds could be due to environmental factors such as very low humidity and extreme temperature (Murdoch & Ellis, 2000) and endogenous factors. Where the germination percentage is low like in the case of *Stevia rebaudiana* Bertoni, carrying out both viability and germination tests on seeds become very essential, both tests will reveal the status of the seeds whether non-viable or dormant (Hidayati, J. Baskin, & C. Baskin, 2002).

1.1 Seed Germination

Germination in seeds require necessary conditions which varies among plant depending on different species. Some of these conditions are stated below.

1.1.1 Light

The seeds of stevia are positively photoblastic (Brandle et al., 1998), thus, light is inevitable as a necessary factor for germinability.

Light plays important roles in plant development, as it is a necessary condition for photosynthesis to occur. Light also influences seed germination and seedling growth. In several plant species, light enhances seed germination, while in some plant species light inhibits seed germination (Jala, 2011). Different light treatments have been reported to affect seed germination of *Nepenthes mirabilis* (Jala, 2011). White and red light positively influenced germination in seeds, while the effect of red light is unparalleled.

In response to light, plants possess photoreceptors which include phytochrome, cryptochromes and one or more unrevealed ultraviolet light receptor(s), which are utilized for detecting and absorbing light (Runkle & Heins, 2001). Phytochrome is protein in nature, and its absorbance value varies in respect of different wavelength of the light spectrum. Peak absorbance are in the red region of the wavelength, ranging from 600 to 700 nm, and far-red region with wavelength ranging from 700 to 800 nm, while the least absorbance is in the blue region with wavelength ranging from 400 to 500 nm (Runkle & Heins, 2001).

1.1.2 Photoperiod Extension

Plant species respond to varying photoperiod for flowering and vegetative development. The use of extended light exposure periods, which varies from 8 h to 24 h length of day light was reported to show different impacts on seed germination in *Rhododendron vaseyi,* however overall germination at 30/20C (86/68F) with a 24-hr photoperiod produced highest percentage of germinated seeds (LeBude et al., 2008, Walker et al., 2006).

1.1.3 Temperature

Optimum temperature for effective seed germination depends on plant species, in most cases optimal germination profiles are found at intermediate temperatures. From previous studies, a temperature regime of 25°C was found to be optimum in *Stevia rebaudiana* seed germination (Carneiro et al., 1992).

1.1.4 Humidity

Seeds need moist condition to germinate (Styer & Koranski, 1997). The amount of water in the air determines its humidity. Optimum humidity is required for seeds to germinate, this is because at extreme low or high humidity, due to soil water content, poor seed germination may result.

1.2 Seed Viability

The most common and reliable method used for seed viability test is tetrazolium chloride (TTC) method (Traveset, 1998). TTC is accepted by the International Rules for Seed Testing (ISTA) as a chemical for testing seed viability (1999). This method has been effectively utilized across a wide range of plant species to predict germination and growth of seedling (Oliveira, Forni-Martins, Magalhães, & Alves, 2004).

1.3 Statement of the Problem

Seed germination is very low making production of plant materials difficult and expensive.

Studying new approaches to enhance seed germination profile in stevia becomes essential since the poor seed germination problem remains unresolved, despite several research works.

1.4 Aim of the Study

The aim of the current study therefore, is to unravel new protocols and innovate a germination apparatus to simulate a conducive climatic conditions for high seed germination profile in *Stevia rebaudiana* Bertoni.

2. Method

2.1 Seed Germination

In this study, new protocols and apparatus were developed and investigated for efficacy on seed germination enhancement in *Stevia rebaudiana* Bertoni.

2.1.1 Prototype: Seed Germination Apparatus

The seed germination apparatus comprised of the following: (i) Planting tray where the seeds are germinated, (ii) peat moss, which served as the sowing medium, (iii) plastic dome to house the planting tray, (iv) light chamber affixed with two red fluorescent tubes to supply the light, (v) watering can for spraying water over the peat moss, (vi) timer switch to regulate period of light exposure, which was 7am to 9pm, (vii) air conditioner, to regulate the temperature of the environment where the apparatus was set up. The whole apparatus was set up in the biological science laboratory of the kulliyyah of science.

2.1.2 Protocols

The protocols were developed to overcome poor seed germination in *Stevia rebaudiana* Bertoni, the details are explained below:

[i] Seed Harvesting, Storage Period and Sowing

Seed was harvested on sunny days as the pappus were dried and easily detached from the parent plants by gentle shaking. Harvesting on wet weather may pose some stress and seeds may decay during storage. Stevia seeds are sown 3-5 days after harvest, but may be stored up to 14 days in a fridge before sowing. Under room temperature with proper ventilation, storage should not exceed one week. Otherwise the small endospermic source of food for the embryo is consumed and seed fertility becomes affected.

[ii] Seed sowing in planting tray

Peat moss was used as the planting medium, it was filled into the holes of a planting tray. There were twenty (Lee et al., 2002) holes per a planting tray, and three seeds were sown in the peat moss per hole, indicating total number of 60 seeds per tray. A sowing depth of 0.3 - 0.5 cm was maintained because light irradiation could easily reach the seeds. The planting tray was then transferred into a thick, transparent, plastic dome. Three replicates of the dome were placed in the light chamber n under light irradiation.

[iii] Light

The plastic dome containing the planting tray was placed under red light irradiation (wavelength 660 nm) in the light chamber. The distance between the affixed light tubes (36 watts) in the light chamber and the surface of the planting tray in the plastic dome is about 16 cm. The light intensity value using data logger was at an average of 423 ± 10 lux. Red light gives better germination rate considering earlier findings (Raji & Osman, 2011; Shyam & David, 1975).

[iv] Extension of light exposure period

The period of light exposure of 14 h was maintained using a timer switch. Red light was switched on between 7.30 am to 9.30 pm daily in the light chamber. This is because *Stevia rebaudiana* Bertoni is a short day plant (Brandle, Starratt, & Gijzen, 2000). In addition, extension of light exposure period had shown positive impact on seed germination profile (LeBude et al., 2008, Walker et al., 2006).

[v] Temperature

Temperature was kept at 24°C using the air conditioner throughout the period of the experiment, based on previous work by Raji and Osman (2011), Sakaguchi and Kan (1992), following standard method (Goettemoeller & Ching, 1999).

[vi] Humidity

Water was sprayed on the peat moss at 2 days interval. Since the planting tray was placed inside the plastic dome, the water loss was controlled, thereby assisting in maintaining the humidity of the environment at an average value RH = $83.8 \pm 3.2\%$, using data logger, throughout the experiment.

2.1.3 Control Experiment

A control experiment was set up as in section 2.1.2.2, except that the domes were placed under natural environmental conditions to observe the seed germination performance.

2.2 Seed Viability

Two hundred black *Stevia rebaudiana* Bertoni seeds were divided into four groups, each containing 50 seeds. Three groups (replicates) were treated with 1 % (w/v) Tetrazolium chloride (TTC) in schott bottles, prepared by dissolving 1 g of TTC in 100 ml of distilled water (Traveset, 1998). For the control remaining one group of seeds was boiled in water inside a beaker placed on a Bunsen burner for five minutes in order to kill the seeds. Seeds were then transferred into a schott bottle containing TTC solution. All the samples were incubated at room temperature for 24 h in dark condition. After 24 h, the seeds were removed from TTC solution and staged on a Carl Zeiss micro imaging dissecting microscope for embryo observation. Seeds were considered viable when 90-100 % of the embryo surface was stained red with TTC (Oliveira et al., 2004; Bhering, Dias, & Barros, 2005). Experiment was conducted three times and data were subjected to one-way analysis using paired sample't' test, with SPSS version 16.

2.3 Statistics

Data collected from experiments were analyzed using ANOVA with the SPSS version 16.

3. Results

3.1 Seed Germination

3.1.1 Seed Germination Apparatus

Seed germination rate and percentage were highly influenced under the controlled environmental conditions using the germination apparatus (Figure 1).

The seed germination rate was 3.5 days (Figure 2), while the percentage of seed germination showed 67.33% (Figure 3) due to effects of the germination apparatus. However, the seed germination rate was 12 days (Figure 2), while the percentage of seed germination was 14% (Figure 3) considering the control experiment.

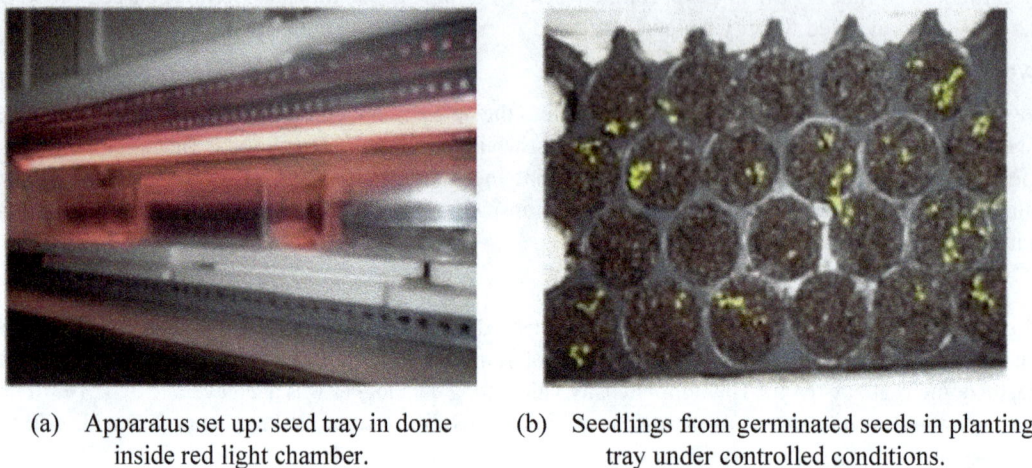

(a) Apparatus set up: seed tray in dome inside red light chamber.

(b) Seedlings from germinated seeds in planting tray under controlled conditions.

Figure 1. Seed germination apparatus

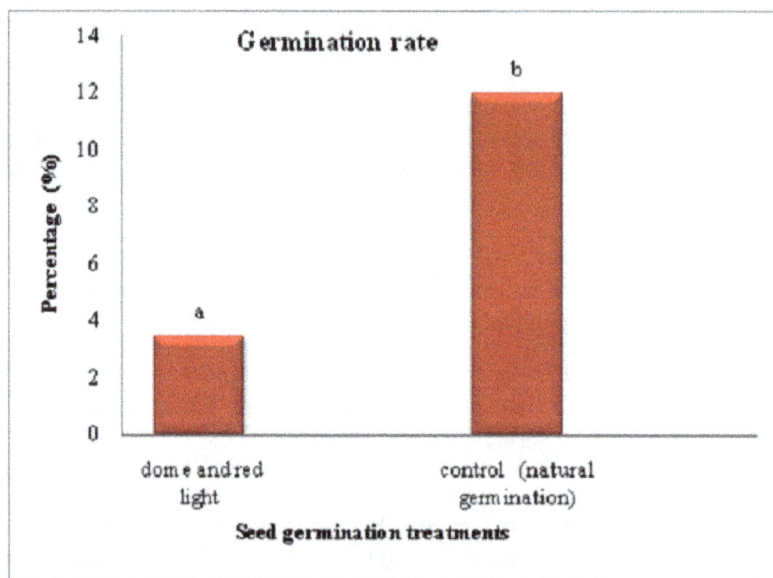

Figure 2. Effects of irradiation and box techniques on germination rate in seed of stevia. Different alphabets denote significant difference at $p < 0.05$

In summary, the seed germination rate and percentage have been successfully improved in *Stevia rebaudiana* Bertoni through the innovation of the irradiated seed germination box and necessary protocols. The 67.33% of germinated seeds observed in this study showed better output over findings from earlier publications which reported 36.3 % (Goettemoeller & Ching, 1999) and 41% (Raji & Osman, 2011) for seed germination in stevia. The efficacy may be attributed to provision of suitable conditions for germination.

3.2 Seed Viability

The viability test showed that stevia seeds are viable. The embryo of viable seeds are stained uniformly red as shown in Figure 4. Results in Figure 5 showed that 68.67 % of the seeds were viable while 31.33% were non-viable. The red colour of embryo was due to reduction of 2, 3, 5- triphenyl tetrazolium chloride in the presence of oxygen as a result of respiration occurring in the living cells of the embryo of viable seeds. The observed yellowish colour on other embryos was due to inability of the embryo to pick stain, possibly because it was dead.

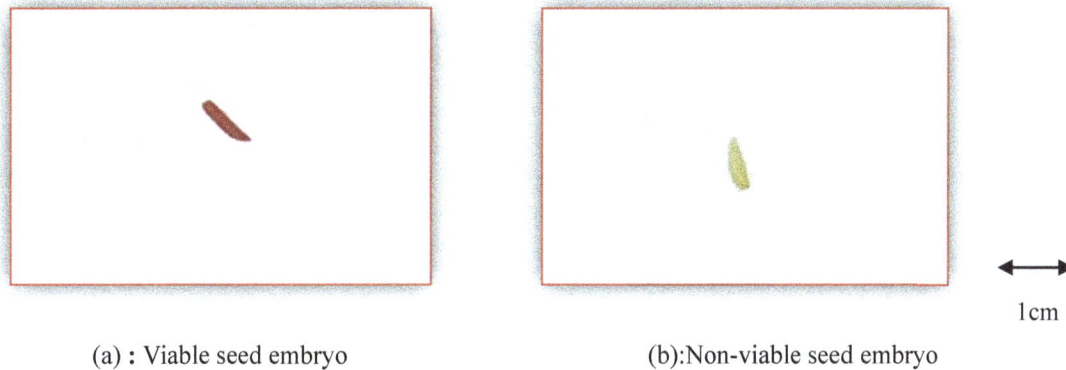

(a) : Viable seed embryo   (b):Non-viable seed embryo

Figure 4. Seed Viability Tests With Tetrazolium Chloride (Ttc)

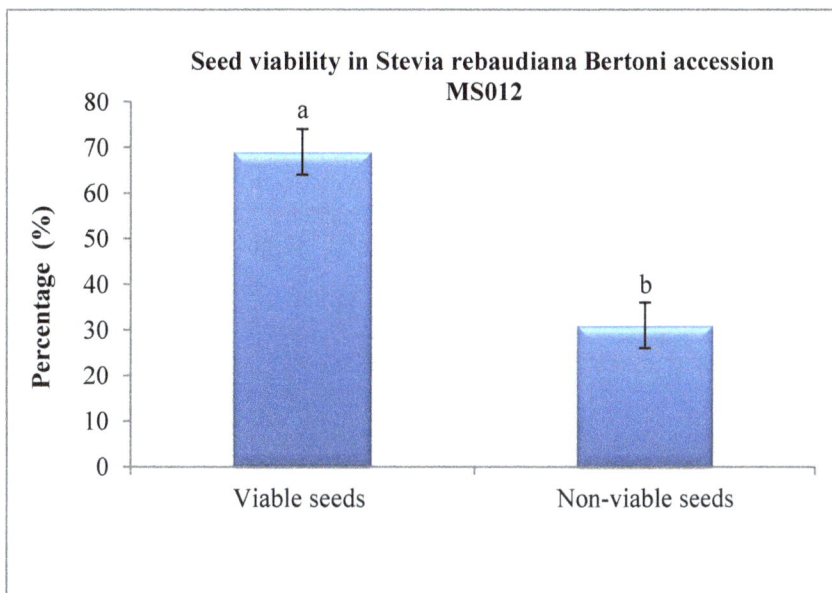

Figure 5. Percentage seed viability. Different alphabets denote significant difference at $p < 0.05$.

4. Discussion

The influence of conducive environmental factors or conditions on seed germination and plant growth and development generally, cannot be over emphasized. In this study, environmental conditions such as light, light extension period, temperature and humidity were controlled for, using invented protocols and the germination apparatus. This approach yielded high level profile in seed germination of *Stevia rebaudiana* Bertoni, a plant species known for poor seed germination (Sakaguchi & Kan, 1992).

The achievement of 3.5 days in rate of seed germination in *Stevia rebaudiana* Bertoni, as compared to 12 days in previous studies by Raji and Osman (2011), and 67.33% in seed germination in comparison with 36.3% and 41% by Goettemoeller and Ching (1999) and Raji and Osman, (2011) respectively, showed the efficacy of the protocols and the apparatus.

Stevia rebaudiana is a short day plant (Brandle et al, 2000) and the seeds are positively photoblastic (Brandle et al., 1998) in nature, meaning therefore, that, exposure to light, and at the same time, the period of exposure, are two inevitable conditions for the seeds to germinate. The use of red light in particular, and the 14 h exposure of the seeds to light are in consonance with previous findings. Dry seeds contain a higher concentration of ABA than GA, thus inhibiting seed germination (Lee et al., 2002). However, the concentration of ABA decreases after water imbibitions, which allow germination to occur. Red light decreases ABA concentration and increased GA levels, this stimulates seed germination. Seeds of *Nepenthes mirabilis* germinated under red light showed the highest speed of emergence, compared to white, green, blue, and yellow lights (Jala, 2011). Similarly, the red light was utilized to increase the germination percentage of *Merremia* sp. (Seo et al., 2006) and *Chromolaena odorata* (Ambika, 2006). Furthermore, extension in period of light exposure of seeds in *Rhododendron vaseyi* species yielded high percentage germination (LeBude et al., 2008, Walker et al., 2006).

Seed germination is temperature dependent (Young et al., 2003) and also vary according to plant species. Sakaguchi and Kan (1992) found an optimum temperature range of 15 – 30 °C to be effective for seed germination in *Stevia rebaudiana* Bertoni. The plastic dome serve to conserve water loss from the peat moss, thereby regulating moisture content and thus, the humidity in the environment within it. Seeds subjected to persistent optimum humidity therefore, stand good chance of high germination profile.

The fact that certain seeds have poor germination rates does not mean they are dead. Non-germinability in seeds could be due to environmental factors such as very low humidity and extreme temperature (Murdoch and Ellis, 2000). The most common and reliable method used for seed viability test is tetrazolium chloride (TTC) method (Traveset, 1998). TTC is accepted by the International Rules for Seed Testing (ISTA) as a chemical for testing seed viability (ISTA, 1999). Oliveira et al. (2004) stated that this method has been effectively utilized across wide range of plant species to predict germination and growth of seedling. The high percentage viability result in this study indicated that the stevia seeds are viable.

5. Conclusion

Stevia rebaudiana seeds are viable. The poor seed germination problem is actually due to inappropriate conditions of germination. Controlled environmental conditions highly improved seed germination in the crop.

Acknowledgement

Our appreciation goes to the research management center of the International Islamic University Malaysia (IIUM) for sponsoring this research through the endowment fund, 'EDW B 10-116-0455' JAZAKUMULLAHU KHAIRAH.

References

Abdullateef, R. A., & Osman, M. B. (2011). Effects of Visible Light Wavelengths on Seed Germinability in Stevia rebaudiana Bertoni. *International Journal of Biology, 3*(4), 83.

Ambika, S. R. (2006). Effect of light quality and intensity on emergence, growth and reproduction in Chromolaena odorata. *Proc. of the Seventh International Workshop on Biological Control and Management of Chromolaena odorata and Mikania micrantha, Taiwan* (pp. 14-17).

Bhering, M. C., Dias, D. C. F. S., & Barros, D. I. (2005). Adequação da metodologia do teste de tetrazólio para avaliação da qualidade fisiológica de sementes de melancia. *Revista Brasileira de Sementes, 27*(1), 176-182. http://dx.doi.org/10.1590/S0101-31222005000100022

Brandle, J. E., Starratt, A. N., & Gijzen, M. (2000). *Stevia rebaudiana. Its biological, chemical and agricultural properties. Agriculture and AgriFood Canada.* Southern Crop Protection and Food Research Centre, 1391 Sandford St., London, Ontario N5V 4T3. Retrieved from www.lni.unipi.it/stevia/stevia/stevia0005.htm

Carneiro, J. W. P., & Guedes, T. A. (1992).Influence of the contact of stevia seeds with the substrate, evaluated by means of the Wiebull function.*Revista Brasileira de Sementes, 14*(1), 65-68.

Goettemoeller, J., & Ching, A. (1999). Seed germination in Stevia rebaudiana. p. 510-511. In: J. Janick (ed.), Perspectives on new crops and new uses. ASHS Press, Alexandria, VA. Retrieved from http://www.lni.unipi.it/stevia/stevia/v4-510.htm

Hidayati, S. N., Baskin J. M., & Baskin C. C. (2002). Effects of dry storage on germination and survivorship of seeds of four Lonicera species (Caprifoliaceae). *Seed Sci. Technol., 30*, 137 –148.

Hsieh, W. P., Hsieh, H. L., & Wu, S. H. (2012). Arabidopsis bZIP16 transcription factor integrates light and hormone signaling pathways to regulate early seedling development.*Plant cell, 24*(10), 3997-4011. http://dx.doi.org/10.1105/tpc.112.105478

International Seed Testing Association. (1999). Biochemical test for viability. *Seed Sci. Technol., 27*(supplement), 201-244.

Jala, A. (2011). Effects of Different Light Treatments on the Germination of Nepenthes mirabilis. *International Transaction Journal of Engineering, Management, & Applied Sciences & Technologies, 2*(1). Retrieved from http://TuEngr.com/V02/083-091.pdf

LeBude, A. V., Blazich, F. A., Walker, L. C., & Robinson, S. M. (2008). Seed germination of two populations of Rhododendron vaseyi: Influence of light and temperature. *J. Environ. Hort., 26*, 217-22

Lee, S., Cheng, H., King, K. E., Wang, W., He, Y., Hussain, A., & Peng, J. (2002). Gibberellin regulates Arabidopsis seed germination via RGL2, a GAI/RGA-like gene whose expression is up-regulated following imbibition. *Genes & development, 16*(5), 646-658. http://dx.doi.org/10.1101/gad.969002

Murdoch, A. J., & Ellis, R. H. (2000). Dormancy, viability and longevity. *Seeds: the ecology of regeneration in plant communities, 2*, 183-214.

Oh, E., Yamaguchi, S., Kamiya, Y., Bae, G., Chung, W. I., & Choi, G. (2006). Light activates the degradation of PIL5 protein to promote seed germination through gibberellin in Arabidopsis. *Plant J.* http://dx.doi.org/ 10.1111/j.1365-313X.2006.02773.x

Oliveira, V. M. de, Forni-Martins, E. R., Magalhães, P. M., & Alves, M. N. (2004). Chromosomal and morphological studies of diploid and polyploid cytotypes of Stevia rebaudiana (Bertoni) Bertoni (Eupatorieae, Asteraceae). *Genetics and Molecular Biology, 27*(2), 215-222. http://dx.doi.org/10.1590/S1415-475720040 0200015

Runkle, E. S., & Heins, R. D. (2001). Specific functions of red, far red, and blue light in flowering and stem extension of long-day plants. *Journal of the American society for horticultural science, 126*(3), 275-282.

Sakaguchi, M., & Kan, T. (1992). Japanese researches on Stevia rebaudiana Bertoni and 335 stevioside. *Ci, Cult. 34*, 235-248.

Seo, M., Hanada, A., Kuwahara, A., Endo, A., Okamoto, M., Yamauchi, Y., ... & Nambara, E. (2006). Regulation of hormone metabolism in Arabidopsis seeds: phytochrome regulation of abscisic acid metabolism and abscisic acid regulation of gibberellin metabolism. *The Plant Journal, 48*(3), 354-366. http://dx.doi.org/10.1111/j.1365-313X.2006.02881.x

Shyam, S. S., & David, N. S. (1975). Effect of light on seed germination and seedling growth of Merremia species. *Folia Geobotanica & Phytotaxonomia, 10*(3), 265-269.

Soejarto, D. D., Compadre, C. M., Medon, P. J., Kamath, S. K., & Kinghorn, A. D. (1983). Potential sweetening agents of plant origin. II. Field search for sweet-tasting Stevia species. *Econ. Bot., 37*, 71-79. http://dx.doi.org/10.1007/BF02859308

Soejarto, D. D., Kinghorn, A. D., & Farnsworth, N. R. (1982). Potential sweetening agents of plant origin. III.Organoleptic evaluation of stevia leaf herbarium samples for sweetness. *J. Nat. Prod., 45*, 590-599.

Strauss, S. (1995). The perfect sweetener? *Technol. Rev, 98*, 18–20.

Traveset, A. (1998). Effect of seed passage through vertebrate frugivores' guts on germination: a review. *Perspectives in Plant ecology, evolution and systematics, 1*(2), 151-190. http://dx.doi.org/10.1078/1433-83 19-00057

Walker, L. C., LeBude, A. V., Blazich, F. A., & Conner, J. E. (2006). Seed germination of pinkshell azalea (Rhododendron vaseyi) as influenced by light and temperature. Proc. SNA Res. Conf. (51st Annu. Rpt. p. 370-373).

Whitaker, J. (1995). Sweet justice: FDA relents on stevia. *Human Events, 51*, 11.

10

Identification of Japanese *Lecanorchis* (Orchidaceae) Species in Fruiting Stage

Hirokazu Fukunaga[1], Yutaka Sawa[2] & Shinichiro Sawa[3]

[1] Tokushima-cho, Tokushima city, Tokushima, Japan

[2] Sawa Orchid Laboratory, Ikku, Kochi city, Kochi, Japan

[3] Kumamoto University, Graduate school of Science and Technology, Kumamoto, Japan

Correspondence: Shinichiro Sawa, Kumamoto University, Graduate school of Science and Technology, Department of Sciences,2-39-1 Kurokami,Kumamoto 860-8555,Japan.E-mail: sawa@sci.kumamoto-u.ac.jp

Abstract

Plants of *Lecanorchis* species are heteromycotrophic and they lack green leaves. Although flowering time is short, plants with fruits can be easily found in the forests. Here we discuss the features of nine taxa namely *L. triloba*, *L. trachycaula*, *L. nigricans*, *L. amethystea*, *L. kiusiana*, *L. suginoana*, *L. japonica*, *L. hokurikuensis*, and *L. flavicans* var. *acutiloba*. From the detailed phenotypes of aerial parts of fruiting plants, we propose a method to identify the Japanese *Lecanorchis* species.

Keywords: *Lecanorchis*, Japanese orchids, Orchidaceae, diagnosis method, fruiting plants

1. Introduction

Lecanorchis Blume (Orchidaceae) comprises a group of mycoparasitic plants with numerous clustered, tuberous roots and an erect, branched or unbranched stem (Blume, 1856). The genus comprises about thirty taxadistributed across a large area between Southeast Asia, Taiwan, New Guinea, and Japan (Garay & Sweet, 1974; Seidenfaden, 1978; Lin, 1987; Hashimoto, 1990; Pearce & Cribb, 1999; Szlachetko & Mytnik, 2000; Averyanov, 2005; Sing-chi, Cribb, & Gale, 2009; Suddee & Pedersen, 2011; Tsukaya & Okada, 2013). In Japan, ten species and/or varieties are reported (Honda, 1931; Tuyama, 1955; Masamune, 1963; Ohwi, 1965; Satomi, 1982; Tuyama, 1982; Hashimoto, 1990; Serizawa, 2005; S. Sawa, Fukunaga, & Y. Sawa, 2006; Fukunaga, S. Sawa, & Y. Sawa, 2008; BG Plants, 2013). Most of Orchidaceaemembers that lack green leaves produce stem, flowers and seeds, and they wither and die in a short time. However, *Lecanorchis* species maintain withered plants at upper ground for a longer period, and the fruiting plants can be easily found in the forests. So, a diagnostic method using fruitingspecimens is needed to identify the species of *Lecanorchis*.

Here we have provided a simple method to identify eight *Lecanorchis* species namely *L. triloba*, *L. trachycaula*, *L. nigricans*, *L. amethystea*, *L. kiusiana*, *L. suginoana*, *L. japonica*, and *L. hokurikuensis*, and one variety, *L. flavicans* var. *acutiloba*. Other Japanese *Lecanorchis* taxa namely *L. flavicans* Fukuy., *L. virella* T. Hashim., *L. japonica* var. *tubufolmis* T. Hashim., *L.nigricans* var. *patipetala* Y. Sawa and *L. nigricans* var. *yakushimensis* T. Hashim. are uncommon and excluded from this diagnosis.

2. Materials and Methods

Field surveys were conducted in Kochi, Tokushima, Ehime, Kagawa, Osaka, Wakayama, Hyogo, Kyoto, Shimane, Shizuoka, Aichi, Gifu, Fukui, Ishikawa, Toyama, Fukuoka, Kagoshima, Miyazaki, and Okinawa prefectures. Herbarium specimens were deposited at MBK, OSA, KANA, TUAT, and mainly at TI herbaria.

Capsule angle was measured between stem and peduncle. Specimens of opened capsules were reconstituted by treating with hot water for a few second.

Obtuse excrescence produced on capsulesand upper part of infructescence is sometimes used as a taxonomic character (Masamune, 1963; Satomi, 1982) (Figure 1A). However, in some cases the phenotype was not observed. Further, the phenotype is induced by sapping by Flatidae (Figure 1B), aphid (Figure 1C), or scale insect (Sawa, 1980), and in some cases, this obtuse excrescence was not considered as a taxonomic character

(Serizawa, 2005). In this article, we exclude this phenotype as a character for classification.

Figure 1. Obtuse excrescence and insects. A: Obtuse excrescence produced in *L. hokurikuensis*. B: Flatidae on the peduncle of *L. hokurikuensis*. C: Aphid on the peduncle of *L. hokurikuensis*. Scale bar: 1 cm

3. Results and Discussion

Stem length, stem width, length of infructescence (stem with fruits), peduncle length, bract length, calyculus length, capsule number, color of capsule and stem, protuberance on capsule and stem, capsule angle, capsule shape, and branching pattern were examined by using eight *Lecanorchis* species, *L. triloba*, *L. nigricans*, *L. amethystea*, *L. kiusiana*, *L. suginoana*, *L. japonica*, and *L. hokurikuensis*, and one variety, *L. flavicans* var. *Acutiloba* (Table 1).

Table 1. Japanese *Lechanorchis* characters

	L. triloba	L.trachycaula	L.nigricans	L.amethystea	L.flavicans var. acutiloba	L.kiusiana	L.suginoana	L. japonica	L.hokurikuensis
stem length (average) (cm)	14.5-42.0 (34.6) n=8	23.5-41.7 (32.1) n=8	11.4-27.7 (19.1) n=38	22.2-42.7 (33.6) n=12	14.0-35.1 (23.3) n=13	3.9-25.4 (17.3) n=246	12.0-29.8 (20.5) n=38	21.2-45.2 (30.9) n=18	17.6-50.4 (30.5) n=59
stem diameter (mm)	0.8-2.2 n=8	0.5-3.5 n=8	1.0-2.2 n=38	1.4-2.5 n=12	0.5-1.3 n=13	0.5-2.2 n=246	1.7-3.5 n=38	1.0-2.6 n=18	1.3-3.6 n=59
inflorescence length (average) (cm)	1.9-5.8 (4.0) n=8	2.8-15.0 (6.1) n=17	1.7-16 (5.2) n=27	8.5-17.0 (13.1) n=9	1.0-6.0 (3.2) n=11	1.5-10 (4.6) n=17	3.5-8.0 (5.6) n=5	2.0-14.0 (7.1) n=11	3.0-8.5 (5.7) n=8
capsule length (average) (cm)	1.3-1.7(1.4) n=24	2.4-3.2(2.7) n=5	2.1-3.6 (2.6) n=66	1.9-3.3 (2.6) n=51	2.3-2.7 (2.5) n=6	1.9-3.0 (2.4) n=381	2.0-3.0 (2.7) n=72	3.1-4.4 (3.7) n=15	3.1-4.8 (3.8) n=111
peduncle length (mm)	0.5-1.5 n=24	0.5-2.5 n=5	1.5-2.5 n=15	1.5-2.0 n=51	1-1.5 n=3	0.5-5.5 n=156	2.0-3.5 n=72	1.0-7.0 n=15	2.0-7.5 n=81
bract length (mm)	1.5-2.2 n=6	1.4-2.5 n=6	1.3-3.0 n=6	1.8-2.5 n=6	1.2-2.2 n=6	1.7-5.2 n=6	2.2-4.2 n=6	2.2-6.0 n=6	3.0-5.5 n=6
calyculus length (mm)	0.4-1.3 n=6	0.5-1.4 n=6	0.5-1.5 n=6	0.3-1.0 n=6	0.3-1.0 n=6	0.5-1.2 n=6	0.5-1.0 n=6	0.5-2.2 n=6	0.7-1.5 n=6
capsule number (average)	3-14 (6.7) n=11	1-11 (5.3) n=10	2-11 (5.7) n=42	3-15 (6.4) n=11	1-7 (3.9) n=13	1-7 (3.6) n=248	3-6 (4.7) n=36	4-10 (6.2) n=18	2-10 (5) n=60
capsule angle to stem	10-45° n=6	20-70° n=6	0-75° n=6	5-60° n=6	5-70° n=6	0-20° n=6	0-30° n=6	0-35° n=6	5-90° n=6

Figure 2. Fruiting plants of *Lecanorchis* spp. A: *L. triloba*, B: *L. trachycaula*, C: *L.nigricans*, D: *L. amethystea*, E: *L. flavicans* var. *acutiloba*. F: *L. kiusiana*, G: *L. suginoana*, H: *L. japonica*, I: *L. hokurikuensis*. Scale bar: 3 cm

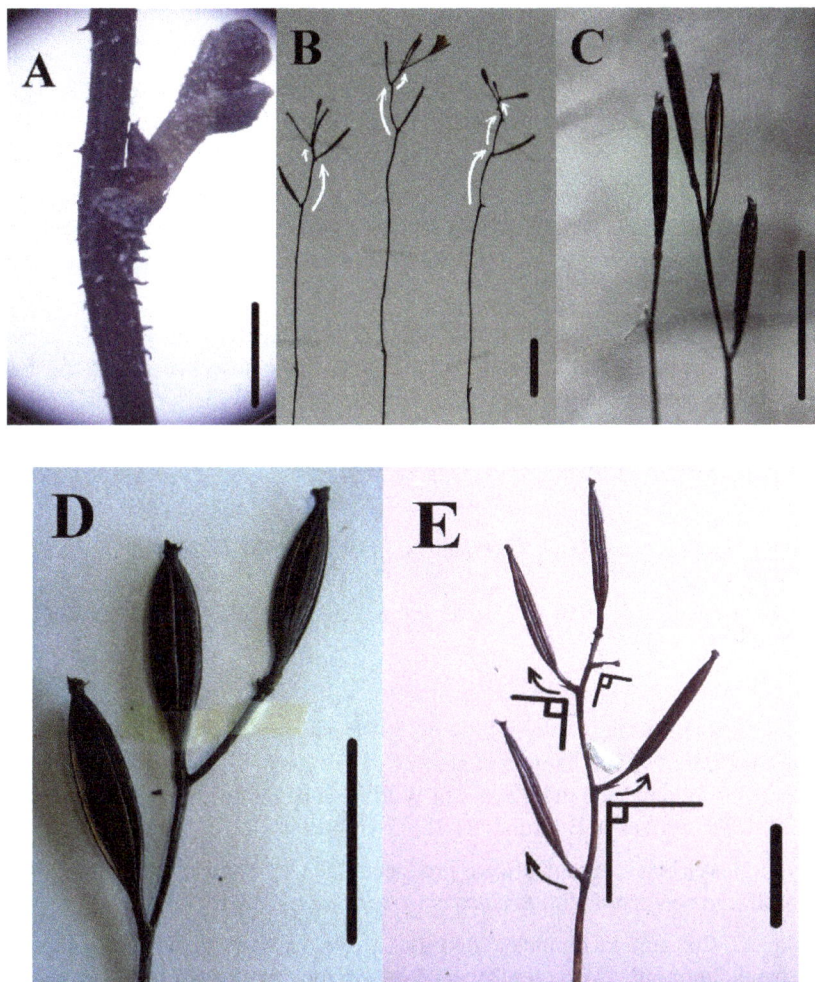

Figure 3. Characteristic phenotypes of Lecanorchis. A: Pointed, short prickle-like protuberances of *L. trachycaula*. B: Curved stem of *L. flavicans* var. *acutiloba*. C: Thin cylindrical to fusiform capsule of *L. kiusiana*. D: Roundish fusiform capsule of *L. suginoana*. E: Curved peduncle with a bract in 45-90° of *L. hokurikuensis*. Scale bar: 5 mm (A) 3 cm (B-E)

3.1 Lecanorchistriloba J. J. Sm.

Lecanorchistriloba J. J. Sm. in Bull. Dep. Agric. Indes Neerl. 19: 26. 1908; Hashimoto in Ann. Tsukuba Bot. Gard. (9): 35. 1990; T. C. Hsu & S. W. Chung in Taiwania 54: 83. 2009; *L. brachycalpa* Ohwi, in Acta Phytotax. Geobot. 7: 35. 1938; Tuyama in J. Jap. Bot. 30: 184. ff. 3 c&d. 1955; Garay & Sweet, Orchids of southern Ryukyu Islands 52. f.3-k. 1974; *L. multiflora* var. *brachycalpa* (Ohwi) T. Hashim., Hashimoto in Ann. Tsukuba Bot. Gard. (8): 6. 1989. (Figure 2A).

Characteristic features: Short infructescence, short peduncles, and small capsules.

Stems 14.5-42 cm long, 0.8-2.2 mm in diameter, solitary, rarely branched, black to brownish black. Scale leaves 3-4 mm long, acute triangle. Infructescence (stem with fruits) 1.9-5.8 cm long, with 3-14 fruits. Bract 1.5-2.2 mm long, acute, round to triangle. Capsule 1.3-1.7 cm long, fusiform, 10-45° in capsule angle, densely produced at the top. Peduncle 0.5-1.5 mm long, Calyculus 0.4-1.3 mm long, irregularly denticulate.Fruits black to brownish black.

Specimen Examined:

Y. Taira 158, Jan. 22, 1938, Type of *L. brachycalpa* Ohwi (KYO); *Y. Kimura s. n.*, Oct. 13, 1940 (TI); *H. Ito s. n.*, Oct. 21, 1936 (TI); *T. Hashimoto s. n.*, Feb. 19, 1990 (TNS-*9504619*); *Y. Hanei s. n.*, May 29, 1989 (TNS-*9504456*); *Y. Hanei s. n.*, Oct. 11, 1990 (TNS-*9504683*); *T. Hashimoto s. n.*, Feb. 20, 1990 (TNS-*9504620*); *T. Hashimoto*, Feb. 20, 1990 (TNS-*9504621*); *T. Amano 7735*, Nov. 27, 1957 (TNS-*139953*); *M. Tawada s. n.*,

Dec. 28, 1941 (TNS-*226051*).

3.2 Lecanorchis trachycaula Ohwi

Lecanorchis trachycaula Ohwi, in Fl. Jap., ed. Rev. 1438. 1965; Hatus., Fl. Ryukyus 808. 1971; Satomi in Satake et al., Wild Fl. Jap. Herb. 1: 206. 1982; Hashimoto in Ann. Tsukuba Bot. Gard. 9: 19. 1990. (Figures 2B & 3A).

Characteristic features: Pointed, short prickle-like protuberances were found on stem and capsules, and the protuberances were obvious in later capsule stages (Figure 3A). Furthermore, non-glossy black capsules were produced laterally except for the uppermost and the lowermost capsules, and the stems were frequently branched.

Stems 23.5-41.7 cm long, 0.5-3.5 mm in diameter, 2-4 order branched, non-glossy black to purplish black, occasionally with pointed protuberances. Scale sheath 3-5 mm long. Infructescence 2.8-15 cm long, with 1-11 fruits. Bract 1.4-2.5 mm long, round to triangle. Capsule 2.4 cm-3.2 cm long, fusiform, 20-70° in capsule angle except for the uppermost and the lowermost capsules, with occasional protuberances. Peduncle 0.5-2.5 mm long, Calyculus 0.5-1.4 mm long, irregularly denticulate.Fruits non-glossy black.

Specimen Examined:

C. Abe 49821, Jun. 11, 1974 (TKPM-*214352*); *S. Takafuzi s. n.*, Oct. 27, 1963 (KANA); *T. Hashimoto s. n.*, Feb. 24, 1990 (TNS-*940462*); *J. Nakayama s. n.*, Aug. 10, 1936 (TUS); *F. Maekawa s. n.*, Jun. 16, 1976 (TUAT-*87954*); *S. Sugaya s. n.*, Aug. 14, 1958 (TUS-*26508*); *S. Sugaya s. n.*, Aug. 15, 1958 (TUS-*26509*); *N. Satomi s. n.*, Jul. 20, 1962 (KANA-*158148*).

3.3 Lecanorchis nigricans Honda

Lecanorchis nigricans Honda, in Bot. Mag. Tokyo 45: 470. 1931; Tuyama, in J. Jpn. Bot. 30: 184. ff. 3A&B. 1955; Ohwi, Fl. Jap., Engl. Ed. 336. 1955; Kitamura et al., in Col. Ill. Herb. Pl. Jap. 3: 28. 1969; F. Maekawa, in Wild Orch. Jap. Col. 239. 1971; Satomi, in Satake et al., in Wild Fl. Jap. Herb. 1: 206.1982; Hashimoto, in Ann. Tsukuba Bot. Gard. (9): 27. 1990; Serizawa, Bunrui 5: 34. 2005. (Figure 2C).

Characteristicfeatures: Non-glossy black capsules were produced sideway except for top and bottom capsules. Protuberances were not produced compared with *Lecanorchistrachycaula* Ohwi.

Stems 11.4-27.7 cm long, 1.0-2.2 mm in diameter, non-glossy black, smooth surface, solitary, occasionally branched basally. Scale sheath 3mm long. Infructescence 1.7-16 cm long, with 2-11 fruits. Bract 1.3-3 mm long, acute triangle. Capsule 2.1 cm-3.6 cm long, thin fusiform, smooth surfaced, 0-75° in capsule angle except for the uppermost and the lowermost capsule. Peduncle 1.5-2.5 mm long, Calyculus 0.5-1.5 mm long, irregularly denticulate.Fruits non-glossy black.

Specimen Examined:

Type-*K. Kashiyama s. n.*, 1931 (TI); *T. Hashimoto s. n.*, Mar. 18, 1990 (TNS-*9504625*); *E. Yamahata s. n.*, Jan. 6, 1976, det. by *Y. Sawa* (OSA), *Y. Hanei s. n.*, Aug. 12, 1989 (TNS-*9504539*); *Y. Kurosawa s. n.*, Aug. 8, 1978 (TUAT); *K. Gunzi s. n.*, Oct. 13, 1963 (IBAR); *T. Hino s. n.*, Oct. 16, 1928 det. by G. Murata (KYO), (OSA-*19436*), (KPM-NA-*102387*), (KPM-NA-*162332*), (KPM-NA-*105066*), (KPM-NA-*73112*), (KPM-NA-*72853*); *Y. Sawa 2277*, 16 Mar. 1999, (TI); *I. Yamashita s. n.*, Oct. 16, 1965 (KANA-*158148*), *T. Ogawa 26679* (KANA); *Y. Sawa 2725*, Aug. 18, 1998. (TI); *Y. Sawa 2726*, Aug. 18, 1998 (TI).

3.4 Lecanorchis amethystea Y. Sawa, Fukunaga & S. Sawa

Lecanorchis amethystea Y. Sawa, Fukunaga & S. Sawa, in Acta Phytotax. Geobot. 57: 123. 2006; Hsu, T.-C. & S.-W. Chung, in Taiwania 55: 366. 2010. (Figure 2D).

Characteristicfeatures: Brown stems and capsules. Infructescences and stems were longer than that of *Lecanorchisnigricans* Honda (Sawa et al., 2006).

Stems 22.2-42.7 cm long, 1.4-2.5 mm in diameter, rarely branched basally, dark brown. Scale sheath 3-5 mm long, acute. Infructescence 8.5-17 cm long, with 3-15 fruits.Bract 1.8-2.5 mm long, acute triangle. Capsule 1.9 cm-3.3 cm long, fusiform, smooth surface, 5-60° in capsule angle.Peduncle 1.5-2.0 mm long, Calyculus 0.3-1.0 mm long, irregular denticulate. Fruits dark brown.

Specimen Examined:

Type-*Y. Sawa 1702*, July 26, 1987 (TI); *H. Fukunaga & Y. Sawa 2*, Aug. 29, 2000 (MBK); *H. Fukunaga & Y. Sawa 4*, Aug. 29, 2000 (MBK); *H. Fukunaga & Y. Sawa 7*, Aug. 29, 2000 (MBK); *Y. Sawa 1864*, Oct. 27, 1987

(MBK); *Y. Sawa 1869*, Oct. 27, 1987 (MBK); *Y. Sawa 1863*, Oct. 27, 1987 (MBK).

3.5 Lecanorchis flavicans Fukuy. var. acutiloba T. Hashim.

Lecanorchis flavicans Fukuy. var. *acutiloba* T. Hashim., in Ann. Tsukuba Bot. Gard.(8): 8. 1989; Hashimoto, in Ann. Tsukuba Bot. Gard. (9): 25. 1990. (Figure 2E).

Characteristic features: Thin stems, short capsules and peduncles. Infructescence produced in zigzag with strong phototropism (Figure 3B).

Stems 14-35.1cm long, 0.5-1.3 mm in diameter, solitary, rarely branched basally, growth in a zigzag way, black to brownish black. Infructescence 1-6 cm long, with 1-7 fruits. Bract 1.2-2.2 mm long. Capsule 2.3-2.7cm long, thin fusiform, smooth surface, 5-70° in capsule angle. Peduncle 1-1.5 mm long, Calyculus 0.3-1 mm long, irregular denticulate. Fruits brownish black.

Specimen Examined:

Y. Hanei s. n., July 26, 1991 (TNS-*9507125*); *Y. Hanei s. n.*, July 26, 1991 (TNS-*9507126*); *Y. Miyagi 5623*, Jun. 25, 1966 (RYU-*29181*); *Y. Miyagi 3394*, Jun. 25, 1967 (RYU-*18494*); *G. Murata et al. 865*, det. by *Y. Sawa* (KYO), *Y. Niiro s. n.* (RYU-*6422*), *Y. Sawa s. n.* (TI).

3.6 Lecanorchis kiusiana Tuyama

Lecanorchis kiusiana Tuyama, in J. Jpn. Bot. 30: 182. 1955; Ohwi, in Fl. Jap., ed. Rev. 335. 1965; Chuma, in J. Jap. Bot. 55: 306. 1980; Satomi, in Satake et al., Wild Fl. Jap. Herb. 1: 205. 1982; Hashimoto, in Ann. Tsukuba Bot. Gard. (9): 13.1990; Serizawa, Bunrui 5: 37. 2005. (Figure 2F).

Characteristicfeatures: This is the smallest *Lecanorchis* species in Japan. Further, it produces narrowly tubular to fusiform capsules produced upward almost in parallel with the slender shoots (Figure 3C).

Stems 3.9-25.4 cm long, 0.5-2.2 mm in diameter, solitary, moderate gloss black to purplish black. Scale sheath 5-20 mm long, egg shaped at the base, acute at the top, smooth surface. Infructescence 1.5-10 cm long, with 1-7 fruits. Bract 1.7-5.2 mm long, triangle. Capsule 1.9-3.0 cm long, thin cylindrical to fusiform, smooth, parallel to stem, 0-20° in capsule angle. Peduncle 0.5-5.5 mm long, Calyculus 0.5-1.2 mm long, irregular denticulate. Fruitsmoderate gloss black to purplish black.

Specimen Examined:

N. Satomi 1010, Oct. 4, 1951 det. by *Y. Sawa* (KANA-*148801*); *T. Kobayashi 11259*, July 27, 1988 det. by *Y. Sawa* (KYO); *N. Satomi 22755*, Mar. 29, 1964 det. by *Y. Sawa* (KANA*148804*); *N. Satomi 16782*, Mar. 28, 1961 det. by *Y. Sawa* (KANA*148800*); *K. Inami s. n.* , Jun. 17, 1975, det. by *Y. Sawa & G. Murata* (KYO); *H. Ono s.n.*, Nov. 17, 1968, det. by *Y. Sawa* (KAG); *G. Ikeda s.n.* , Jul. 30, 1965, det. by *Y. Sawa* (KAG); *I. Yamashita 3*, Jun. 1979 (TUAT-*87990*), *Y. Sawa 891* (OSA-*47829, 167462*); *Y. Sawa & H. Fukunaga 2572*, Feb. 4 1999 (TI); *Y. Sawa & H. Fukunaga 2722*, Mar. 16, 1999 (TI); *Y. Sawa & H. Fukunaga 2721*, Mar. 16 1999 (TI); *Y. Sawa & H. Fukunaga 2729*, Mar. 16 1999 (TI); *Y. Sawa & H. Fukunaga 2730*, Mar. 16 1999 (TI); *Y. Sawa & H. Fukunaga 2732*, Mar. 16 1999 (TI); *Y. Sawa & H. Fukunaga 2724*, Mar. 16 1999 (TI); *Y. Sawa & H. Fukunaga 2731*, Mar. 16 1999 (TI); *Y. Sawa & H. Fukunaga 2747*, Mar. 17 1999 (TI); *Y. Sawa & H. Fukunaga 2748*, Mar. 17 1999 (TI); *Y. Sawa & H. Fukunaga 2749*, Mar. 17 1999 (TI); *Y. Sawa & H. Fukunaga 2279*, Jan. 23 1999 (TI); *Y. Sawa & H. Fukunaga 2719*, Aug. 6 1998 (TI); *N. Satomi 22755* Mar. 29, 1964 (KANA-*148804*); *N. Satomi 25033*, Apr. 2, 1965 (KANA-*147795*); *E. Ikeda 21674*, Jun. 2, 1963 (KANA-*134468*); *N. Satomi 16782*, Mar. 28, 1961 (KANA-*148800*); *N. Satomi 26845*, Jul. 16, 1973 (KANA-*148805*); *N. Satomi 148801*, Oct. 4, 1951 (KANA-*148801*).

3.7 Lecanorchis suginoana (Tuyama) Seriz

Lecanorchis suginoana (Tuyama) Seriz, in Bunrui 5: 38. 2005; *L. japonica* var. *sugioana* Tuyama, in J. Jpn. Bot. 57: 211. 1982; *L. kiusiana* Tuyama var. *suginoana* Hashim, In Ann. Tsukuba Bot. Gard. (9): 18. 1990. (Figure 2G).

Characteristic features: Small plants with non-glossy brown to black, and round capsules (Figure 3D). Lower part of the shoot is thicker than that of *L. kiusiana*.

Stems 12-29.8 cm long, 1.7-3.5 mm in diameter, solitary, smooth surface, brown to black. Scale sheath 5-10 mm long. Infructescence 3.5-8 cm long, with 3-6 fruits.Bract 2.2-4.2 mm long, acute triangle.Capsule 2-3 cm long, roundish fusiform, 0-30° in capsule angle. Peduncle 2-3.5 mm long. Calyculus 0.5-1 mm long, irregular denticulate, 0.5-1 mm long.Fruitsnon-glossy brownish black to black.

Specimen Examined:

M. Wadas. n. Jul. 5, 1983 (TI), *M. Tanaka 3856* (OSA), *T. Umehara 4836* (OSA-91409); *Y. Sawa 2524*, Feb. 20, 1998 (TI); *Y. Sawa 2524*, Feb. 20, 1999 (TI); *Y. Sawa 2524*, Feb. 20, *2526*, (TI); *Y. Sawa 2527*, Feb. 20, 1999, (TI); *Y. Sawa 2533*, Feb. 20, 1999 (TI); *Y. Sawa 2763*, Feb. 20, 1999 (TI); *Y. Sawa 2766*, Feb. 20, 1999 (TI); *Y. Sawa 2538*, Feb. 20, 1999 (TI); *Y. Sawa 1439*, May 18, 1987 (TI); *Y. Sawa 1584*, Jun. 12, 1987 (TI); *Y. Sawa 872*, May 23, 1987 (TI); *Y. Sawa 875*, May 18, 1987 (TI); *Y. Sawa 871*, May 18, 1987 (TI); *Y. Sawa 945*, Apr. 5, 1984 (TI); *Y. Sawa 948*, Apr. 5, 1984 (TI); *Y. Sawa 949*, Apr. 5, 1984 (TI); *Y. Sawa 882*, Jun. 15, 1984 (TI).

3.8 Lecanorchis japonica Blume

Lecanorchis japonica Blume, in Mus. Bot. Ludg. Bat. 2: 188. 1856; Kitamura et al., in Col. Ill. Herb. Pl. Jap. 3: 28. 1969; F. Maekawa, in Wild Orch. Jap. Col. 236. 1971; Satomi in Satake et al., in Wild Fl. Jap. Herb. 1: 206. 1982; Hashimoto, in Ann. Tsukuba Bot. Gard. (9): 3. 1990; Serizawa Bunrui 5: 35. 2005. (Figure 2H).

Characteristic features: This is the largest *Lecanorchis* species in Japan. Further it produces capsules slant upward with straight peduncle and capsules.

Stems 21.2-45.2 cm long, 1-2.6 mm in diameter, solitary, black to brownish black. Scale Sheath 6-12 mm long. Infructescence 2-14 cm long, with 4-10 fruits. Bract 2.2-6 mm long, acute triangle. Capsule 3.1-4.4 cm long, fusiform, 0-35° in capsule angle. Peduncle 1-7 mm long, straight, Calyculus 0.5-2.2 mm long, irregular denticulate. Fruits black to brownish black.

Specimen Examined:

K.Kawanabe 4705, Dec. 7, 1958 det. By *Hatushima* (KAG); *Y. Sawa 1383*, Dec. 26, 1986 (TUAT); *S. Hatusima 20766*, May 14, 1957 (KAG); *Y. Kashiyama s. n.* Jan. 29, 1933 (KYO); *K. Seto s. n.*, Jun. 28, 1962 (OSA); *Y. Sawa & H. Fukunaga 2718*, Jan. 12, 1999 (TI); *N. Satomi 25091*, Apr. 25, 1965 (KANA-*158145*); *N. Satomi 15232*, Aug. 2, 1960 (KANA-*158144*); *S. Kato s. n.*, Sep. 5, 1976 (KANA-*145571*); *K. Aoki s. n.*, Jul. 1, 1970, (KANA-*145570*); *N. Satomi 7269*, Dec. 31, 1955 (KANA-*148807*); *T. Omura s. n.*, (KANA-*147707*).

3.9 Lecanorchis hokurikuensis Masam.

Lecanorchis hokurikuensis Masam., in J. Geobot. 12: 69. 1963; Satomi in Satake et al., Wild Fl. Jap. Herb. 1: 205. 1982; Bunrui 5: 36. 2005; *L. japonica* Blume var. *hokurikuensis* (Masam.); Hashimoto, in Ann. Tsukuba Bot. Gard. (9): 7. 1990. (Figure 2I).

Characteristic features: The capsule angle (45-90°) to the shoot (Figure 3E), and peduncle is curved to upward (Figure 3E).

Stems 17.6-50.4 cm long, 1.3-3.6 mm in diameter, solitary, light brown to black. Scale Sheath 5-12 mm. Infructescence 3-8.5 cm long, with 2-10 fruits. Bract 3-5.5 mm long, acute triangle. 45-90° in bract angle. Capsule 3.1-4.8 cm long, fusiform, 5-90° in capsule angle. Peduncle 2-7.5 mm long, upward. Calyculus 0.7-1.5 mm long, irregular denticulate. Fruits brown to brownish black.

Specimen Examined:

I. Yamashita s. n., Jun. 16, 1981, det. by *Y. Sawa*, (TUAT-*85827*); *Suzuki, Nakamura, Moriguchi & Tasaki s. n.*, Oct. 15, 1975, det. by *Y. Sawa* (IBAR); *N. Satomi 27163*, Nov. 19, 1972 (KANA-*103715*); *A. Kimura s. n.*, Oct.3, 1931, det. by *Y. Sawa* (TUS-*26500*); *F. Maekawa s. n.*, Jan. 1972, det. by *Y. Sawa* (TUAT-*87864*); *F. Maekawa s.n.*, Mar. 4, 1972, det. by *Y. Sawa* (TUAT-*87863*); *S. Tamaki s. n.*, Oct. 1970, det. by *Y. Sawa* (RYU-*6832*); *Y. Miyagi 1763*, Aug. 1972, det. by *Y. Sawa* (RYU-*9966*); *N. Satomi s. n.*, Oct. 22, 1980 (KANA-*94944*); *N. Satomi 27088*, Oct. 22, 1978, (KANA-*92137*); *N. Satomi 27028*, Jul. 10, 1977 (KANA-*146686*); *N. Satomi 27030*, Jul. 10, 1977 (KANA-*146687*); *S. Okuda s. n.*, May 31, 1970 (KANA-*145579*); *G. Masamune 14169*, Oct. 16, 1961 (KANA-*44546*); *N. Satomi 15430*, Sep. 8, 1960 (KANA-*146680*); *N. Satomi 27163*, Nov. 19, 1972 (KANA-*103715*); *N. Satomi 26774*, Nov. 3, 1972 (KANA-*146685*); *N. Satomi 26941*, Nov. 2, 1975 (KANA-*146688*); *Y. Shirosaki 27024*, (KANA-*147792*); *N. Satomi s. n.*, Oct. 21, 1980 (KANA-*97123*); *N. Satomi s.n.*, Sep. 20, 1953 (KANA-*13921*), *N. Satomi 26942*, Oct. 15, 1972 (KANA-*146689, 81904*), *N. Satomi 27630*, Jul. 22, 1986 (KANA-*119847*); *Y. Sawa 2645*, Jun. 12, 1998 (TI); *Y. Sawa 2285*, Jun. 12, 1998 (TI); *Y. Sawa 2286*, Jun. 12, 1998 (TI); *Y. Sawa 2287*, Jun. 12, 1998 (TI); *N. Satomi 15181*, Jul.1, 1960 (KANA-*147790*); *N. Satomi 15661*, Sep. 23, 1960 (KANA-*146681*); *N. Satomi 20144*, Jun. 25, 1962 (KANA-*146682*); *N. Satomi 15167*, Jul. 3, 1960 (KANA-*146683*); *N. Kurosaki, 2009*, Jul. 2, 1966 (KANA-*668290*); *M. Ihara s. n.*, Jul. 1, 1957 (KANA-*146684*); *Uzihara s. n.*, Sep. 4, 1955 (KANA-*27649*); *N. Satomi 26846*, Jul. 22, 1973 (KANA-*145578*); *S. Sugaya s. n.*, Jun. 12, 1956 (KANA-*145572*); *H. Kitami s. n.*, Jul. 5, 1968 (KANA-*145575*); *Y. Ueno 12856*, (KANA-*145575*).

Key to the species of Japanese *Lecanorchis* (fruiting specimens)

1. Capsules 1.3-1.7 cm long ………. *L. triloba*

1. Capsules 1.9-4.8 cm long ………. 2

2. Stems branched at base ………. 3

2. Stems unbranched ………. 6

3. Capsule formation laterally ………. 4

3. Capsule formation upward ………. 5

4. Stems prickled ………. *L. trachycaula*

4. Stems smooth ………. *L. nigricans*

5. Infructescence straight, 8.5 -17 cm long ………. *L. amethystea*

5. Infructescencecurved, 1-6 cmlong ………. *L. flavicans*var. *actiloba*

6. Capsules 1.9-3.0 cm long ………. 7

6. Capsules 3.1-4.8 cm long ………. 8

7. Capsulesnarrowly cylindrical to fusiform ………. *L. kiusiana*

7. Capsules roundish fusiform ………. *L. suginoana*

8. Peduncles curved, with a bract in 45-90°………. *L. hokurikuensis*

8. Peduncles straight, with a bract in -45°………. *L. japonica*

References

Averyanov, L. V. (2005). New Orchids from Vietnam. *Rheedea, 15*, 83-101.

BG Plants. (2013). In *Web BG Plants, Y-list.* Retrieved December 25, 2013, from http://bean.bio.chiba-u.jp/bgplants/ylist_main.html

Blume, C. L. (1856). *Lecanorchis Bl. Mus. Bot. Ludg. Bat. 2*, 188. Brill: Leiden.

Fukunaga, H., Sawa, S., & Sawa, Y. (2008). A new form of *Lecanorchiskiusiana. Orchid Rev., 116*, 106-108.

Garay, L. A., & Sweet, H. R. (1974). *Orchids of southern Ryukyu Islands.* Massachusetts, MA: Bot. Mus. Harvard Univ. Cambridge-Mass.

Hashimoto, T. (1990). A taxonomic review of the Japanese *Lecanorchis* (Orchidaceae). *Ann. Tsukuba Bot. Gard., 9*, 1-40.

Honda, M. (1931). Nuntia ad FloramJaponiae XIV. *Bot. Mag. Tokyo, 45*, 469-471, 493-494.

Lin, T. P. (1987). *Native Orchids of Taiwan 3* (pp. 1-300). Taipei: Southern Materials Center.

Masamune, G. (1963). *Lecanorchis hokurikuensis. J. Geobot., 12*, 69.

Ohwi, J. (1965). *Flora of Japan* (Rev. ed.). Tokyo, Japan: Shibundo.

Pearce, N., & Cribb, P. (1999). Notes Relating to the Flora of Bhutan: XXXVII. New species and records of Orchidaceae from Bhutan and India (Sikkim). *Edinb. J. Bot., 56*, 273-284. http://dx.doi.org/10.1017/S096042860000113X

Satomi, N. (1982). Ranka (Orchidaceae). In Y. Satake et al. (Eds.), *Wild flowers of Japan: Herbaceous Plants I* (pp. 187-235, pls. 170-208). Tokyo, Japan: Heibonsya Ltd., Pub.

Sawa Y. (1980). Reconsideration of obtuse excrescence as a taxonomic character. *Annual meeting of the botanical society of Japan, 45*, 54.

Sawa, S., Fukunaga, H., & Sawa, Y. (2006). *Lecanorchis amethystea* (Orchidaceae), A new species from Kochi. *Acta Phytotax. Geobot., 57*, 123-128.

Seidenfaden, G. (1978). Orchid genera in Thailand VI. NeottioideaeLindl. *Dansk Bot. Arkiv, 32*(2), 1-195.

Serizawa, S. (2005). The genus *Lecanorchis* (Orchidaceae) in Aichi Prefecture, Central Honshu. *Bunrui, 5*, 33-38.

Sing-chi, C., Cribb, P. J., & Gale, S. W. (2009). Lecanorchis. In Z. Y. Wu, P. H. Raven & D. Y. Hong (Eds.), *Flora of China* (Orchidaceae) (Vol. 25, pp. 171-172). Beijing, Chinaand St. Louis, Missouri: Science Press and Missouri Botanical Garden Press.

Suddee, S., & Pedersen, H. Æ. (2011). A new species of *Lecanorchis* (Orchidaceae) from Thailand. *Taiwania, 56*, 37-41.

Szlachetko, D. L., & Mytnik, J. (2000). *Lecanorchis seidenfadeni* (Orchidaceae, Vanilloideae), a new orchid species from Malaya. *Ann. Bot. Fennici, 37*, 227-230.

Tsukaya, H., & Okada, H. (2013). A new species of Lecanorchis (Orchidaceae, Vanilloideae) from Kalimantan, Borneo. *Systematic Botany, 38*, 69-74. http://www.bioone.org/doi/full/10.1600/036364413X662079

Tuyama, T. (1955). A new saprophytic orchid, *Lecanorchis kiusiana. J. Jap. Bot., 30*, 181-187.

Tuyama, T. (1982). *Lecanorchis japonica* Bl. var. *suginoana* Tuyama, a new variety, with comments on the other taxa of the genus. *J. Jap. Bot., 57*, 205-211.

11

Potential Yield of Soybean Lines Are Higher Than Their Parent Indonesian Lowland Popular Variety

Heru Kuswantoro[1]

[1] Indonesian Legume and Tuber Crops Research Institute, Indonesian Agency for Agricultural Research and Development, Indonesia

Correspondence: Heru Kuswantoro, Indonesian Legume and Tuber Crops Research Institute, Indonesian Agency for Agricultural Research and Development, Malang, Indonesia. E-mail: herukusw@yahoo.com

Abstract

Breeding materials should be evaluated under a number of different environments, to ensure their genetic value. The objective of the research was to study the potential yield of soybean lines compared to the parent in two environments. A total of 15 genotypes (including one Indonesian lowland popular variety, 'Anjasmoro') were grown in Jambegede (Entisol-Inceptisol association) and Ngale (Vertisol) Research Stations in June-September 2011. The design was a randomized completely blocks design with three replications. The results showed that G × E interaction highly affected the grain yield causing different rank for the genotypes in different environment. Tgm/Anj-795 and Tgm/Anj-790 lines consistently yielded the highest grain in both environments. Tgm/Anj-790 also had large seed size. Tgm/Anj-789 and Tgm/Anj-796 consistently showed the earliest days to maturity in both environments. 'Anjasmoro' had longer days to maturity but lower grain yield. Tgm/Anj-795 and Tgm/Anj-790 lines had potentially high yield stability that need to be tested further in many different environments to gain knowledge of their actual yield stability.

Keywords: agronomic characters, G × E interaction, soybean lines, yield

1. Introduction

The main goal of growing crops is to maximize net profit through increasing grain yield (Alghamdi, 2004). Hence, the primary goal of most soybean breeding programs is high grain yield (Toledo et al., 2000). To increase soybean growth area and production, it is also important to develop the high-yielding early-maturing cultivars under a wide range of different environments (Alghamdi, 2004). For sustainable agriculture, the use of stable genotypes as a mean of high grain yield is very important (Carpenter & Board, 1997). In Indonesia the soybean production always decreases due to the decresing planting area (Statistic Indonesia, 2013). Therefore, high grain yield soybean variety is needed to maintain or increase soybean production.

In developing an improved variety, the genotype vs environment interaction (G × E) is of major importance (Sharrifmoghaddassi & Omiditabrizi, 2010). Trait stability parameters are estimated to determine the superiority of individual genotypes across the range of environments when the G × E is the present (Ulker et al., 2006); but in the absence of information on G × E, estimation of heritability and prediction of genetic advance become biased. Hence, breeding material should be evaluated under different environments (Duzdemir, 2011). Information on G × E is very important in selecting and developing variety that will be recommended in a certain area. The G × E occurs when variability relative or rank of a genotype change with the environment change. The interaction of genotype and environment can also be described as the interplay between genetic and environmental factors on the growth and development of plants (Cucolotto et al., 2007). Thus, a high-yielding variety in an area does not necessarily have high yields in other areas and vice versa.

Yield stability of a plant in certain area will be different from other area and dependent on the environmental conditions of its both abiotic and biotic environment. Comparatively, the abiotic/physical environmental conditions such as soil type, rainfall, temperature, and humidity play a greater role on the stability of crop yields than that of biotic environmental conditions; because abiotic environmental condition always exist in longterm period than biotic condition. Beside, biotic condition, such as disease and pest infestation, is an incidental condition. Therefore, plant breeders always test their promising lines in various environments, to determine the yield stability of the candidate varieties to be released. Promising lines that were assessed stable in various environments are released as varieties for broad areas. Promising lines having the highest yield potential in a particular area are classified as unstable lines, and can be recommended for narrow adaptation (Gurmu et al., 2009).

Stability is defined as the ability of plants to maintain its yield potential under the changing of environmental conditions, so the stability is dynamic in character and always changes based on a specific range of different environments (Hidayat, 2002). From the agronomic point of view, stability follows the homeostatic processes of living organisms in the short term to maintain productivity under environmental changes. In this sense, stability is characterized by high sustainability and equitability (Conway, 1982). In this experiment, soybean lines were tested in two different environments to study their potential yield compared to the Indonesian lowland popular variety ('Anjasmoro').

2. Materials and Method

A total of 14 soybean (*Glycine max* (L.) Merrill) lines which derived of 'Tanggamus' and 'Anjasmoro' crossing, and one of the parents (Indonesian lowland popular variety, 'Anjasmoro') were grown on lowland with "Rice-Rice-Soybean" cropping pattern at Jambegede Research Station and Ngale Research Station from June to September 2011. Crossing was conducted at Indonesian Legume and Tuber Crops Institute, Malang, in 2005 by using 'Tanggamus' as female parent and 'Anjasmoro' as male parent. Jambegede Research Station is located on Malang Regency, East Java Province, while Ngale Research Station is located on Ngawi Regency, East Java Province. Soil type of Jambegede Research Station is Entisol-Inceptisol association, while Ngale Research Station is Vertisol. Climate type of both environments is C3 according to Oldeman classification (Oldeman, 1975), where there are 5-6 wet months and 4-6 dry months. The altitude for Jambegede and Ngale Research Station are 335 m and 168 m above sea level, respectively.

The design was a randomized complete blocks design with three replications. The plot size was 1.6 m × 3.0 m with plant spacing of 0.4 m × 0.15 m, two plants per hill. Fertilization was done by applying 50 kg Urea, 75 kg SP36, and 75 kg of KCl per hectare at sowing time. Weeding was conducted manually at 14 and 28 day after planting (dap). Watering was conducted by technical irrigation. Pest control was intensively done by applying insecticides with 5 days interval. Harvesting was carried out after the plant was physiologically matured, that was shown by pods having turned to yellow/brown and the leaves fallen. Data were collected for 50% days to flowering, days to maturity, plant height, number of branches per plant, number of pods per plant, 100 grains weight, and grain yield per hectare.

3. Results and Discusion

Results showed that there were interactions between the genotypes and environment regarding the grain yield, flowering and days to maturity, and number of branches (Table 1). Similar results were reported by Ashraf et al. (2010) and Jandong et al. (2011). This indicates that there was a difference in the response of the lines tested against the growth environment. Comparatively, there was no statistical siginificant G × E regarding plant height, number of pods per plant, and weight of 100 grains, but significant differences among the tested genotypes were acquired for three characters. This indicates that there existed diversity among the tested lines.

Table 1. Genetic × environment (G × E) analysis of soybean lines at Jambegede and Ngale Research Stations. Dry season 2011

Source	Degree of freedom	Flower	Maturity	Height	Branches	Pod-f	100SW	Yield
Environments	1	405.34**	182.04**	1,418.22**	0.78	422.50	16.73	1.57
Genotypes	14	37.95**	91.82**	236.02**	2.48**	248.25**	11.51**	0.55**
G × E	14	3.80**	36.62**	32.93	0.69**	48.97	1.20	0.22*
Error	1	0.68	60.09	18.59	0.27	65.27	1.18	0.10
Coefficient of variation (%)		2.42	1.30	7.74	13.11	14.84	8.49	16.16

**Significant at 1%, *Significant at 5%, Flower = flowering indices, Maturity = maturity indices, Height = plant height [cm], Branches = number of branches per plant, Pod-f = number of pods per plant, 100 GW = 100 grains weight, Yield = yield per hectare.

Tgm/Anj-790 line had the highest grain yield (2.53 t/ha) in Jambegede, whereas Tgm/Anj-795 line had the highest grain yield (2.50 t/ha) at Ngale (Table 2). 'Anjasmoro' as control variety yielded much lower yield, in Jambegede and Ngale i.e. 1.50 and 1.71 t/ha respectively, than either of those two best-yielding lines. Lines that had lower grain than 'Anjasmoro' were Tgm/Anj-777, Tgm/Anj-789, Tgm/Anj-803 and Tgm/Anj-824 grown at Jambegede (Table 2). At Ngale, no line had lower yield than 'Anjasmoro'. Furthermore, Tgm/Anj-789 had low grain yield (1.28 t/ha) at Jambegede, but at Ngale was quite high yield (2.13 t/ha) (Table 2). Average grain yield of Ngale was higher than that of Jambegede, i.e. 2.09 and 1.83 t/ha respectively. It may be due to the varying level of availability of the soil moisture, since at Ngale, the soil moisture was presumed higher than at Jambegede. Indeed the soil type of Ngale is Vertisol, able to retain more water than soil type of Jambegede (Entisol-Inceptisol association). On the other hand, the average grain yield obtained in this experiment was higher than these in analogical studies in tropical areas (Aremu et al., 2006), but lower than in subtropical areas (De Bruin & Pedersen, 2009; Pedersen & Lauer, 2004; Wilhelm & Wortmann, 2004).

Table 2. Grain yield of soybean lines at Jambegede and Ngale Research Stations. Dry season 2011

Genotypes	Yield [t/ha]	
	Jambegede	Ngale
Tgm/Anj-743	2.25 a-d	2.17 a-f
Tgm/Anj-744	1.95 c-i	2.15 a-f
Tgm/Anj-764	2.01 b-i	2.19 a-f
Tgm/Anj-773	2.32 a-d	2.34 a-d
Tgm/Anj-777	1.22 j	1.88 c-i
Tgm/Anj-778	2.21 a-e	1.96 c-i
Tgm/Anj-780	1.63 g-j	2.37 a-c
Tgm/Anj-789	1.28 j	2.13 a-g
Tgm/Anj-790	2.53 a	2.31 a-d
Tgm/Anj-795	2.24 a-d	2.50 ab
Tgm/Anj-796	2.07 a-h	1.84 d-i
Tgm/Anj-799	1.60 h-j	1.99 b-i
Tgm/Anj-803	1.31 j	1.67 f-j
Tgm/Anj-824	1.29 j	2.17 a-f
'Anjasmoro'	1.50 ij	1.71 e-j
LSD 5%	0.52	

Values followed by the same letter were not significantly different at Least Significant Different (LSD) 5%.

Tgm/Anj-744 and Tgm/Anj-824 lines scored the longest days to flowering periods both at Jambegede and at Ngale (Table 3). Line with the shortest days to flowering planted at Ngale was Tgm/Anj-789 (29 days). At Jambegede, line with the shortest days to flowering was Tgm/Anj-795 (32 days). 'Anjasmoro' scored days to flowering of 33 days and 31 days at Jambegede and Ngale, respectively. In general, days to flowering counts scored for soybean lines grown at Jambegede were higher than for those grown at Ngale. In this study, days to flowering values did not correlate with yield (data not shown), but in another study (Egli & Bruening, 2002) with the time of flower development and pollination being possibly an important determinant of seed number.

Tgm/Anj-773 and Tgm/Anj-790 lines scored the highest values of days to maturity indices at Jambegede (Table 3). At Ngale, the tested lines had a generally relatively lower values of days to flowering indices, while 'Anjasmoro' had the highest value of day to maturity at this location (Table 3). Tgm/Anj-789 and Tgm/Anj-796 consistently showed the lowest value of days to maturity at both environments. Similar to days to flowering indices' value, the days to maturity indices of soybean lines grown at Jambegede scored higher than these at Ngale. Variety 'Anjasmoro', with the highest value of days to maturity, did not give the highest yield. It might be due to the higher use of photosyntates for vegetative growth for this higher scoring days to maturity indices. The higher vegetative growth may decrease grain yield because higher vegetative growth will affect leaf area index. Similar result also showed by Mellendorf (2011) which reported that in higher plant density lead decreasing grain yield.

The highest number of branches per plant was developed by Tgm/Anj-803 line grown at Jambegede, and equivalent to Tgm/Anj-778 line grown at Ngale Research Station. At the same time, the lowest branches per plant index was shown by 'Anjasmoro' in both locations, and equivalent to Tgm/Anj-780 at Jambegede and Tgm/Anj-789 at Ngale (Table 4). The number of branches per plant was related to grain yield, similar to report by Machikowa and Laosuwan (2011), who stated that number of branches per plant gives the highest positive direct effect on grain yield per plant after number of pods per plant.

Table 3. Days to flowering and maturity of soybean lines at Jambegede and Ngale Research Stations. Dry season 2011

Genotypes	Days to flowering [days]		Days to maturity [days]	
	Jambegede	Ngale	Jambegede	Ngale
Tgm/Anj-743	38.6 b	33.0 gh	82.7 ab	78.7 g-i
Tgm/Anj-744	40.3 a	38.0 bc	81.3 b-d	78.0 h-j
Tgm/Anj-764	38.0 bc	34.0 fg	79.0 f-i	76.7 j
Tgm/Anj-773	35.0 ef	31.0 ij	84.0 a	79.3 e-h
Tgm/Anj-777	37.0 cd	31.0 ij	80.7 c-f	76.7 j
Tgm/Anj-778	36.7 cd	30.0 jk	81.7 bc	79.3 e-h
Tgm/Anj-780	33.0 gh	30.0 jk	81.0 b-e	79.3 e-h
Tgm/Anj-789	33.3 gh	29.0 k	79.7 d-h	77.3 ij
Tgm/Anj-790	35.7 de	31.0 ij	83.7 a	78.0 hij
Tgm/Anj-795	32.7 gh	30.0 jk	80.7 c-f	79.3 e-h
Tgm/Anj-796	35.7 de	32.0 hi	79.7 d-h	77.3 ij
Tgm/Anj-799	37.3 bc	31.0 ij	81.7 bc	78.0 h-j
Tgm/Anj-803	38.0 bc	32.0 hi	80.7 c-f	79.3 e-h
Tgm/Anj-824	40.3 a	38.0 bc	81.0 b-e	78.7 ghi
'Anjasmoro'	33.0 gh	31.0 ij	81.3 b-d	80.0 c-g
LSD 5%	1.35		1.69	

Values followed by the same letter were not significantly different at Least Significant Different (LSD) 5%.

Table 4. Number of branches per plant of soybean lines at the Jambegede and the Ngale Research Stations. Dry season 2011

Genotypes	Number of branches/plant	
	Jambegede	Ngale
Tgm/Anj-743	3.63 g-m	4.77 a-d
Tgm/Anj-744	4.43 a-g	3.17 j-n
Tgm/Anj-764	4.70 a-d	4.57 a-e
Tgm/Anj-773	3.83 e-k	3.67 f-m
Tgm/Anj-777	3.97 d-j	4.03 d-i
Tgm/Anj-778	4.03 d-i	5.13 ab
Tgm/Anj-780	2.97 l-n	3.73 e-l
Tgm/Anj-789	3.27 i-m	2.93 l-n
Tgm/Anj-790	3.77 e-l	4.30 b-g
Tgm/Anj-795	3.00 k-n	4.20 c-h
Tgm/Anj-796	3.60 g-m	3.37 h-m
Tgm/Anj-799	4.23 c-g	4.30 b-g
Tgm/Anj-803	5.23 a	4.90 a-c
Tgm/Anj-824	4.50 a-f	4.40 a-g
'Anjasmoro'	2.37 n	2.87 mn
LSD 5%	0.84	

Values followed by the same letter were not significantly different at Least Significant Different (LSD) 5%.

No G × E was found for plant height (Table 1). In Jambegede and Ngale the highest plant height was shown by Tgm/Anj-744 (60.1 cm and 78.3 cm respectively). The shortest plant height was shown by Tgm/Anj-780 and Tgm/Anj-789 in Jambegede and Tgm/Anj-789 in Ngale (Table 5). 'Anjasmoro' had the plant height slightly below the average of the tested lines, i.e. 50 cm in Jambegede and 58.9 cm in Ngale. Plant height is important for other agronomic traits, because it correlates with those agronomic traits such as number of branches per plant, number of pods per plant, and grain yield of soybean (Kuswantoro et al., 2006). Moreover, Board (2002) suggested a regression model to predict the high-yielding cultivars and correlated it with yield per plot based on plant height, in addition to total dry matter at beginning seed (reproductive stage 5 - R5) and pod filling period.

Table 5. Plant height of per plant of soybean lines at the Jambegede and the Ngale Research Stations. Dry season 2011

Genotypes	Plant height [cm]	
	Jambegede	Ngale
Tgm/Anj-743	53.8 b-d	56.0 bd
Tgm/Anj-744	60.1 a	78.3 a
Tgm/Anj-764	55.8 a-d	60.1 b
Tgm/Anj-773	45.5 ef	60.2 b
Tgm/Anj-777	45.0 ef	56.7 b-d
Tgm/Anj-778	53.0 cd	60.3 b
Tgm/Anj-780	40.0 f	51.1 cd
Tgm/Anj-789	41.4 f	49.0 d
Tgm/Anj-790	56.9 a-c	61.5 b
Tgm/Anj-795	51.7 cd	57.0 b-d
Tgm/Anj-796	52.7 cd	54.6 b-d
Tgm/Anj-799	59.5 ab	70.2 a
Tgm/Anj-803	54.9 a-d	61.5 b
Tgm/Anj-824	56.2 a-c	59.9 b
'Anjasmoro'	50.0 de	58.9 bc
LSD 5%	5.96	8.28

Values in the same column followed by the same letter were not significantly different at Least Significant Different (LSD) 5%.

The highest number of pods per plant developed the Tgm/Anj-773, while the lowest the Tgm/Anj-789 (Table 6). 'Anjasmoro' also classified as genotype with the lowest number of pods after Tgm/Anj-790 (Table 6). Genetic factors play a greater role in expression of these traits because there was no interaction with the environment (Table 1). Inadequate supply of assimilates to flowers is a dominant factor in flower abortion (Yashima et al., 2005), causing pod number reduction. Machikowa and Laosuwan (2011) reported that pods per plant count had the highest positive direct effect on grain yield. The number of pods per plant was reported as largely dependent on the number of floral buds that initiate pods (Desclaux et al., 2000). There was a significant correlation found between the number of pods per plant and grain yield (Malik et al., 2011). Furthermore, number of pods per plant was reported as having a positive direct effect on grain yield (Machikowa & Laosuwan, 2011; Sudaric & Vrataric, 2002).

Grain size was expressed as the weight of 100 grains and presented in Table 6. The largest seed size was achieved by Tgm/Anj-790, while the smallest seed size by Tgm/Anj-764. Environment markedly influences seed size during the seed development period of growth (Table 1). At the seed development period, grain size reduction caused by drought or other stresses can substantially reduce yield (Kuswantoro & Zen, 2013). Nevertheles, several smaller-seeded soybean lines with high yields have been reported and vice versa, which indicates that seed size does not directly correlate with yield potential (Klein et al., 2005). Some researchers obtained different results with a significant correlation (Malik et al., 2011) and a genetic correlation (Arshad et al. 2006) was found between grain yield and 100 grains weight. Also, 100 grains weight had a positive direct effect on grain yield (El-Badawy & Mehasen, 2012).

Table 6. Average of number of filled pods per plant and 100 grain weight of soybean lines at the Jambegede and the Ngale Research Stations. Dry season 2011

Genotypes	Number of filled pods/plant	Grain weight [g/100 grains]
Tgm/Anj-743	62.62 ab	11.17 fg
Tgm/Anj-744	56.02 a-d	12.51 c-e
Tgm/Anj-764	54.17 b-e	10.58 g
Tgm/Anj-773	63.80 a	11.72 e-g
Tgm/Anj-777	50.52 c-e	13.71 bc
Tgm/Anj-778	60.28 ab	11.84 d-f
Tgm/Anj-780	49.48 c-e	12.87 b-e
Tgm/Anj-789	45.10 e	13.02 b-d
Tgm/Anj-790	47.30 de	14.04 b
Tgm/Anj-795	49.15 c-e	13.34 bc
Tgm/Anj-796	49.20 c-e	12.71 c-e
Tgm/Anj-799	62.60 ab	13.40 bc
Tgm/Anj-803	60.88 ab	12.00 d-f
Tgm/Anj-824	57.38 a-c	12.57 c-e
'Anjasmoro'	48.30 c-e	16.45 a
LSD 5%	9.34	1.26

Values in the same column and followed by the same letter were not significantly different at Least Significant Different (LSD) 5%.

4. Conclusion

The grain yield was highly affected by G × E interaction. In the different environment, ranking of the genotypes were also different. However, Tgm/Anj-795 and Tgm/Anj-790 lines consistently had the highest grain yield in both locations. Tgm/Anj-790 high yield was defined by the large seed size. The G × E interaction also affected other traits, such as days to flowering indices, days to maturity indices, and number of branches per plant. Tgm/Anj-789 and Tgm/Anj-796 consistently showed the lowest values of days to maturity in both locations. 'Anjasmoro' had higher value of days to maturity but lower grain yield. Tgm/Anj-795 and Tgm/Anj-790 lines had potentially high yield stability that need to be further tested in many different locations and seasons to know the actual stability.

Acknowledgements

This study was funded by the State Ministry of Research and Technology of Republic of Indonesia through SINTA Project 2011. Thank Mr. Agus Supeno for the assistance in the research.

References

Alghamdi, S. S. (2004). Yield stability of some soybean genotypes across diverse environments. *Pak. J. Biol. Sci., 7*, 2109-2114. http://dx.doi.org/10.3923/pjbs.2004.2109-2114

Aremu, C. O., Ojo, D. K., Oduwaye, O. A., & Amira, J. O. (2006). Comparison of joint regression analysis (JRA) and additive main effect and multiplicative interaction (AMMI) model in the study of G × E interaction in soybean. *Nigerian J. Genet., 20*, 74-83.

Arshad, M., Ali, N., & Ghafoor, A. (2006). Character correlation and path coefficient in soybean *Glycine max* (L.) Merrill. *Pak. J. Bot., 38*, 121-130.

Ashraf, M., Iqbal, Z., Arshad, M., Waheed, A., Gufran, M. A., Chaudhry, Z., & Baig, D. (2010). Multi-environment response in seed yield of soybean [*Glycine max* (L.) Merrill], genotypes through GGE biplot technique. *Pak. J. Bot., 42*, 3899-3905.

Board, J. E. (2002). A regression model to predict soybean cultivar yield performance at late planting dates. *Agron. J., 94*, 483-492. http://dx.doi.org/10.2134/agronj2002.0483

Carpenter, A. C., & Board, J. E. (1997). Branch yield components controlling soybean yield stability across plant populations. *Crop Sci., 37*, 885-891. http://dx.doi.org/10.2135/cropsci1997.0011183X003700030031x

Conway, G. (1982). A Guide to Agroecosystem Analysis. *Multiple Croping Project.* Thailand: Faculty of Agriculture. University of Chiang Mai.

Cucolotto, M., Pípolo, V. C., Garbuglio, D. D., da S. F. Junior, N., Destro, D., & Kamikoga, M. K. (2007). Genotype x environment interaction in soybean: evaluation through three methodologies. *Crop Breeding and Applied Biotechnology, 7*, 270-277.

De Bruin, J. L., & Pedersen, P. (2009). Growth, yield, and yield component changes among old and new soybean cultivars. *Agron. J., 101*, 124-130. http://dx.doi.org/10.2134/agronj2008.1087

Desclaux, D., Huynh, T. T., & Roumet, P. (2000). Identification of soybean plant characteristics that indicate the timing of drought stress. *Crop Sci., 40*, 716-722. http://dx.doi.org/ 10.2135/cropsi2000.403716x

Duzdemir, O. (2011). Stability analysis for phenological characteristics in chickpea. *Afr. J. Agric. Res., 6*, 1682-1685. http://dx.doi.org/ 10.5897/AJAR10.1138

Egli, D. P., & Bruening, W. P. (2000). Potential of early-maturing soybean cultivars in late plantings. *Agron. J., 92*, 532-537. http://dx.doi.org/10.2134/agronj2000.923532x

Egli, D. P., & Bruening, W. P. (2002). Synchronous flowering and fruit set at phloem-isolated nodes in soybean. *Crop Sci., 42*, 1535-1540. http://dx.doi.org/ 10.2135/cropsci2002.1535

El-Badawy, M. El. M., & Mehasen, S. A. S. (2012). Correlation and path coefficient analysis for yield and yield components of soybean genotypes under different planting density. *Asian J. Crop Sci., 4*, 150-158. http://dx.doi.org/10.3923/ajcs.2012.150.158

Gurmu, F., Mohammed, H., & Alemaw, G. (2009). Genotype x environment interactions and stability of soybean for grain yield and nutrition quality. *African Crop Sci. Journal, 17*, 87-99.

Hidayat. (2002). Analisis Interaksi Galur Lingkungan Beberapa Galur Padi Di Lahan Pasang Surut Berjenis Tanah Gambut Kalimantan Barat. Disertasi Doktor. *Program Pascasarjana Universitas Gadjah Mada.* Yogyakarta.

Jandong, E. A., Uguru, M. I., & Oyiga, B. C. (2011). Determination of yield stability of seven soybean (*Glycine max*) genotypes across diverse soil pH levels using GGE biplot analysis. *J. of Appl. Biosci., 43*, 2924-2941.

Klein, R., Elmore, R. W., & Nelson, L. A. (2005). *Using Soybean Yield Data to Improve Variety Selection – Part I.* University of Nebraska – Lincoln Extension educational programs abide with the nondiscrimination policies of the University of Nebraska – Lincoln and the United States Department of Agriculture.

Kuswantoro, H., Basuki, N., & Arsyad., D. M. (2006). Identifikasi plasma nutfah kedelai toleran terhadap tanah masam berdasarkan keragaman genetik dan fenotipik. *Agrivita, 28*, 54-63.

Kuswantoro, H., & Zen, S. (2013). Performance of acid-tolerant soybean promising lines in two planting seasons. *Inter. J. Biol., 5*, 49-56. http://dx.doi.org/10.5539/ijb.v5n3p49

Machikowa, T. & Laosuwan, P. (2011). Path coefficient analysis for yield of early maturing soybean. *Songklanakarin J. Sci. Technol., 33*, 365-368.

Malik, M. F. A., Ashraf, M., Qureshi, A. S., & Khan, M. R. (2011). Investigation and comparison of some morphological traits of the soybean populations using cluster analysis. *Pak. J. Bot., 43*, 1249-1255.

Oldeman, L. R. (1975). *An agro-climatic map of Java.* Central Research Institute for Agriculture. Bogor, Indonesia.

Pedersen, P., & Lauer, J. G. (2004). Response of soybean yield components to management system and planting date. *Agron. J., 96*, 1372-1381. http://dx.doi.org/10.2134/agronj2004.1372

Sharrifmoghaddassi, M., & Omiditabrizi, A. H. (2010). Stability analysis of seven Iranian Winter safflower cultivars. *World Applied Sci. J., 8*, 1366-1369.

Statistic Indonesia. (2013). Trends of Selected Socio-Economic Indicators of Indonesia. *BPS – Statistic Indonesia.* Jakarta.

Sudaric, A., & Vrataric, M. (2002). Variability and interrelationships of grain quantity and quality characteristics in soybean. *Die Bodenkultur, 53*, 137-142.

Toledo, J. F. F. de, Arias, C. A. A., Oliveira, M. F. de, Triller, C., & Miranda, Z. de F. S. (2000). Genetical and environmental analyses of yield in six biparental soybean crosses. *Pesq. agropec. bras., Brasília, 35*, 1783-1796. http://dx.doi.org/10.1590/S0100-204X2000000900011

Ülker, M., Sönmez, F., Çiftçi, V., Yilmaz, N., & Apak, R. (2006). Adaptation and stability analysis in the selected lines of tir wheat. *Pak. J. Bot., 38*, 1177-1183.

Wilhelm, W. W., & Wortmann, C. S. (2004). Tillage and rotation interactions for corn and soybean grain yield as affected by precipitation and air temperature. *Agron. J., 96*, 425-432. http://dx.doi.org/10.2134/agronj2004.4250

Yashima, Y., Kaihatsu, A., Nakajima, T., & Kokobun, M. (2005). Effect of source/sink ratio and cytokinin application on pod set in soybean. *Plant Prod. Sci., 8*, 139-144.

12

Response of Some Soybean Germplasm to Mangan Toxicity

Heru Kuswantoro[1]

[1] Indonesian Legume and Tuber Crops Research Institute, Indonesian Agency for Agricultural Research and Development, Malang, Indonesia

Correspondence: Heru Kuswantoro, Indonesian Legume and Tuber Crops Research Institute, Indonesian Agency for Agricultural Research and Development, Malang, Indonesia. E-mail: herukusw@yahoo.com

Abstract

In excessive manner, mangan (Mn) as an essential nutrient can be toxic to the plant. This phenomenon often occurs in acid soil. Hence, it is needed gene resources to develop plant in acid soil. The soybean germplasm tolerance to Mn toxicity was tested in seed laboratory, using two factors experimental design. The first factor was Mn toxicity containing two treatments (1) 0 ppm Mn in pH 7 as control, and (2) solution concentration of 75 ppm Mn in pH 4. The second factor was 14 accessions of soybean germplasm. Results showed that generally root characters decreases while shoot characters increased in Mn toxicity condition. However, some genotypes showed different performance. There was one genotype having the highest root dry weight in Mn toxicity condition, i.e. MLGG 0091. The highest root dry weight in this genotype was also supported by the root length and number of roots. MLGG 0091 was also capable to increase the length epicotyle that contribute to the increase in seedling dry weight. Therefore MLGG 0091 can be used as a gene source for tolerance to Mn toxicity.

Keywords: tolerance of acidity and Mn, characterization, germplasm, soybean

1. Introduction

Mangan is an essential nutrient that can be toxic to the plant in excessive manner (Marschner, 1995). Mn level in a soil is controlled by Mn availability in soil, pH and electron availability (Adams, 1981; Sparow and Uren, 1987). Soil with high Mn availability can cause Mn toxicity in soil pH below 6.0 (Hue et al., 1987). In high electron environment, caused by excessive watering, low drainage, or high organic manure application, Mn toxicity is able to arise even in alkaline pH (Hue, 1988). It is because some organic molecules are capable to split Mn oxide through electron transfer in reductive process (Stone & Morgan, 1984).

Mn toxicity in acid soil is a complex trait and involving multiple physiological and biochemical mechanisms (Millaleo et al., 2010). Beside, at different times during the growing and in the same soil season, manganese can be deficient and toxic to plants (Johansen, 2005). One of the reasons is Mn toxicity relates to other nutrients availability. Mn toxicity also decreases due to high availability of other nutrients in the soil such as Ca (Horst, 1988), Mg (Goss et al., 1991), and Si (Horst & Maschner, 1978).

Plant species differences or even variety differences in a species have different response level to Mn toxicity (Foy et al., 1988; Reddy et al., 1991). Mn toxicity in *Phaseolus vulgaris* is identified at 0.5 uM if the nutrients solution free of Si (Horst & Marschner, 1978). In soybean, Mn toxicity is identified at 15 ppm (Heenan & Carter, 1976). In this research, response of some germplasm to Mn toxicity was studied.

2. Materials and Method

Response of some soybean germplasm to mangan toxicity was conducted in Seed Laboratory of Indonesian Legume and Tuber Crops Research Institute, Malang-Indonesia. Research was carried out by using factorial design of randomized complete block design (RCBD). The first factor was Mn toxicity containing two treatments (1) 0 ppm Mn in pH 7 as control, and (2) solution concentration of 75 ppm Mn in pH 4. The second factor was 14 accessions of soybean germplasm.

A total of 25 sterilized soybean seeds were germinated in a petridish. A gauze was put inside the petridish to ensure the germinated seeds to be able to stand straight up. Solution was pour in the petridish up to half of seed size to maintain seeds respiration. The decreasing solution due to the absorption by the seeds was overcome by adding aquadest up to the specified limit. Germination was conducted at 25°C. The observation was recorded on root length, root dry weight, hypocotile length, number of lateral roots, seedling dry weight and epicotyle length at 6 days after germinating.

3. Results and Discussion

The results showed that genotype MLGG 0147 and MLGG 0025 had the longest root length among the 14 tested genotypes, where the roots length of those two genotypes were more than 10 cm. In Mn toxicity condition, the two genotypes also had relatively long roots. Other genotype having relatively long roots was MLGG 0091, with narrow standard deviation. There were two genotypes with root length longer or equal to the previous three genotypes, but the standard deviation were wider, i.e. MLGG 0096 and MLGG 0118 (Figure 1). High Mn concentration affected the roots becoming shorter (Abou et al., 2002), but the five genotypes were capable to maintain roots length than other ten genotypes.

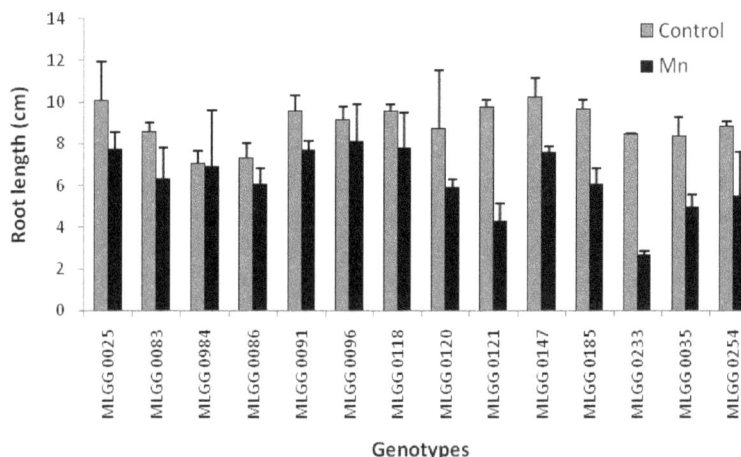

Figure 1. Root length of soybean germplasm in control and Mn toxicity conditions

Genotype MLGG 0233 was the genotype with shortest root length in Mn toxicity condition. However, in condition without Mn toxicity (control), this genotype was able to grow its roots up to 8 cm long. Other genotypes having the shortest root length in Mn toxicity condition was MLGG 0121. In normal conditions, this genotype was remain able to perform roots length almost 10 cm. Therefore, based on root length these two genotypes were classified as Mn toxicity sensitive genotypes (Figure 1). The highest percentage decrease in root length was shown by MLGG 0233 and MLGG 0121, i.e. 68.65% and 55.71% respectively. Genotype with the lowest of percentage of root length decrease was MLGG 0984, where the percentage of root length decrease was 3.45% (Table 1). Root resistance to Mn toxicity was the environmentally sensitive quantitative traits (Kaseem et al., 2004), that affected by the environments.

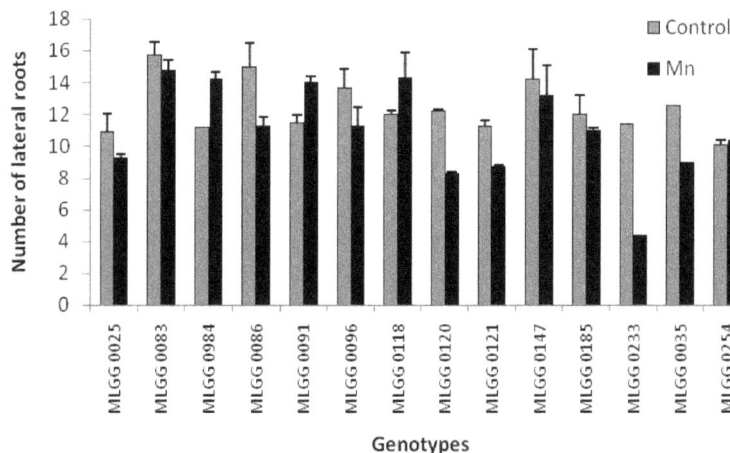

Figure 2. Number of lateral roots of soybean germplasm in control and Mn toxicity conditions

Genotypes with the highest number of lateral roots was MLGG 0083 and MLGG 0086 in the control condition, while the least number of lateral roots was shown by MLGG 0233. In Mn stress conditions, MLGG 0083 also showed the highest amount of the root followed by MLGG 0984, MLGG 0091 and MLGG 0118 (Figure 2). Similar to the root length, number of lateral roots decreased in Mn toxicity condition. The highest decrease of the root number was shown by MLGG 0233. Other genotypes showed a decrease in relatively high number of lateral roots were MLGG 0035, MLGG 0120 and MLGG 0086. Even though MLGG 0086 having a relatively high reduction in number of roots, but in stress condition the number of lateral roots was still relatively high. The interesting thing about this study is that there were several genotypes which have more number of lateral roots in stress conditions, three of them were MLGG 0984, MLGG 0091 and MLGG 0118 where the number of lateral roots increased 27.59%, 28.90% and 15.33% respectively (Table 1).

The highest root dry weight was shown by MLGG 0091 in Mn toxicity followed by MLGG 0035 also in Mn toxicity condition. It mean, root dry weight of some genotypes increased in Mn toxicity condition (Figure 3). Genotype having the lowest root dry weight was MLGG 0233, occurred in both of control and Mn toxicity conditions. For this genotype, it seemed that root dry weight was more affected by genetic factor than environmental factor, because root dry weight in different solution conditions were not significantly different. Other genotype having equivalent root dry weight in both two conditions was MLGG 0120. However, generally root dry weight decreased. The highest root dry weight decreasing were shown by MLGG 0025 and MLGG 0086, i.e. 23.69% and 23.38% (Table 1). Izaguirre-Mayoral and Sinclair (2005) also reported the decreasing of root dry weight when soybean grown in Mn solution from 60uM/L. Khabaz-Saberi et al., (2009) also reported that the increasing solution Mn concentration decreased root dry weight. Increasing root dry weight of *Heliantus anuus* in lower Mn concentration and then decreasing in higher concentration was reported by Hadjiboland et al. (2008). Soybean tolerance to Mn toxicity related with roots necrotic marker (Kaseem et al., 2004). It indicated the dammage of the roots caused by Mn toxicity. The increasing of root dry weight presumably was caused by the increasing root diameter such as root dry weight of cultivar IAC-15 increased by 63 and 116 % with the increase of Mn (Junior et al., 2009).

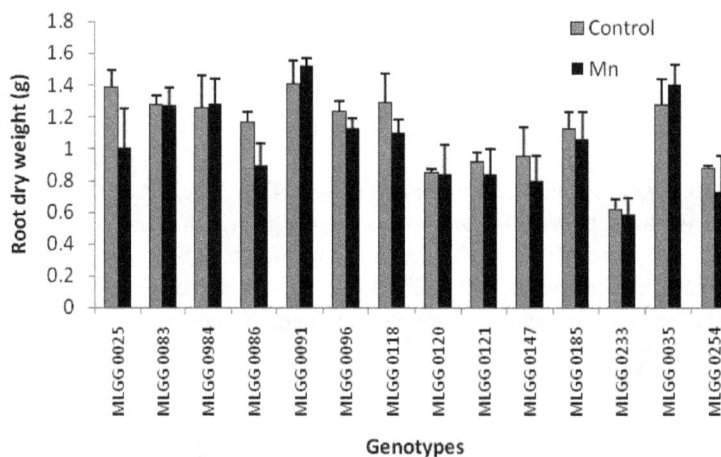

Figure 3. Root dry weight of soybean germplasm in control and Mn toxicity conditions

Hipocotyle length in control condition was higher than in Mn toxicity condition. In control condition, the highest hipocotyle length was achieved by MLGG 0121, followed by MLGG 0233, MLGG 0185 and MLGG 0086 (Figure 4). The decreasing hipocotyle length was achieved by MLGG 0233, while the lowest was achieved by MLGG 0096. However, there were some genotypes which had hipocotyle length higher in Mn toxicity condition than in control, i.e. MLGG 0086, MLGG 0984, MLGG 0083 and MLGG 0096. Therefore, the hipocotyle length increased in those four genotypes, with the highest increasing was achieved by MLGG 0984 as 17.17% (Tabel 2). In shortening of shoot, the high Mn concentration was also involved (Abou et al., 2002).

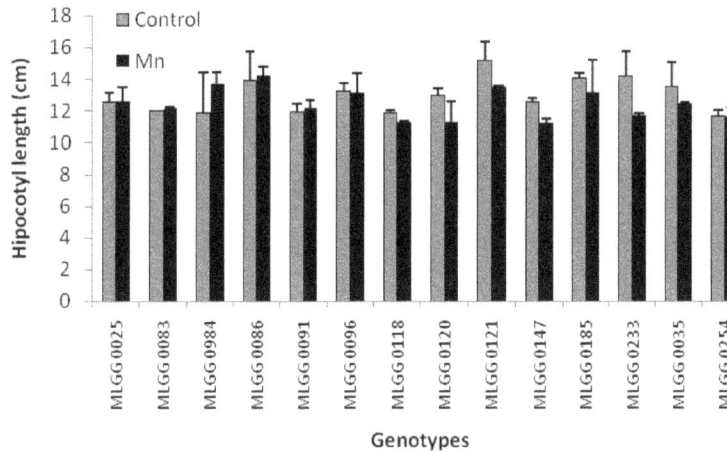

Figure 4. Hipocotyle length of soybean germplasm in control and Mn toxicity conditions

Unlike root length and hipocotyle length, epicotyle length in Mn toxicity condition was higher than in control condition. The highest epicotyle length was achieved by MLGG 0096 in Mn toxicity condition, followed by MLGG 0096 in control condition. Genetic factor also more affected in this genotype than environmental factor, because the difference of epicotyle length between Mn toxicity and control conditions was relatively low. It was different to the genotypes of MLGG 0035 and MLGG 0254, where in these two genoytpes epicotyle length in Mn toxicity condition was relativlely higher than in control condition (Figure 5). In these two genoytpes increasing epicotyle length achieved up to 163.89% and 113.16% respectively (Table 2).

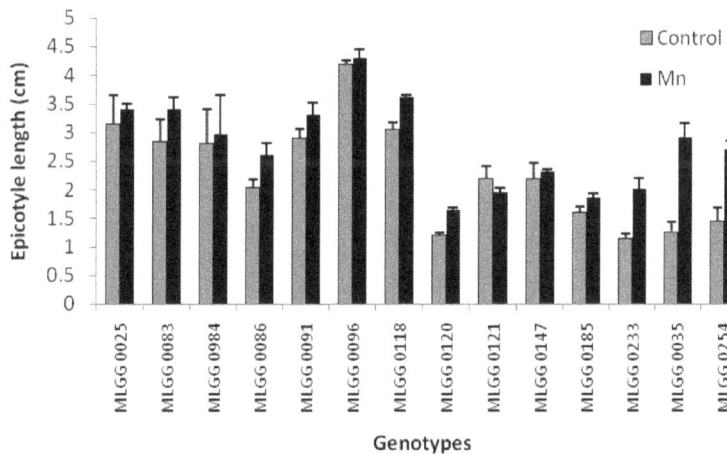

Figure 5. Epicotyle length of soybean germplasm in control and Mn toxicity conditions

The highest seedling dry weight was shown by MLGG 0091 in Mn toxicity condition, while the lowest was shown by MLGG 0233 in both Mn toxicity and control conditions (Figure 6). Generally, seedling dry weight decreases in Mn toxicity condition. The highest decreasing was shown by MLGG 0025 that achieved 21.87%, while other genotypes had relatively low decreasing percentage. Some researcher also reported similar results, where increasing solution Mn concentration decreased shoot dry weight (Izaguirre-Mayoral & Sinclair, 2005; Khabaz-Saberi et al., 2009). Some genotypes did not experience seedling dry weigth decreasing. Even, there were three genotypes that experienced seedling dry weight increasing in Mn toxicity conditions, i.e. MLGG 0091 and MLGG 0035 up to 7.32% and 8.14% respectively (Table 2). Reis and Junior (2011) reported the differences among the genotypes in leaves ultrastructural parts. Presumably, it also affected in whole seedling dry weight in addition to the difference due to seedling dry weight.

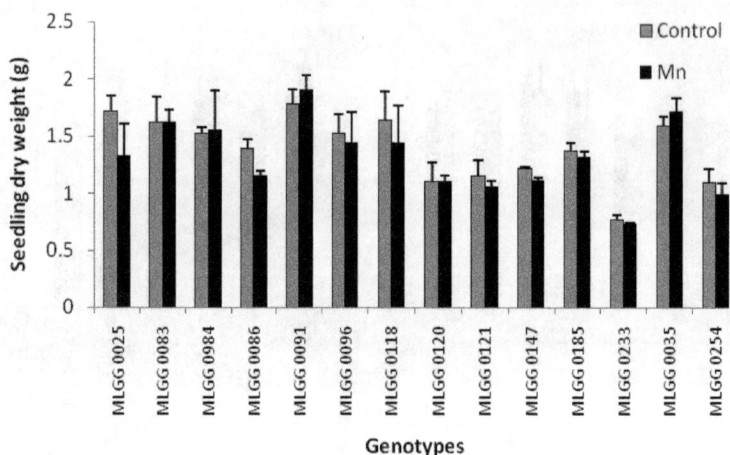

Figure 6. Seedling dry weight of soybean germplasm in control and Mn toxicity conditions

Table 1. Decrease percentage of roots length, number of roots, and root dry weight of soybean germplasm in Mn toxicity condition

Genotypes	Root length (cm)	Number of lateral roots	Root dry weight (g)
MLGG 0025	22.69	10.04	28.69
MLGG 0083	26.11	-0.52	0.88
MLGG 0984	3.45	-27.59	-4.04
MLGG 0086	16.52	15.35	23.38
MLGG 0091	19.28	-28.90	-7.81
MLGG 0096	11.39	6.28	8.67
MLGG 0118	18.60	-15.33	14.33
MLGG 0120	27.56	30.90	1.46
MLGG 0121	55.71	19.73	9.61
MLGG 0147	25.43	-11.46	16.91
MLGG 0185	36.52	-2.75	6.55
MLGG 0233	68.65	61.48	5.26
MLGG 0035	40.13	28.48	-10.06
MLGG 0254	38.12	-5.97	17.82

Table 2. Decrease percentage of hipocotyle length, epicotyle length, and seedling dry weight of soybean germplasm in Mn toxicity condition

Genotypes	Hipocotyle length (cm)	Epicotyle length (cm)	Seedling dry weight (g)
MLGG 0025	0.05	-20.84	21.87
MLGG 0083	-1.25	-32.14	-0.84
MLGG 0984	-17.17	5.68	-1.65
MLGG 0086	-2.36	-31.16	17.21
MLGG 0091	-1.85	-16.47	-7.32
MLGG 0096	0.88	-2.61	5.98
MLGG 0118	5.45	-18.56	10.98
MLGG 0120	12.87	-38.81	-1.84
MLGG 0121	10.96	9.62	7.78
MLGG 0147	10.76	-11.33	8.98
MLGG 0185	6.56	-18.65	4.18
MLGG 0233	17.39	-72.31	5.04
MLGG 0035	7.13	-163.89	-8.14
MLGG 0254	0.46	-113.16	8.54

4. Conclusion

In Mn toxicity condition, the highest root dry weight was achieved by MLGG 0091. The high root dry weight in this genotype was also supported by the root length and number of roots. Decrease percentage of the root length in MLGG 0091 was classified as low if compared to other genotypes. In addition, MLGG 0091 was also capable to increase the length epicotyle that contribute to the increase in seedling dry weight. Therefore MLGG 0091 can be used as a source of Mn tolerant genes.

References

Abou M., Symeonidis L., Hatzistavrou E., & Yupsanis T. (2002). Nucleolytic activities and appearance of a new DNase in relation to nickel and manganese accumulation in Alyssum murale. *J. Plant Physiol., 159*, 1087-1095. http://dx.doi.org/10.1078/0176-1617-00667

Adams F. (1981). Nutritional imbalances and consntraints to plant growth on acid soils. *J. Plant Nutr., 4*, 81-87. http://dx.doi.org/10.1080/01904168109362905

Foy C. D., Scott B. J., & Fisher J. A. (1988). Genetics differences in plant tolerance to manganese toxicity. p. 293-307. In R. D. Graham et al. (Eds.), *Manganese in soils and plants*. Kluwer Acad. Publ. Dordrecht, Netherland.

Goss M. J., Carvalho M. J. G. P. R., & Kirby E. A. (1991). Predicting toxic concentration in acid soils. p. 729-732. In R. J. Wright et al. (Eds.), *Plant-soils interactions at low pH*. Kluwer Acad. Publ. Dordrecht, Netherland.

Hajiboland R., Aliasgharpour M., Dashtbani F., Movafeghi A., & Dadpour M. R. (2008). Localization and Study of Histochemical Effects of Excess Mn in Sunflower (*Helianthus annuus* L. cv. Azarghol) Plants. *Journal of Sciences, Islamic Republic of Iran, 19*, 305-315.

Heenan D. P., & Carter O. G. (1976). Tolerance of soybean cultivars to manganese toxicity. *Crops Sci., 16*, 389-391. http://dx.doi.org/10.2135/cropsci1976.0011183X001600030018x

Horst W. J. (1988). The physiology of manganese toxicity. p. 175-188. In R. D. Graham et al. (Eds.), *Manganese in soils and plants*. Kluwer Acad. Publ. Dordrecht, Netherland.

Horst W. J., & Marschner H. (1978). Effect of silicone on manganese tolerance on bean plant (*Phaseolus vulgaris*). *Plant and Soil, 50*, 287-303. http://dx.doi.org/10.1007/BF02107179

Hue N. V. (1988). A possible mechanism for manganese toxicity in Hawaii soils amended with a low-sewage sludge. *J. Environ. Qual., 17*, 473-479. http://dx.doi.org/10.2134/jeq1988.00472425001700030022x

Hue N. V., Fox R. L., & McCall W. W. (1987). Aluminum, Ca, and Mn concentrations in macadamia seedlings as affected by soil acidity and liming. *Commun. Soil Sci. Plant Anal., 18*, 1253-1267. http://dx.doi.org/10.1080/00103628709367897

Izaguirre-Mayoral M. L., & Sinclair T. R. (2005). Soybean genotypic difference in growth, nutrient accumulation and ultrastructure in response to manganese and iron supply in solution culture. *Annals of Botany, 96*, 149–158. http://dx.doi.org/10.1093/aob/mci160

Johansson J. (2005). Manganese solubility due to compaction in soils under corn and soybean. Institutionen för markvetenskap. Uppsala.

Junior J. L., Malavolta E., Nogueira Nd. L., Moraes M. F., Reis A. R., Rossi M. L., & Cabral C. P. (2009). Changes in anatomy and root cell ultrastructure of soybean genotypes under manganese stress. *R. Bras. Ci. Solo., 33*, 395-403. http://dx.doi.org/10.1590/S0100-06832009000200017

Kassem, My. A., Meksem, K., Kang, C. H., Njiti, V., Kilo, V., Wood, A. J., & Lightfoot, D. A. (2004). Loci underlying resistance to manganese toxicity mapped in a soybean recombinant inbred line population of "Essex" x "Forrest". *Plant and Soil, 260*, 197-204. http://dx.doi.org/10.1023/B:PLSO.0000030189.96115.21

Khabaz-Saberi H., Rengel Z., Wilson R., & Setter T. L. (2009). Variation of tolerance to manganese toxicity in Australian hexaploid wheat. *J. of Plant Nut. and Soil Sci., 173*, 103-112. http://dx.doi.org/10.1002/jpln. 200900063

Marschner, H. (1995). *Mineral nutrition of higher plants* (2nd ed.). Academic Press. San Diego, CA.

Millaleo R., Reyes-Díaz M., Ivanov A. G., Mora M. L., & Alberdi M. (2010). Manganese as essential and toxic element for plants: transport, accumulation and resistance mechanisms. *J. Soil Sci. Plant Nutr., 10*, 476 – 494. http://dx.doi.org/10.4067/S0718-95162010000200008

Reddy M. R, Ronaghi A., & Bryant J. A. (1991). Differential responses of soybean genotypes to excess manganese in an acid soil. *Plant and Soil, 134*, 221-226. http://dx.doi.org/10.1007/BF00012039

Reis dos A. R., & Junior J. L. (2011). Genotypic influence on the absorption, use and toxicity of manganese by soybean. Soybean - Molecular Aspects of Breeding. *InTech.* Retrieved from http://www.intechopen.com/books/soybean-molecularaspects-of-breeding/genotypic-influence-on-the-absorption-use-and-toxicity-of-manganese-by-soybean

Sparow L. A., & Uren N. C. (1987). Oxidation and reduction of Mn in acidic soils: Effect of temperature and soil pH. *Soil Biol. Biochem., 19*, 143-148. http://dx.doi.org/10.1016/0038-0717(87)90073-3

Stone A. T., & Morgan J. J. (1984). Reduction and distribution of manganese (III) and manganese (IV) oxides by organics: 2. Survey of the reactivity of organics. *Environ. Sci. Technol., 18*, 617-624. http://dx.doi.org/10.1021/es00124a011

13

Detection of *Beet Necrotic Yellow Vein Virus* by Double Stranded RNA Analysis

Handan Çulal Kılıç[1], Nejla Yardımcı[1] & Gözde Urgen[1]

[1] Department of Plant Protection, Faculty of Agriculture, Süleyman Demirel University, Turkey

Correspondence: Handan Çulal Kılıç, Department of Plant Protection, Faculty of Agriculture, Süleyman Demirel University, Isparta, Turkey. E-mail: handankilic@sdu.edu.tr

Abstract

Beet necrotic yellow vein virus (BNYVV), the type member of the *Benyvirus* genus, has a multipartite, positive-sense single-stranded RNA genome, which consists generally of four, or in some isolates five, distinct RNA species. In this study, 108 BNYVV infected soil samples were collected from Isparta province, Turkey. Sugar beet plants cv Kasandra were grown in these soil samples using bait plant techniques and root samples were then analyzed by dsRNA analysis. The RNA was purified by CF-11 cellulose chromatography and gel electrophoresis. In 108 samples tested, dsRNA profiles were observed in 53 samples. No dsRNA bands were observed in negative control used in the analysis.

Keywords: sugar beet, BNYVV, rhizomania, dsRNA analysis

1. Introduction

Rhizomania is caused by *Beet necrotic yellow vein virus* (BNYVV). Rhizomania causes serious disease of sugar beet. It was first reported in Italy in the 1950s (Canova, 1959) but now it is present in sugar beet areas all over the world (Chiba et al., 2011). BNYVV is transmitted in soil by zoospores of plasmodiophorid, *Polymyxa betae* (Keskin, 1964). BNYVV is member of the genus *Benyvirus* (Tamada, 1989). BNYVV is characterized by rod-shaped particles, 20 nm in diameter and four different model lengths 85, 100, 265 and 390 nm (Putz, 1977) containing four seperate single stranded genomic RNAs of 1467, 1774, 4612 and 6746 base pairs, respectively.

In some Asian, French and English isolates, 5th RNA, which is 1349 nucleotide long has been described (Saito et al., 1996; Koenig et al., 1997; Harju et al., 2002; Ward et al., 2007). RNA 1 and RNA 2 have "housekeeping" genes involved in replication, assembly and cell to cell movement, whereas RNA 3, RNA 4 and RNA 5 are associated with vector-mediated infection and disease development in sugar beet roots (Tamada, 2007).

The first sign of rhizomania disease in a sugar beet crop appears as light green or yellow irregularly shaped patches in the field. Individual plants show the characteristic proliferation of fibrous roots around the tap root, "the root madness symptoms" of rhizomania. In severely infected plants, the tap root and lateral roots become necrotic and die then and the vascular tissue develops a pale brown coloration (Brunt & Richards, 1989).

BNYVV leads to serious decreases in root yield and quality of sugar. Virus reduces sugar content in the roots by 3-4% and yields of sugar beet more than 50-60% (Henry, 1996).

In Turkey, Rhizomania was first detected in Alpulu Sugar Refinery area. Later the presence of the disease was reported in different beet growing areas of Turkey (Vardar & Erkan, 1992; Kıymaz & Ertunç, 1996; Ertunç et al., 1998; Kutluk-Yılmaz & Yanar, 2001; Kaya, 2009).

More than 90% of plant viruses have single or double stranded RNA genomes. During the replication of single-stranded RNA viruses, a complementary strand of viral RNA is synthesized. An annealed dsRNA can be isolated by phenol extraction (Zaitlin & Hull, 1987). Morris & Dodds (1979) developed methods for the isolation and analysis of viral dsRNAs in diseased plant tissue.

Analysis of dsRNA has been used as a means of virus detection in various crops (Rezaian & Krake, 1987; Monette et al., 1989; Yardımcı & Açıkgöz, 1997; Bostan & Açıkgöz, 2000; Yardımcı &Korkmaz, 2004; Yardımcı & Eryiğit, 2006) including sugar beet (Hutchinson et al., 1992; Ilhan & Ertunç, 2001; Ertunç & Ilhan, 2002).

This study aimed to identify BNYVV on sugar beet plants by dsRNA analysis.

2. Materials and Methods

2.1 Sampling

Soil samples were collected in August and September 2011 from soils used in sugar beet culture of Isparta Isparta province. The samples were selected considering the visual indications for the presence of rhizomania in field-grown sugar beet plants, such as yellow coloration of leaves and beard-like appearance of the roots (Figure 1). Each sample consists on a mixture of 5 sub-samples collected from different parts of the same fields.

Figure 1. Typical root symptoms of Rhizomania in field-grown sugar beet plants in Gonen region, Isparta Province

2.2 Bait Plant Technique

The soil samples were air dried for 3-4 weeks placed in sterilized pots and 10 seeds of *Beta vulgaris* cv. Kasandra were sown in each pot. Sugar beet plants were grown for 9 weeks in the greenhouse at 23 °C then roots from each pot were harvested separately. Pots containing sterilized soil were used as negative control. Roots were placed into polyethylene bags labeled and stored at -20 °C until dsRNA analysis was performed.

2.3 dsRNA Analysis

The method for dsRNA extraction was performed according to Morris and Dodds (1979). 20 g of bait plant roots was homogenized with a mortar and pestle. It was added 1 ml of 10% SDS, 1 ml of 2% Bentonit, 10 ml of 1XSTE (0.1 M NaCl, 0.05 M Tris- HCl, 1mM EDTA, pH: 6.9), 10 ml of water-saturated phenol and 5 ml of Chloroform: pentanol (25:1). The homogenate was mixed for 60 min. The homogenate was centrifuged at 8000 g for 20 min. The upper aqueous phase was withdrawed and placed it in a 50 ml centrifuge tube and 2.1 ml of 96% Ethanol were added. The samples were stored overnight at 4 °C. 1 g portion of Whatman CF-11 cellulose per sample was added. Cellulose colons were prepared with 20 ml plastic syringe. The samples were added to one colon and let it drain completely. The colon was washed with 60 ml of 1XSTE containing ethanol (16%). Double-stranded RNA was then eluted with STE buffer and precipitated by adding 0.5 ml of 3 M Sodium acetate (pH 5.5) and 20 ml of 96% Ethanol to each sample.

The dsRNA was stored at -20 °C overnight. The precipitated dsRNAs were collected by centrifuged at 8000 g for 25 min. The pellet was dried and resuspended in 200 µl TBE (Tris, Boric Acid, EDTA). The samples were mixed with 30 µl 3M Sodium acetate and 0.9 ml of 96% Ethanol then incubated at -20 °C overnight. After centrifugation at 5000 g for 20 min. The pellet was resuspended in sterile distilled water and used in electrophoresis.

Double- stranded RNA was analysed by electrophoresis in agarose gels. The gels were electrophoresed at 100 V for 1 hour, stained with ethidium bromide and then visualized and photographed by Doc-It gel imaging and documentation system (UVP, England).

3. Results and Discussion

Typical rhizomania symptoms on leaves and roots were observed on plants growing under the greenhouse conditions after 9 months (Figure 2). Typical symptoms of BNYVV which are beard-like apperance of the roots, light green coloration of leaves.

dsRNA profiles typical of BNYVV were observed in 53 of 108 samples tested. However, healthy sugar beet did not show any dsRNA bands onto agarose gel. In 28 samples were observed RNA 1+2+3+4, in 9 samples were observed RNA 3, in 7 samples were observed RNA 1+2+3, in 9 samples were observed RNA 1+3 (Figures 3-4). In this research, RNA 5 has not been identified in the analyzed samples. Profiles of dsRNA obtained in this study are shown in Table 1.

The dsRNA profile detected in this study was same as previously reported BNYVV dsRNA profiles (Hutchinson et al., 1992; Ilhan & Ertunç, 2001; Ertunç & Ilhan, 2002). Non-specific bands were observed in agarose gel. The main reason is thought to be the lack of enzymatic treatment.

Variable dsRNA profiles were observed in BNYVV isolates. These alterations in the deletion mutations of smaller RNAs were associated with changes in symptoms expression of BNYVV, molecular differences between the strain of BNYVV or low virus concentration in the isolates (Henry et al., 1986; Koenig et al., 1986). The yield low amount of dsRNA in infected plants is related specifically to the time of infection and incubation temperature (Valverde et al., 1990).

Figure 2. Beard like appaerance of bait plant roots

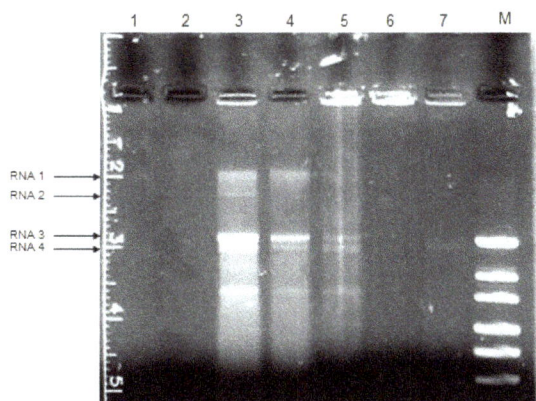

Figure 3. Agarose gel electrophoresis of dsRNA from BNYVV- Infected roots (M: Marker (100 bp-1.5 kb DNA Ladder, Biobasic); 1: Atabey-3 isolate; 2: Gonen-44 isolate; 3: Islamkoy-40 isolate; 4: Yalvac-11 isolate; 5: Keciborlu-17 isolate; 6: Kuleonu-60 isolate; 7: Negative control

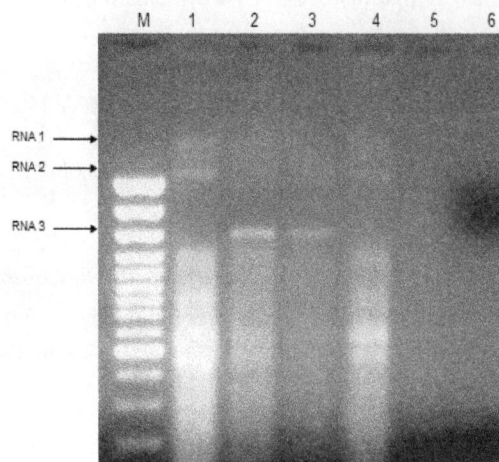

Figure 4. Agarose gel electrophoresis of dsRNA from BNYVV- Infected roots (M: Marker (100 bp DNA Ladder, Biobasic)); 1: Gonen-33 isolate; 2: Kuleonu-14 isolate; 3: Keciborlu-7 isolate; 4: Yalvac-22 isolate; 5: Senirkent 20 isolate; 6: Negative control

dsRNA analysis is simple, quick, efficient and relatively inexpensive. The purified dsRNA can be used as template for cDNA synthesis and subsequent PCR, molecular cloning, prob preparation and a reagent for mechanical inoculations (Valverde et al., 1990). dsRNA analysis is more sensitive when compared to serological tests (Ertunç & Ilhan, 2002). A disadvantage is that this procedure can generally not be used to process a large number of samples at a time. Also, since dsRNA is related to replication of the virus the titre of the dsRNA molecule may increase or decrease in the plant at certain times of the year, thereby influencing detection.

4. Conclusions

We can say that practicality of using analysis of dsRNA as an alternative or complementary method for diagnosis of BNYVV. At the very least it provides a jumping-off place in the diagnostic process before proceeding to more specific techniques.

Table 1. dsRNA Profile of BNYVV on samples in different districts of Isparta Province

Samples No	dsRNA Profile
Atabey-5	RNA 1+2+3+4
Atabey-24	RNA 1+2+3+4
Atabey-25	RNA 1+2+3+4
Atabey-78	RNA 1+2+3+4
Islamkoy-1	RNA 1+2+3+4
Islamkoy-21	RNA 1+2+3+4
Islamkoy-40	RNA 1+2+3+4
Islamkoy-48	RNA 1+2+3+4
Yalvac-2	RNA 1+2+3+4
Yalvac-11	RNA 1+2+3+4
Yalvac-41	RNA 1+2+3+4
Yalvac-23	RNA 1+2+3+4
Keciborlu-8	RNA 1+2+3+4
Keciborlu-22	RNA 1+2+3+4

Keciborlu-20	RNA 1+2+3+4
Keciborlu-17	RNA 1+2+3+4
Senirkent-5	RNA 1+2+3+4
Kuleonu-66	RNA 1+2+3+4
Kuleonu-23	RNA 1+2+3+4
Kuleonu-88	RNA 1+2+3+4
Kuleonu-29	RNA 1+2+3+4
Kuleonu-67	RNA 1+2+3+4
Sarkikaraagac-77	RNA 1+2+3+4
Sarkikaraagac -1	RNA 1+2+3+4
Sarkikaraagac -95	RNA 1+2+3+4
Sarkikaraağac-15	RNA 1+2+3+4
Sarkikaraağac-58	RNA 1+2+3+4
Gonen-3	RNA 1+2+3+4
Gonen-33	RNA 1+2+3
Gonen-70	RNA 1+2+3
Gonen-2	RNA 1+2+3
Gonen-56	RNA 1+2+3
Gonen-68	RNA 1+2+3
Gonen-23	RNA 1+2+3
Gonen-99	RNA 1+2+3
Islamkoy-43	RNA1+3
Islamkoy-9	RNA1+3
Kuleonu-77	RNA1+3
Kuleonu-78	RNA1+3
Kuleonu-86	RNA1+3
Kuleonu-95	RNA1+3
Atabey-33	RNA1+3
Atabey-11	RNA1+3
Atabey-10	RNA1+3
Keciborlu-7	RNA 3
Keciborlu-12	RNA 3
Islamkoy-12	RNA 3
Islamkoy-61	RNA 3
Islamkoy-10	RNA 3
Islamkoy-88	RNA 3
Kuleonu-14	RNA 3
Gonen-28	RNA 3
Gonen-81	RNA 3

References

Bostan, H., & Açıkgöz, S. (2000). Determination of PVX and PVS symptoms on some test plants and identification of these viruses using dsRNA analysis. *The Journal of Turkish Phytopathology, 29*(1), 41-49.

Brunt, A. A., & Richards, K. E. (1989). Biology and molecular biology of furoviruses. *Advances in Virus Research, 36,* 1-32.

Canova, A. (1959). On the pathology of sugar beet. *Inf Fitopatology, 9,* 390-396.

Chiba, S., Kondo, H., Miyanishi, M., Andika, I. B., Han, C., & Tamada, T. (2011). The evolutionary history of *Beet necrotic yellow vein virus* deduced from genetic variation, Geographical origin and spread, and the breaking of host resistance. *Molecular Plant- Microbe Interaction, 24*(2), 207-218. http://dx.doi.org/10.1094/MPMI-10-10-0241

Ertunç, F., Erzurum, K., Karakaya, A., Ilhan, D., & Maden, S. (1998). Incidence of Rhizomania disease on sugar beet in Çorum, Kastamonu and Turhal Sugar Refinery Regions. *The Journal of Turkish Phytopathology, 27*(1), 39-46.

Ertunç, F., & Ilhan, D. (2002). dsRNA analysis of Turkish *Beet Necrotic Yellow Vein Virus* (BNYVV) isolates. *The Journal of Turkish Phytopathology, 31*(3), 173-183.

Harju, V. A., Mumford, R. A., Blockley, A., Boonham, N., Clovert, G. R. G., Weekes, R., & Henry, C. M. (2002). Occurence in the United Kingdom of *Beet Necrotic Yellow Vein Virus* isolates which contain RNA-5. *Plant Pathology, 51,* 811. http://dx.doi.org/10.1046/j.1365-3059.2002.00781.x

Henry, C. M., Jones, R. A. C., & Coutts, R. H. A. (1986). Occurrence of a soil-borne virus of sugarbeet in England. *Plant Pathology, 38,* 585-591.

Henry, C. (1996). Rhizomania-its effect on sugar beet yield in the UK. *British Sugar Beet Review, 64,* 224-26.

Hutchinson, P. J., Henry, C. M., & Coutts, R. H. A. (1992). A comparison, using dsRNA analysis, between beet soil-borne virus and some other tubular viruses isolated from sugar beet. *Journal of General Virology, 73,* 1317-1320. http://dx.doi.org/10.1099/0022-1317-73-5-1317.

Ilhan, D., & Ertunç, F. (2001). Investigation of some Furoviruses by dsRNA analysis method. *The Journal of Turkish Phytopathology, 30*(1), 27-34.

Kaya, R. (2009). Distribution of Rhizomania disease in sugar beet growing areas of Turkey. *Journal of Agricultural Science, 15*(4), 332-340.

Keskin, B. (1964). *Polymyxa betae* n.sp., ein Parasit in den Wurzeln von *Beta vulgaris* Tournefort, besonders während der Jugendent wicklung der Zuckerrübe. *Archivi für Mikrobiologie, 49,* 348-374.

Kıymaz, B., & Ertunç, F. (1996). Research on the detection of virus diseases in sugar beet in Ankara. *The Journal of Turkish Phytopathology, 25*(1-2), 55-63.

Koenig, R., Haeberle, A. M., & Commandeur, U. (1997). Detection and characterization of a distinct type of Beet Necrotic Yellow Vein Virus RNA 5 in a sugar beet growing area in Europe. *Archives of Virology, 142,* 1499-1504.

Koenig, R., Burgermeister, W., Weich, Sebald, H. W., & Kothe, C. (1986). Uniform RNA patterns of *Beet necrotic yellow vein virus* in sugarbeet roots, but not in leaves from several plant species. *Journal of General Virology, 67,* 2043-2046.

Kutluk-Yılmaz, N. D., & Yanar, Y. (2001). Study on the distribution of *Beet necrotic yellow vein virus* (BNYVV) in sugar beet growing area of Tokat –Turkey. *The Journal of Turkish Phytopathology, 30*(1), 21-25.

Monette, P. L., James, D., & Godkin, S. E. (1989). Double-stranded RNA from rupestris stem pitting affected grapevine. *Vitis, 28,* 137-144.

Morris, T. J., & Dodds, J. A. (1979). Isolation and analysis of double-stranded RNA from virus-infected plant and fungal tissue. *Phytopathology, 69,* 854-858.

Putz, C. (1977). Composition and structure of *Beet necrotic yellow vein virus*. *Journal of General Virology, 35,* 397-401.

Rezaian, M. A., & Krake, L. R. (1987). Nucleic acid extraction and virus detection in grapevine. *Journal of Virological Methods, 17,* 277-285.

Saito, M., Kiguchi, T., Kusume, T., & Tamada, T. (1996). Complete nucleotide sequence of the Japanese isolate

S of *Beet necrotic yellow vein virus* RNA and comparison with European isolates. *Archives of Virology, 141,* 2163-2175.

Tamada, T. (1989). Production and pathogenicity of isolates of *Beet necrotic yellow vein virus* with different numbers of RNA components. *Journal of General Virology, 70,* 3399-3409.

Tamada, T. (2007). Susceptibility and resistance of Beta vulgaris *subsp.*maritima to foliar rub-inoculation with Beet necrotic yellow vein virus. *Journal of General Plant Pathology, 73, 76*-80.

Vardar, B., & Erkan, S. (1992). The first studies on detection of *Beet necrotic yellow vein benyvirus* in sugar beet in Turkey. *The Journal of Turkish Phytopathology, 21*(2-3), 74-76.

Valverde, R. A., Nameth, S. T., & Jordan, R. L. (1990). Analysis of double stranded RNA for plant virus diagnosis. *Plant Disease, 74*(3), 255-258.

Ward, L, Koenig, R., Budge, G., Garrido, C., Mcgrath, C., Stubbley, H., & Boonham, N. (2007). Occurrence of two different types of RNA-5-containing *Beet necrotic yellow vein virus* in the UK. *Archives of Virology, 152,* 59-73. http://dx.doi.org/10.1007/s00705-006-0832-x

Yardımcı, N., & Açıkgöz, S. (1997). Studies of *Alfalfa mosaic virus* of alfalfa growing areas in Erzurum. *The Journal of Turkish Phytopathology, 26*(1), 23-30.

Yardımcı, N., & Korkmaz, S. (2004). Studies on spread and identification of *Zucchini yellow mosaic virus* disease in the North-West Mediterranean Region of Turkey by biological indexing and double-stranded RNA analysis. *Plant Pathology Journal, 3*(1), 1-4.

Yardımcı, N., & Eryiğit, H. (2006). Identification of *Cucumber Mosaic Virus* in tomato (*Lycopersicon esculentum*) growing areas in the North-West Mediterranean Region of Turkey. *New Zealand Journal of Crop and Horticulture Science, 34,* 173-175. http://dx.doi.org/10.1080/01140671.2006.9514403

Zaitlin, M., & Hull, R. (1987). Plant virus–host interactions. *Annual Review of Plant Physiology, 38,* 291-315. http://dx.doi.org/10.1146/annurev.pp.38.060187.001451

14

Studies on The Allelopathic Effects of *Tithonia rotundifolia* on the Germination and Seedling Growth of Some Legumes and Cereals

Olutobi Otusanya[1] & Olasupo Ilori[2]

[1] Botany Department, Obafemi Awolowo University, Ile-Ife, Nigeria

[2] Biology Department Adeyemi College of Education, Ondo, Nigeria

Correspondence: Olasupo Ilori, Biology Department Adeyemi College of Education, Ondo, Nigeria. E-mail: olasupoilori@yahoo.com

Abstract

The study investigated the allelopathic effects of *Tithonia rotundifolia* on the germination and growth of two legumes (*Vigna unguiculata* and *Glycine max*) and two cereals (*Zea mays* and *Sorghum bicolor*). This was with a view to determining the susceptibility of these test crops to allelochemicals. The germination studies were carried out by raising seedlings in Petri-dishes which had been lined with Whatman No. 1 filter paper. Ten millilitres of 100%, 75%, 50% and 25% concentrations of the methanolic or water extract solutions were used for the treatments while distilled water served as control. Germination and growth analyses were carried out according to standard methods. The data obtained were analysed by Factorial Analysis of Variance (ANOVA) to determine significant ($P < 0.05$) effects. The germination and growth of the juvenile seedlings of all the test crops were significantly inhibited by the methanolic and water extracts dose dependently. However, the methanolic extracts had a more pronounced inhibitory effect on these parameters. The study concluded that the methanolic extracts were more phytotoxic and had higher inhibitory effects on the parameters than the water extracts. Also, it was observed that the response of plants to allelochemical toxicity was dependent on plant species.

Keywords: allelopathic, methanolic extract, water extract, *Tithonia rotundifolia*, legumes, cereals

1. Introduction

The phenomenon of allelopathy has received increasing attention as a means of explaining vegetation patterns in plant communities (Miller, 1996). According to Inderjit et al. (1999), allelopathy may occur in all environments and should be considered as a part of community interaction. Kohli et al. (1998) and Singh et al. (2001) opined that allelopathy plays an important role in many agro-ecosystems. A large number of plants impose inhibitory effects on the germination and growth of neighbouring or successional plants by releasing allelopathic chemicals into the soil, either as exudates from living tissues or by decomposition of plant residues (Rice, 1984; Narwal, 1999; Alam & Islam, 2002; Khan et al., 2009). Bendall (1975) studied water and ethanol extracts and residues in soil and concluded that an allelopathic mechanism might be involved in the exclusion of some annual thistle (*Carduus crispus* L.), pasture and crop species in areas infested with *Cirsium arvense* (L) Scop. According to Stachon and Zimdal (1980), *C. arvense* litter reduced the growth of *Amaranthus retroflexus* L. and *Setaria viridus* L. more than that of cucumber (*Cucumis sativus* L.) or barley (*Hordeum vulgare* L.) in greenhouse experiments. They also observed that high densities of *C. arvense* reduced the incidence of annual weeds growing in the vicinity of *C. arvense*. Khan et al. (2009) stated that aqueous extracts of *Eucalyptus camaldulensis* L. inhibited seed germination, fresh and dry weight of wheat seedlings. Rawat et al. (2002) reported that aqueous extract of the root of *Helianthus annus* delayed and inhibited the germination and seedling growth of linseed (*Linum usitatissium* L.) and mustard (*Brassica juncea* L.) Aqueous extracts from the leaves of *Helianthus tuberosus* L. *Xanthium occidentale*, *Lactuca sativa* and *Cirsum japonica* all in the Asteraceae family inhibited the root growth of Lucerne (Chon et al., 2003). Ilori et al. (2007) observed that the radical growth of *Oryza sativa* was inhibited by aqueous extract of *T. diversifolia*. Otusanya et al. (2007) reported that the growth of *Amaranthus cruentus* was inhibited by aqueous extract of *T. diversifolia*. Javed and Asghari (2008) reported that the leaf extract of *Helianthus annus* inhibited the rate of germination of wheat seedlings.

In Nigeria, *Tithonia rotundifolia* is a widespread species having colonized roadsides, waste places, fallow land

and disturbed open spaces like abandoned construction sites etc. and displacing traditional weedy species like *Chromolaena odorata and Panicum maximum* (Adebowale & Olorode, 2005). The plant associates with common crops like vegetables, cassava, yam, rice, sorghum, soyabean e.t.c. and becomes a dominant plant where it is present (Tongma et al., 1998). Cowpea (*Vigna unguiculata* (L.) Walpers) and Soybean (*Glycine max* (L.) Merr.) which belong to the family *Fabaceae* are economically significant legumes in the tropics. Maize (*Zea mays* L.) and Sorghum (*Sorghum bicolor* (L.) Moench) are annual grasses belonging to the family Poacea. *Z. mays* L. is one of the most important cereal crops growing in the world. It is used as food for human consumption as well as food grain for animals (Moussa, 2001). *S. bicolor* (L.) Moench is a drought resistant cereal important for grain, forage and bioethanol production (Aishah et al., 2011). Considering the effects of *Tithonia* species on associated crops, the objectives of this work was to determine the effects of water and methanolic extracts of fresh shoots of *T. rotundifolia* on the germination, growth parameters (plumule and radicle lengths) and yield parameters (fresh and dry weights of plumule and radicle) of juvenile seedlings of *V. unguiculata*, *G. max*, *Z. mays* and *S. bicolor*.

2. Materials and Methods

2.1 Study Area

This study was conducted at the Botany Department of the Obafemi Awolowo University (O. A. U.), Ile-Ife, Osun State, Nigeria, Latitude 07°30'N - 07°35'N and Longitude 04 °30' - 04°40'E.

2.2 Plant Materials

The plant materials that were utilized in this study are the seeds of the following plants *Tithonia rotundifolia* (Miller) S. F. Blake, *Vigna unguiculata* L.Walp, *Glycine max* L. Merr., *Zea mays* L. and *Sorghum bicolor* (L.) Moench. The seeds of the test crops (*Vigna unguiculata*, *Glycine max*, *Zea mays*, and *Sorghum bicolor*) were collected from IITA (International Institute of Tropical Agriculture) Ibadan. *T. rotundifolia* seeds were collected along Road 20 at the Senior Staff Quarters of O. A. U., Ile Ife.

2.3 Germination Experiment

Preparation of extracts for the different treatments was carried out according to the modified method of Qasem and Abu - Irmaileh (1985). The extract solution (100%) was diluted appropriately with water to give 75%, 50%, and 25% concentrations of the aqueous extracts while distilled water served as control. Petri-dishes were thoroughly washed and oven dried. The seeds of the different test plants were selected randomly on the basis of uniformity of size and the seeds were then soaked for five minutes separately in 5% sodium hypochlorite to prevent fungal infection. Thereafter they were rinsed for about five minutes in running tap water. Ten of the seeds were placed in each of the clean oven dried Petri-dish which had been lined with a Whatman No. 1 filter paper. The filter paper in each of the Petri-dishes allocated to the control was moistened with ten millilitres of distilled water while that of the Petri-dishes allocated to the other treatments were moistened with ten millilitres of the appropriate concentration of the extracts. The Petri-dishes were incubated at room temperature for two weeks. Emergence of one millimetre of the radicle was used as the criterion for germination. Measurements of germination, plumule and radicle lengths, fresh and dry weights were carried out using standard methods.

2.4 Statistical Analysis

The data obtained were analysed by factorial Analysis of Variance (ANOVA) to determine significant ($P < 0.05$) effects.

3. Results

The percentage germination of the control seeds of the test crops was higher than that of the seeds treated with the different extracts (Figure 1). In most cases the percentage germination increased as extracts concentration decreased. There was significant reduction of the germination of the seeds by all the concentrations of FME and FWE at $P < 0.05$. Seedlings of the test crops in the control had plumule and radicle lengths that were significantly higher than those of the seedlings in all the extract regimes and these plumule and radicle lengths reduced with increase in the concentration of the methanolic extracts and water extracts (Figures 2 & 3). The control seedlings of all the test crops had plumule fresh weight that was significantly higher than that of the seedlings in both the FME and FWE regimes (Figure 4). The plumule fresh weights of the seedlings in all the extract regimes increased with decrease in the concentration of the extracts. The radicle fresh weight of *V. unguiculata* and *S. bicolor* seedlings in the 25% and 50% FWE was almost equivalent or slightly higher than that of the control seedlings while the other extracts inhibited the radicle fresh weight of these seedlings. In the case of the *G. max* and *Z. mays* seedlings, the methanolic extract was more phytotoxic than the water extracts and the radicle fresh weight was inhibited by all the extracts (Figure 5). The plumule dry weight of the control *V. unguiculata* seedlings was higher than that of the seedlings in all the extract regimes while that of the seedlings in the 75% and 100% extract regimes were almost equivalent and lower than that of the 25% extracts regime. Also, the 25% FWE seedlings of *Z. mays* had the same plumule dry weight with that of the control while that of the *S. bicolor* seedlings in the 25% FME was lower than that of the seedlings in the 50% FME regime. In the case of the *G. max*, the plumule dry weight of the seedlings in the 100% FME and FWE regimes was much lower than that of the control, FME and FWE seedlings (Figure 6). *V. unguiculata*, *G. max* and *Z. mays* seedlings in the control regime had radicle dry weight that were higher than those of the seedlings in all the extract regimes. However, the radicle dry weight of the control *S. bicolor* seedlings was equivalent to that of the 25% FWE seedlings (Figure 7). Interactions of extracts x crops, crop x extract concentrations were found significant for all the parameters except plumule dry weight (Table 1).

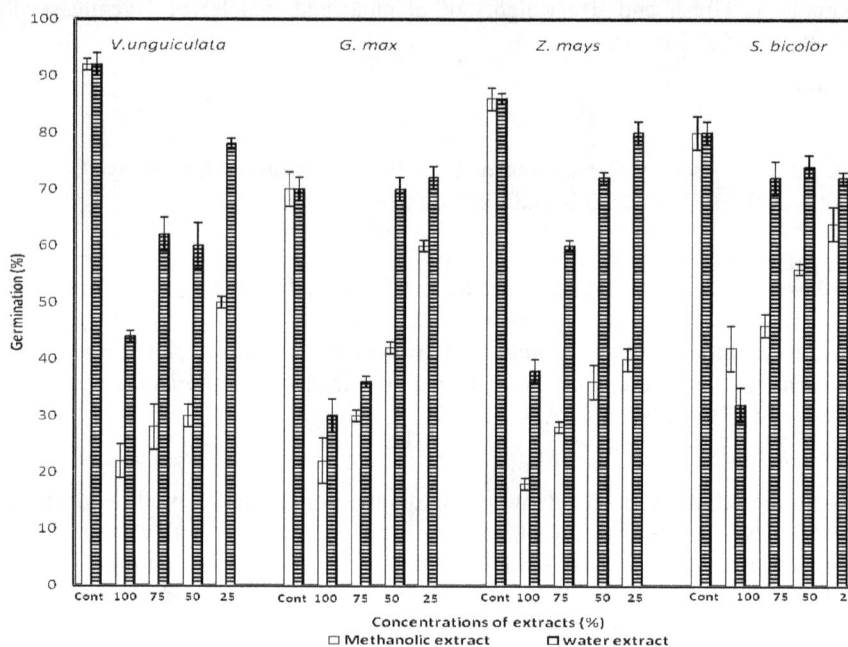

Figure 1. Effect of the methanolic extracts and water extracts of the fresh shoots of *T. rotundifolia* on the germination of the test crops. Capped bars indicate standard errors

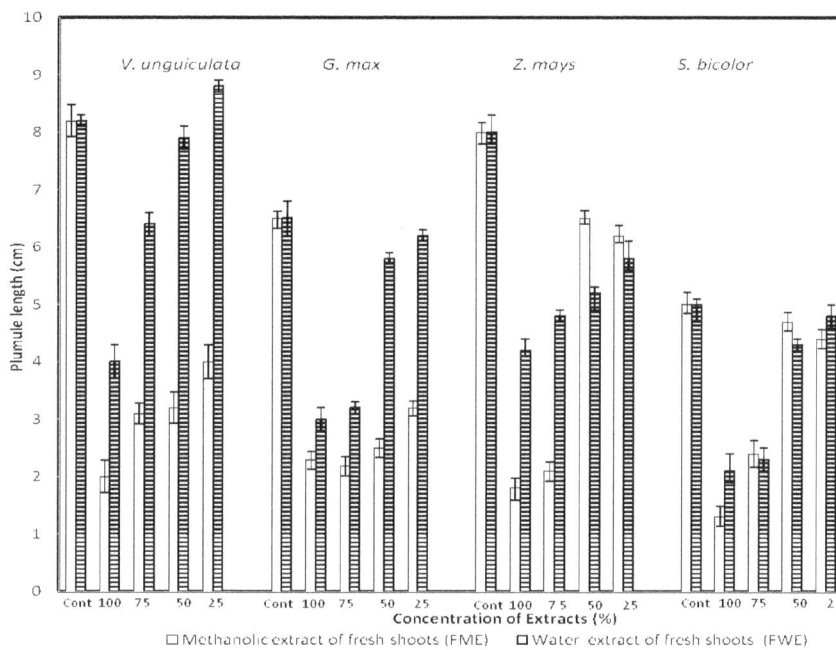

Figure 2. Effect of the methanolic and water extracts of the fresh shoots of *T. rotundifolia* on the plumule length of the test crops. Capped bars indicate standard errors

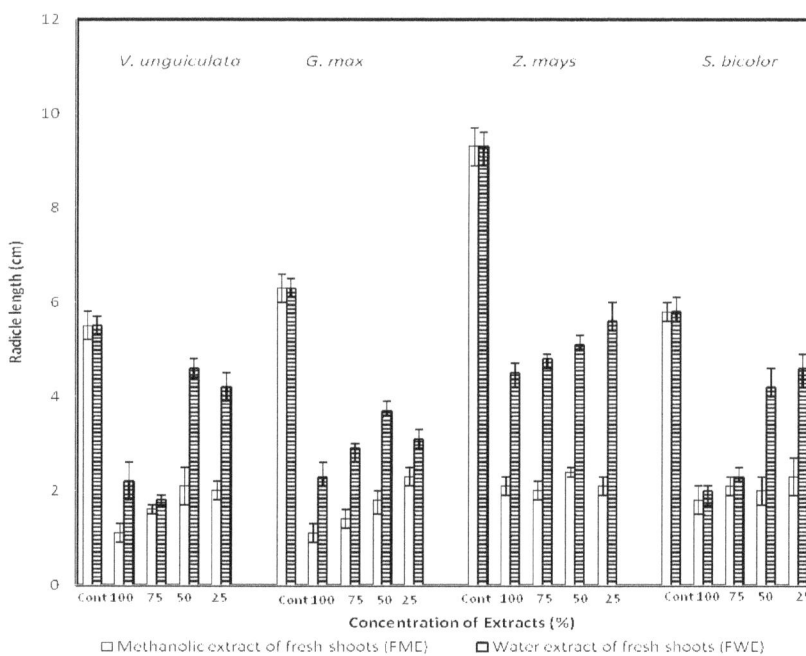

Figure 3. Variation in the radicle length of the test crops treated with the methanolic extracts and water extracts of the fresh shoots of *T. rotundifolia*. Capped bars indicate standard errors

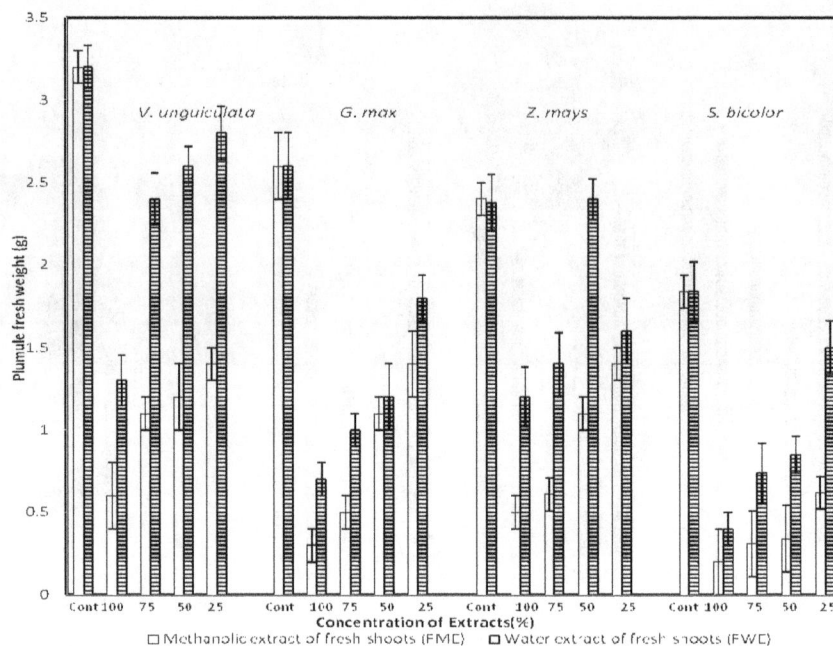

Figure 4. Variation in the plumule fresh weight of the test crops treated with the methanolic extracts and water extracts of the fresh shoots of *T. rotundifolia*. Capped bars indicate standard errors

4. Discussion

According to Leu et al. (2002) and Inderjit and Duke (2003), allelopathy in natural and agricultural ecosystems is receiving increasing attention because allelochemicals significantly reduce the growth of other plants and the yields of crop plants. Allelochemicals are secondary plant products or waste products generated by the plant's main metabolic pathways which are released into the environment in appreciable quantities via root exudates, leaf leachates, roots and other degrading plant residues (Putnam, 1988). These chemicals have harmful effects on crops in the ecosystem resulting in the reduction and delayed germination, seedling mortality and reduction in growth and yield (Herro & Callaway, 2003). The process of seed germination is a crucial stage in plant growth. During germination, biochemical changes take place, which provides the basic framework for subsequent growth and development (Khan et al., 2009). According to Bhownmik and Inderjit (2003), allelochemicals can affect the establishment or regeneration of population by affecting seed germination. These authors were of the opinion that increasing germination can enhance the competitive ability of a plant species for both above-ground and underground resources.

The water and methanolic extracts from *T. rotundifolia* had significant inhibitory effect on the germination of the seeds of all the test crops in this study. This observation agreed with the findings of Inderjit and Dakshini (1994) who reported that the water extracts from the roots of *Pluchea lanceolata* in the family Asteraceae inhibited the germination of tomato and mustard. The water extracts from tissues of *Helianthus annus* were also observed to inhibit germination of *Solanum nigrum* (Sedigheh et al., 2010). Rawat et al. (2002) found that the aqueous extract of the root of *Helianthus annus* delayed and inhibited the germination and seedling growth of linseed (*Linuna usitatissium* L.) and mustard (*Brassia Juncia* L.). Nandal and Dhillon (2005) reported that the aqueous extracts of poplar leaves adversely affected the germination and seedling growth of some wheat varieties at high extract concentrations. Mulatu et al. (2006) reported that aqueous extract of *Parthenium hysterophorus* leaves and flower inhibited seed germination of lettuce. Preliminary investigations have revealed that the aqueous extract from the leaves of *T. diversifolia* retarded the germination and the radicle growth of *Oryza sativa*, *Amaranthus cruentus*, *Capsicum annum* and *Lycopersicon esculentum* (Ilori et al., 2007; Otusanya et al., 2007; Otusanya et al., 2008). Khan et al. (2009) reported that the reduction in germination counts of wheat became more pronounced with increasing levels of *Eucalyptus camaldulensis* aqueous extract concentration. Javed and Asghari (2008) also found that the leaf extract of *Helianthus annus* inhibited the rate of germination of wheat seedlings. A related work by Arshad (2011) showed that the water and methanolic extracts of *Withania somnifera* markedly suppressed the germination, root and shoot growth of *Parthenium hysterophorus*.

The growth of the plumule and radicle of the water and methanolic extracts treated seedlings of the test crops were significantly inhibited at P < 0.05. The inhibition of the growth of the radicle of *G. max* and *Z. mays* were more pronounced than that of the plumule growth. These results corroborates the earlier findings of several other workers such as Chou and Kuo (1986), Alam (1990), Zackrisson and Nilsson (1992) and Munir and Tawaha (2002) who all asserted that root growth was more sensitive to the increasing concentration of plant aqueous extracts in comparison to the shoot growth. The more accentuated effect of the allelochemical on the roots might be due to their closer contact with the leachates or extracts especially when maintained on filter/germination paper (Chung et al., 2001). Rahman (1998) reported that aqueous extract derived from the inflorescence, stem, and leaves of *Parthenium hysterophorus* L. inhibited the growth of radicle and plumule of *Cassia sophera* Linn. Florentine et al. (2006) observed that allelopathy is characterized by reduction in plants emergence or growth, reducing their performance in the association. A similar result was reported by Kushima (1998) who stated that there was an inhibition of the growth of the plumule length of tomato seedlings by the application of leachate from water melon seeds. James and Bala (2003) found that dried mango leaf powder significantly inhibited the sprouting of purple nutsedge tubers while Yang et al. (2006) reported that its aqueous extract inhibited the germination and growth of some crops. Also, the results in this study was consistent with the finding of Ilori et al. (2007) who stated that the radicle growth of *Oryza sativa* was inhibited by the aqueous extract of *T. diversifolia* (a close relative of the donor plant). This retardation of the juvenile seedling growth of the target crops was observed to increase significantly with increasing extract concentrations. This was consistent with the work of Khan et al. (2009) who reported that the inhibitory effects of aqueous extracts of *Eucalyptus camaldulensis* L. on germination and seedling growth (fresh and dry weight) of wheat were increased as the extract concentration increased. A similar result was obtained by Swapnal and Badruzzaman (2010) on the allelopathic effect of *Croton bonplandianum* Baill. weed on seed germination and seedling growth of crop plants. They reported that the root length, shoot length of *Melilotus alba* Medik., *Vicia sativa* L. and *Medicago hispida* Gaertn. decreased progressively when the plants were exposed to increasing concentration of the extract of *Croton bonplandianum* Baill. Khan et al. (2009) from their study of the effect of Eucalyptus extracts on twelve varieties of wheat concluded that the variation in germination of different varieties might be due to variation of the genetics of these twelve varieties. Likewise, it was observed in this study that the response of plants to allelochemicals toxicity was found to be dependent on plant species. The most affected crops were *G. max* and *Z. mays* for radicle fresh and dry weights and *V. unguiculata* for germination.

5. Conclusion

It can be summarized from the results of this study that both the water and methanolic extracts at any concentration inhibited the germination, growth and ultimately the yield of the test crops. The methanolic extract was more phytotoxic than the water extracts. The extent of the inhibition by the water and methanolic extracts followed this order: 100% > 75% > 50% > 25%. This affirmed the fact that the response of the target crops was extract concentration dependent. In conclusion, the effectiveness of these extracts on the germination and growth of the crops in this study showed that the presence of *T. rotundifolia* would negatively affect the neighboring or successional crop plants.

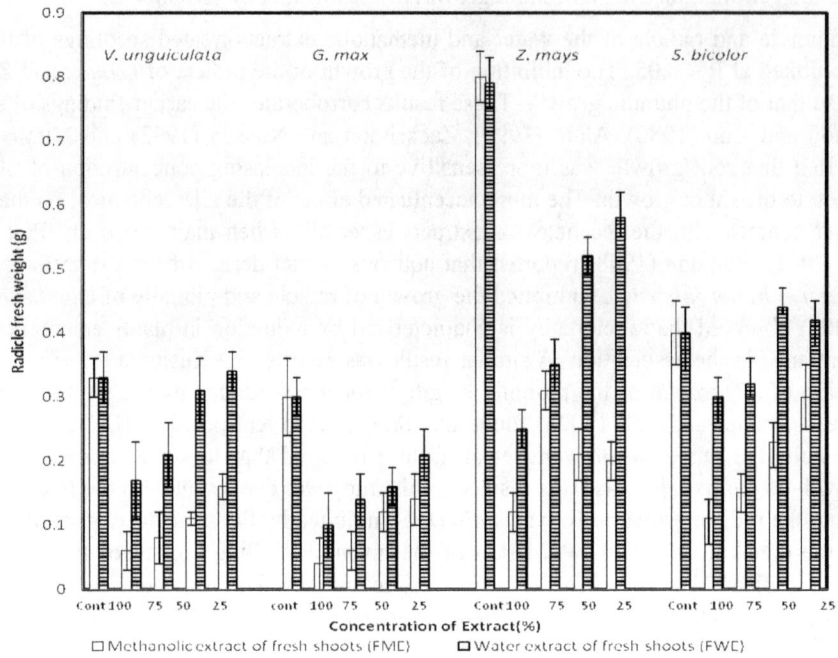

Figure 5. Radicle fresh weight of the test crops as affected by the methanolic and water extracts of the fresh shoots of *T. rotundifolia*. Capped bars indicate standard errors

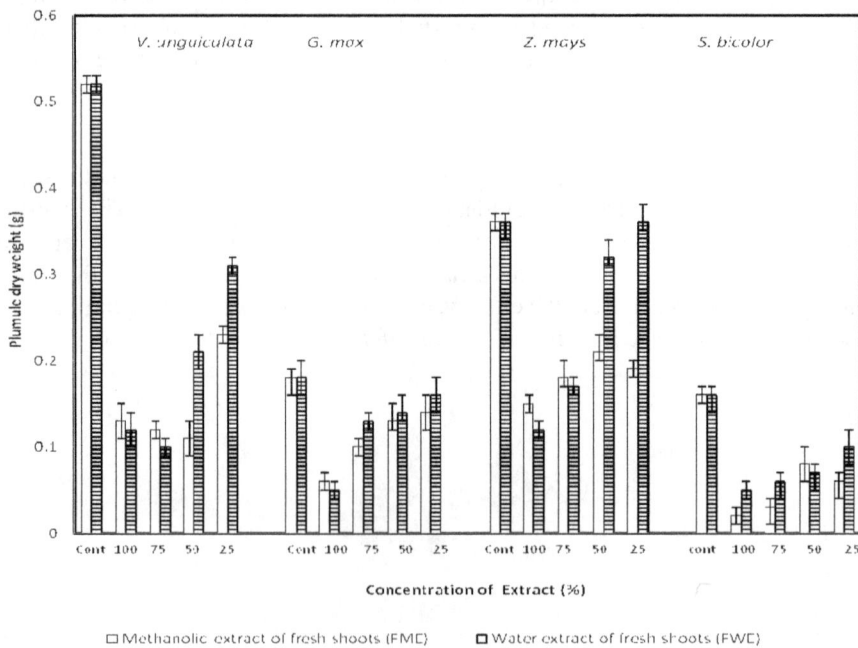

Figure 6. Plumule dry weight of the test crops as affected by the methanolic and water extracts of the fresh shoots of *T. rotundifolia*. Capped bars indicate standard errors

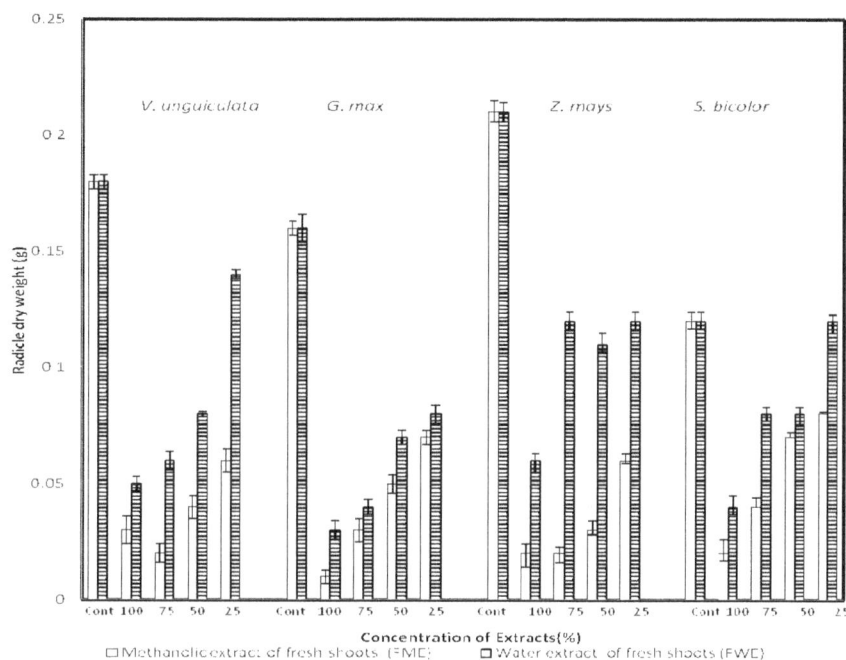

Figure 7. Radicle dry weight of the test crops as affected by the application of the methanolic extracts and water extracts of the fresh shoots of *T. rotundifolia*. Capped bars indicate standard errors

Table 1. Results of analysis of variance of the traits determined

Source	df	Germination percentage	Plumule length	Radicle length	Plumule fresh weight	Radicle fresh weight	Plumule dry weight	Radicle dry weight
Crops(C)	3	**	**	**	**	**	**	**
Extracts(E)	1	**	**	**	**	**	ns	**
Conc.	4	**	**	**	**	**	**	**
C x E	3	**	**	**	**	**	ns	**
C x Conc.	12	**	**	**	**	**	**	**
E x Conc.	4	**	**	**	**	**	ns	**
C x E x Conc	12	**	**	**	**	**	ns	**

** P < 0.05, df, degrees of freedom. Conc., concentration

References

Adebowale, A., & Olorode, O. (2005). An overview of the invasive potential of *Tithonia* species (Asteraceae) in Nigeria. *Science Focus, 10*(3), 65-69.

Aishah, S., Saberi, H. A. R., Halim, R. A., & Zaharah, A. R. (2011). Yield responses of forage sorghums to salinity and irrigation frequency. *Africa Journal of Biotechnology, 10*(20), 4114-4120. Retrieved from http://www.ajol.info/index.php/ajb/article/view/93588

Alam, S. M., & Islam, E. U. (2002). Effect of aqueous extract of leaf stem and root of nettleleaf goosefoot and NaCl on germination and seedling growth of rice. *Pacific Journal of Science and Technology, 1*(2), 47-52.

Alam, S. M. (1990). Effect of wild plant extract on germination and seedling growth of wheat. *Rachis, 9*, 12-13. Retrieved from http://www.cabdirect.org/abstracts/19916776706.html

Arshad, J., Shazia, S., & Sobiya, S. (2011). Management of *Parthenium hysterophorus* (Asteraceae) by *Withania somnifera* (Solanaceae). *Natural Product Research,* *25*(4), 407-416. http://dx.doi.org/10.1080/14786419.2010.483230

Bendall, G. M. (1975). The allelopathic activity of California thistle (*Cirsium arvense*) in Tasmaria. *Weed Research, 15,* 17-81.

Bhownmik, P. C., & Inderjit, C. (2003). Challenges and opportunities in implementing Allelopathy for natural weed management. *Crop Protection, 22,* 661-671. http://dx.doi.org/10.1016/S0261-2194(02)00242-9

Chon, S. U., Kin, Y., & Kee, J. C. (2003). Herbicidal potential and quantification of causative allelochemicals from several compositae weeds. *European Weed Research Society. Weed Research, 43,* 444-450. http://dx.doi.org/10.1046/j.0043-1737.2003.00361.x

Chou, C. H., & Kuo, Y. L. (1986). Allelopathic research in subtropical vegetation in Taiwan HI. Allelopathic exclusion of understory species by *Leucaena lencocephala. Journal of Chemical Ecology, 12,* 1431-1448. http://dx.doi.org/10.1007/BF01012362

Chung, I. M., Ahn, J. K., & Yun, S. J. (2001). Identification of allelopathic compounds from rice (*Oryza sativa* L.) straw and their biological activity. *Canadian Journal of Plant Science, 81,* 815-819. http://dx.doi.org/10.4141/P00-191

Florentine, S. K., Westbrooke, M. E., Gosney, K., Ambrose, G., & O' Keefe, M. (2006). The arid lands invasive weed *Nicotiana glauca* R. Graham (Solanaceae): Population and soil seed bank dynamics, seed germination patterns and seedling response to flood and drought. *Journal of Arid Environment, 66,* 218-230. http://dx.doi.org/10.1016/j.jaridenv.2005.10.017

Herro, J. L., & Callaway, R. M. (2003). Allelopathy and exotic plant invasion. *Plant and soil, 256,* 29-39. http://dx.doi.org/10.1023/A:1026208327014

Ilori, O. J., Otusanya, O. O., & Adelusi, A. A. (2007). Phytotoxic effects of *Tithonia diversifolia* on germination and growth of *Oryza sativa. Research Journal of Botany, 2*(1), 23-32. http://dx.doi.org/10.3923/rjb.2007.23.32

Indergit, M., & Darkshini, K. M. M. (1994). Allelopathic effect of *Pluchea lanceolata* (Asteraceae) on characteristics of four soils and tomato and mustard growth. *American Journal of Botany, 81,* 799-804. http://dx.doi.org/10.2307/2445760

Indergit, M., Darkshini, K. M. M., & Chester, L. (1999). Principle and practices in plant ecology. In Foy (Eds.), *Allelochemical Interactions.*

Inderjit, M., & Duke, S. O. (2003). Ecophysiological aspects of allelopathy. *Planta, 217,* 529-539. http://dx.doi.org/10.1007/s00425-003-1054-z

James, J. F., & Bala, R. (2003). *Allelopathy: How plants suppress other plants.* The Hort. Sc. Depart. Inst. Food Agric. Sci. Univ. Florida.

Javed, K., & Asghari, B. (2008). Effects of sunflower (*Helianthus annuus* L.) extracts on wheat (*Triticum aestivum* L.) and physicochemical characteristics of soil. *African Journal of Biotechnology, 7*(22), 4130-4135.

Khan, M. A., Hussain, I., & Khan, E. A. (2009). Allelopathic effects of Eucalyptus (*Eucalyptus Camaldulensis* L.) on germination and seedling growth of wheat (*Triticum aestivum* L.) *Pakistan Journal of Weed Science Research, 15*(2-3), 131-143.

Kohli, R. K., Batish, D., & Singh, H. P. (1998). Allelopathy and its implications in agroecosystems. *Journal of Crop Production, 1,* 169-202. http://dx.doi.org/10.1300/J144v01n01_08

Kushima, M., Hideo, K., Seiji, K., Shosuke, Y., Yamada, K., Yokotani, T., & Koji, H. (1998). An allelopathic substance exuded from germinating water melon seeds. *Plants Growth Regulation, 25,* 1-4. http://dx.doi.org/10.1023/A:1005907101778

Leu, E., Krieger-Liszkay, A., Goussias, C., & Gross, E. M. (2002). Polyphenolic allelochemicals from the aquatic angiosperm *Myriophyllum spicatum* inhibit photosystem II. *Plant Physiology, 130,* 2011-2018. http://dx.doi.org/10.1104/pp.011593

Miller, D. A. (1996). Allelopathy in forage crop systems. *Agronomy Journal, 88,* 854-859. http://dx.doi.org/10.2134/agronj1996.00021962003600060003x

Moussa, H. R. (2001). *Physiological and biochemical studies on the herbicide (Dual) by using radiolabelled technique. Ph.D. Thesis.* Faculty of Science Ain-Shams University.

Mulatu, W., Gezahegn, B., & Befekadu, B. (2006). Allelopathic effect of *Parthenium hysterophorus* extract on seed germination and seedling growth of lettuce. *Tropical Science, 45*(4), 159-162.

Munir, A. T., & Tawaha, A. R. M. (2002). Inhibitory effects of aqueous extracts of black mustard on germination and growth of lentil. *Pakistan Journal of Agronomy, 1*(1), 28-30. http://dx.doi.org/10.3923/ja.2002.28.30

Nandal, & Dhillon, A. (2005). Allelopathic effects of poplar (*Populus deltoides* Bartr Ex Marsh): an assessment on the response of wheat varieties under laboratory and field conditions. *Proc. Fourth World Congress on Allelopathy* 21-26 August 2005. Charles Sturt University, Wagga, NSW, Australia.

Narwal, S. S. (1999). In Enfield (Ed.), *Allelopathy update, basic and applied aspects* (Vol. 2, pp. 203-54). Science Publishers Inc: New Hampshire.

Otusanya, O. O., Ikonoh, O. W., & Ilori, O. J. (2008). Allelopathic potentials of *Tithonia diversifolia* (Hemsl) A. Gray: Effect on the germination, growth and chlorophyll accumulation of of *Capsicum annum* L. and *Lycopersicon esculentum* Mill. *International Journal of Botany, 4*(4), 471-475. http://dx.doi.org/10.3923/ijb.2008.471.475

Otusanya, O. O., Ilori, O. J., & Adelusi, A. A. (2007). Allelopathic effect of *Tithonia diversifolia* on germination and growth of *Amaranthus cruentus* Linn. *Research Journal of Environmental Sciences, 1*(6), 285-293. http://dx.doi.org/10.3923/rjes.2007.285.293

Putnam, A. R. (1988). Allelochemicals from plants as herbicides. *Weed Technology, 2*, 510-518. Retrieved from http://www.jstor.org/stable/3987390

Rahman, A. (1998). Allelopathic potential of *Pathenium hysterophorus* L on germination, growth, and dry matter production in *Cassia sopheras* L. *Bionature, 18*(1), 17-20.

Rawat, L. S., Rawat, D. S. K., Narwal, S. S, Palaniraj, R., & Sati, S. C. (2002). Allelopathic effects of aqueous extracts of sunflower (*Helianthus annus* L.) root on some winter oil seed crop. *Geobies (Jodhpur), 29*(4), 225-228.

Rice, E. L. (1984). *Allelopathy* (2nd Ed.) (p. 422). New York, U.S.A.: Academic Press.

Sedigheh, S., Aptin, R., & Zoheir, Y. A. (2010). Allelopathic effect of *Helianthus annus* (sunflower) on *Solanum nigrum* (black nightshade) seed germination and growth in laboratory condition. *Journal of Horticultural Science and Ornamental Plants, 2*(1), 32-37.

Singh, H. P., Kohli, R. K., & Batish, D. R. (2001). Allelopathy in agroecosystems: an overview. In R. K. Kohli, H. P. Singh, & D. R.Batish (Eds.), *Allelopathy in agroecosystems* (pp. 1-44). New York: Food Products Press.

Stachon, W. J., & Zimdal, R. L. (1980). Allelopathic activity of Canada thistle (*Cirsium arvense*) in Colarado. *Weed Science, 28*, 83-86.

Swapnal, S., & Badruzzaman, M. S. (2010). Allelopathic effect by aqueous extracts of different parts of *Croton bonplandianum* Baill. on some crop and weed plants. *Journal of Agricultural Extension and Rural Development, 2*(1), 22-28. Retrieved from http://www.researchgate.net/publication/228512918

Tongma, S., Katsuichiro, K. & Kenji, U. (1998). Allelopathic activity of Mexican sunflower (*Tithonia diversifolia)* in soil. *Weed Science, 46*(4), 432-437.

Yang, G., Zhu, C., Luo, Y., Yang, Y., & Wei, J. (2006). Potential allelopathic effect of *Piper nigrum, Magnifera indica and Clausena lansium. Ying Yong Sheng Taixue. Bao., 17*(9), 1633-1636. Retrieved from http://www.ncbi.nlm.nih.gov/pubmed/17147171

Zackrisson, O., & Nilsson, M. C. (1992) Allelopathic effects of *Empetrumherma phrodilbium* on seed germination of two boreal tree species. *Canadian Journal of Research, 22*, 1310-1319. http://dx.doi.org/10.1139/x92-174

15

Phytoplankton Ability to Physiological Acclimatization and Genetic Adaptation to Global Warming

Eduardo Costas[1], Emma Huertas[2], Beatriz Baselga-Cervera[1], Camino García-Balboa[1] & Victoria López-Rodas[1]

[1] Genética, Facultad de Veterinaria, Universidad Complutense, Madrid, Spain

[2] Instituto de Ciencias Marinas de Andalucía (CSIC), Cádiz, Spain

Correspondence: Victoria López-Rodas, Genética, Facultad de Veterinaria, Universidad Complutense, Av. Puerta de Hierro s/n 28040 Madrid, Spain. E-mail: vlrodas@ucm.es

Abstract

Global warming represents a challenge to the survival of phytoplankton organisms, which are the basis of the aquatic food web and drive essential biogeochemical cycles. We propose a direct experimental research mimicking temperature-increasing scenarios as a novel way to explore the adaptation of phytoplankton to predicted future thermal scenarios. This vulnerability analysis of individual phytoplankton species to increased temperature is key to understand the impact of global warming on aquatic ecosystems. Considering the polyphyletic complexity of the phytoplankton community, we compare the adaptation ability of diverse phytoplankton species from oceanic, coastal and inland waters to global warming, evaluating the role played by physiological acclimatization and genetic adaptation. We found that physiological acclimatization allows survival under the lowest temperature increase. Afterwards pre-existing genetic variability allow some genotypes to survive. Finally, when the temperature rises to a certain threshold, only the occurrence of new mutations that confer thermal resistance assures adaptation. Our results also reveal diverse degrees of tolerance to temperature increase among the different functional groups of phytoplankton, with great inter-specific capacity for genetic adaptation.

Keywords: genetic adaptation, global warming, mutation, physiological acclimation, phytoplankton

1. Introduction

There is a broad consensus that global warming is a huge challenge for life worldwide if the present trend of temperature rise continues (Wrona et al., 2006). The majority of studies aimed at addressing the effects of climate change on ecosystems (e.g. Stenseth et al., 2002; Walther et al., 2002; Parmesan & Yohe, 2003). Unfortunately, the microbial world has been scarcely analyzed. Consequently, the effects of climate change are best understood for the animals and plants that for organisms as abundant and important as microbes.

The adaptation studies of microbes to global warming are indispensable, as biogeochemical cycles depend of microbial activity. In particular, the tolerance of phytoplankton to global warming is extremely relevant as these organisms are the primary producers of aquatic ecosystems (Kirk, 1994; Falkowski, & Raven, 1997), responsible for approximately half of the global net primary production (Behrenfeld et al., 2001). Phytoplankton also has a key role in biogeochemical cycles of carbon, nitrogen and phosphorus (Falkowski, Barber, & Smetacek, 1998).

Although the lower trophic levels, such as the primary producers (phytoplankton) are usually considered less sensitive to environmental changes than their consumers or predators, alarming data suggest that phytoplankton biomass and productivity have declined since 1950 due to global warming (Behrenfeld et al., 2006; Boyce, Lewis, & Worm, 2010).

In aquatic environments, phytoplankton experience temperature fluctuations on many different time-scales, but their ability to cope temperature changes could be compromised by the current rise in global temperature. The global temperature has increased around 0.6 °C during the last 100 years (Houghton et al., 2001), and current climate change models predict a further increase of 1–7 °C by the year 2100, relying on the CO_2 emission (Magnuson et al., 1997; Meehl et al., 2007). However, greater changes are observed at higher northern latitudes, and most land areas are warming more rapidly than oceans (Intergovernmental Panel on Climate Change [IPCC], 2007). Under this rapid temperature increase phytoplankton mainly relies on mechanisms to cope with local environmental alterations, such as phenotypic plasticity and genetic adaptation, since the water mass circulation controls its spatial distribution.

A recent work investigated, at the level of individual species, the maximum capacity of adaptation to a gradual warming process in species belonging to various ecological niches, as well as comparing the inter-specific response as a function of the natural habitat and taxonomic group (Huertas, Rouco, López-Rodas, & Costas, 2011). This work has given us a first insight into the response of phytoplankton communities to an expected environment alteration, for instance the temperature increase, that shows there is a high variability in the maximum capacity of adaptation associated with the taxonomic group and natural habitat. The experimental design used by Huertas et al. (2011), the so-called ratchet protocol, is the perfect tool to investigate maximum adaptation capability. However, this technique does not accurately discern between the mechanisms underlying the adaptation process (i.e. physiological acclimatization, pre-selective mutations or adaptive mutations) in waters whose temperature is increasing.

In this work, we carried out a mechanistic approach to disentangling causes of adaptation of phytoplankton to temperature increase. Based on classic concepts on adaptation to environmental change (Fisher, 1930; Lewontin, 1974; Sniegowski & Lensky, 1995), and recent experimental work on adaptation of phytoplankton to global change (Garcia-Villada et al., 2004; López-Rodas, Maneiro, Lanzarot, Perdigones, & Costas, 2008; Marvá, García-Balboa, Baselga-Cervera, & Costas, 2014; García-Balboa et al., 2013), and warming (Flores-Moya et al., 2005; Huertas et al., 2011; Rouco, López-Rodas, Flores-Moya, & Costas, 2011; Romero, López-Rodas, & Costas, 2012), we hypothesise a conceptual model to explain phytoplankton adaptation to temperature increase:

i)   under low levels of environmental stress, physiological acclimatization would be able to allow adaptation;

ii) when the temperature increases beyond the level of physiological adaptation, the pre-existing genetic variability could allow adaptation of some genotypes;

iii) finally, after a certain temperature threshold, only the occurrence of new thermal-resistant mutants can assure adaptation.

We analysed this conceptual model in various common phytoplankton species belonging to different taxonomic groups and isolated from continental water bodies, coastal waters and the Open Ocean. Our experimental model provides evidence for assessing how phytoplankters might respond and evolve to the temperatures envisaged for the near future. In addition, the nature of these resistant cells was investigated in an attempt to distinguish between temperature-resistant cells arising by direct and specific adaptation in response to thermal stress, versus temperature-resistant cells arising by rare spontaneous mutations occurring randomly before thermal stress.

2. Methods

2.1 Species Used

Twenty six different clonal cultures (i.e. strains) of *Microcystis aeruginosa* (Kützing) Lemmermann (Cyanobacteria), isolated from Doñana National Park (southern Spain), eighteen strains of *Prorocentrum triestinum* Schiller (Dinophyceae) isolated from the Atlantic cost (western Spain), twenty strains of *Tetraselmis suecica* (Kylin) Butcher (Prasinophyceae) isolated from coastal waters of east Sardinia (Italy), twenty four strains of *Nitzschia closterium* (Ehrenberg) Smith (Bacillariophyceae), isolated from Galicia (northwest Spain) and twelve strains of *Isochrysis galbana* Parke (Haptophyceae) isolated from the north-central Atlantic from the Algal Culture Collection of the Universidad Complutense (Madrid, Spain) were used. Natural thermal variation range during the year is summarized in Table 1.

Table 1. Main characteristics of the species studied. Temperature causing 100% growth inhibition was measured in ancestral populations before the fluctuation analysis. Growth inhibition was measured in triplicates of each species inoculated with 5×10^5 cells from mid-log exponentially growing cultures and exposed to increasing temperature (increase interval = 2 °C)

Isolation Site	Location (Lat/long)	Species	Isolation temperature (°C)	Annual temperature range (°C)	Temperature causing 100% growth inhibition (°C)	N° of strains isolated	Micro-photographs
CWB	037° 005' N 006° 029'W	*Microcystis aeruginosa*	23	14-31	32	26	
COA	036° 007'N 006° 023'W	*Prorocentrum triestinum*	21	15-24	26	18	
COA	038° 059'N 008° 022'E	*Tetraselmis suecica*	22	13-25	28	20	
COA	036° 007'N 006° 001'W	*Nitzschia closterium*	21	15-24	28	24	
OPE	043° 041'N 011° 013'W	*Isochrysis galbana*	14	13-19	28	12	

Note. CWB= continental waterbodies; COA = coastal waters; OPE = open ocean.

2.2 Disentangle the Causes of Phytoplankton Adaptation to Temperature Increase

In order to assess the adaptation process that takes place in each of the species as a response to an increase of temperature, two different experiments were performed:

2.2.1 Experiment 1

The effect of warming was estimated by monitoring growth of different strain under increasing temperatures. We inoculated triplicates of each strain with 5×10^5 cells (N_0) from mid-log exponentially growing cultures. Each strain was grown axenically in cell-culture flasks (Greiner; Bio-One Inc., Longwood, NJ, USA) with 20 ml of f/2 medium (marine species) (Sigma-Aldrich, Germany) or BG11 medium (freshwater species) (Sigma-Aldrich, Germany) at 15, 20, 25, 30 and 35 °C under 30 µmol m^{-2} s^{-1} light. Each strain was maintained for 20 days under these conditions. Afterwards, the final number of cells was blind counted (i.e. the person counting the test did not know the identity of the tested sample), using a Beckman (Brea, CA, USA) Z2 particle counter, except *M. aeruginosa*, which were counted using a Uriglass settling chambers (Biosiga, Cona, Italy) in an inverted microscope (Axiovert 35, Zeiss Oberkóchen, Germany) because cell aggregation prevents reliability of results using a particle counter in this species. We consider that a strain grow under a given temperature when Nt > 5.5×10^5 cells.

Two different results can be found, each of them being interpreted as the independent consequence of two different phenomena of adaptation:

i)   If all the strains (genotypes) of a species are able to grow after a temperature increase, then a physiological acclimatization to temperature increase is enough to assures adaptation to that temperature (without effect among strains of genetic individual differences);

ii)   In contrast, if only some strains (genotypes) of a species are able to growth after a temperature increase and others do not, then there are preexisting genetic variability among strains of this species in adaptation to increasing temperature.

2.2.2 Experiment 2

A temperature increase above a certain limit, the pre-existing variability is not capable to cope with high temperatures (i.e. all the strains of a species are unable to growth under this temperature). Once reached this limit temperature value, massive cell destruction takes place. However, some microalgae could survive at such high temperature as a result of two different mechanisms:

i)   Adaptation induced directly by the temperature increases (i.e. as a result of post-selective adaptive mutations that occur in response to temperature increase).

ii)   Adaptation not induced by temperature (i.e. as a result of selection of rare pre-selective spontaneous mutations that occur prior to temperature increase).

A modified Luria and Delbrück (1943) fluctuation analysis adapted for liquid cultures of microalgae (Lopez-Rodas et al., 2001; Costas et al., 2001) was performed using temperature as the selective agent (Figure 1). The fluctuation analysis combined set of experiments and statistical methodology to differentiate between adaptation as a result of post-selective mechanisms and adaptation as a result of pre-selective mechanisms. It consists of two different sets of experiments for each temperature and each strain analyzed (Figure 1). A clone of each species randomly chosen was used to perform the fluctuation analysis.

In set 1 experiment, around 90 culture flasks per species and temperature were used. Each culture was inoculated with a small enough number of cells to ensure the absence of pre-existing mutants in the culture ($N_0 \approx 100$ cells) and growth in 5 ml medium at 22 °C (non-selective conditions) until a final population of $Nt \approx 10^5$ cells was reached. The cultures were subsequently exposed to temperatures of 30 °C (*P. triestinum*, *N. closterium*) or 35 °C (*M. aeruginosa*, *T. suecica*, *I. galbana*) as selective conditions. For set 2 (control) around 40 culture flasks containing 5 ml of medium were inoculated with aliquots of $Nt \approx 10^5$ cells from the same parental populations (growing at 22 °C) and exposed to selective conditions (temperatures of 30, 35 and 40 °C) simultaneously with set 1. Set 2 served as control group for the experimental error (Figure 1) and tracked the variance within the parental population (the inter-culture flask-to-flask variation), which was expected to be low.

Experiments and controls were maintained under the selective temperature value for 75 days prior to observation (to assure that only one mutant cell could generate enough progeny to be detected). The resistant cells in each culture were then counted.

According to Luria and Delbrück (1943), two different results can be found in the set 1 experiment, each of them being interpreted as the independent consequence of two distinct adaptation mechanisms (i.e. post-selective or pre-selective mechanisms, respectively):

i)   If the occurrence of resistant cells is induced by the temperature increase (post-selective mechanisms), then every cell is likely to have the same possibility of developing resistance. Consequently, inter-culture (flask-to-flask) variation will be low (i.e. following the Poisson model, variance/mean ≈ 1) and the variance in the number of cells per culture of set 1 should be similar to variance of set 2 (no fluctuation, Fig. 1).

ii)   If the resistant cells arise by random mutations that occur spontaneously during the period in which the cultures grow from N_0 to Nt (before the exposure to an increased temperature), then the inter-culture (flask-to-flask) variation would be high (i.e. not consistent with the Poisson model, variance/mean >1) and the variation in the number of cells per culture of set 1 should be significantly higher than those of set 2 (fluctuation, Figure 1).

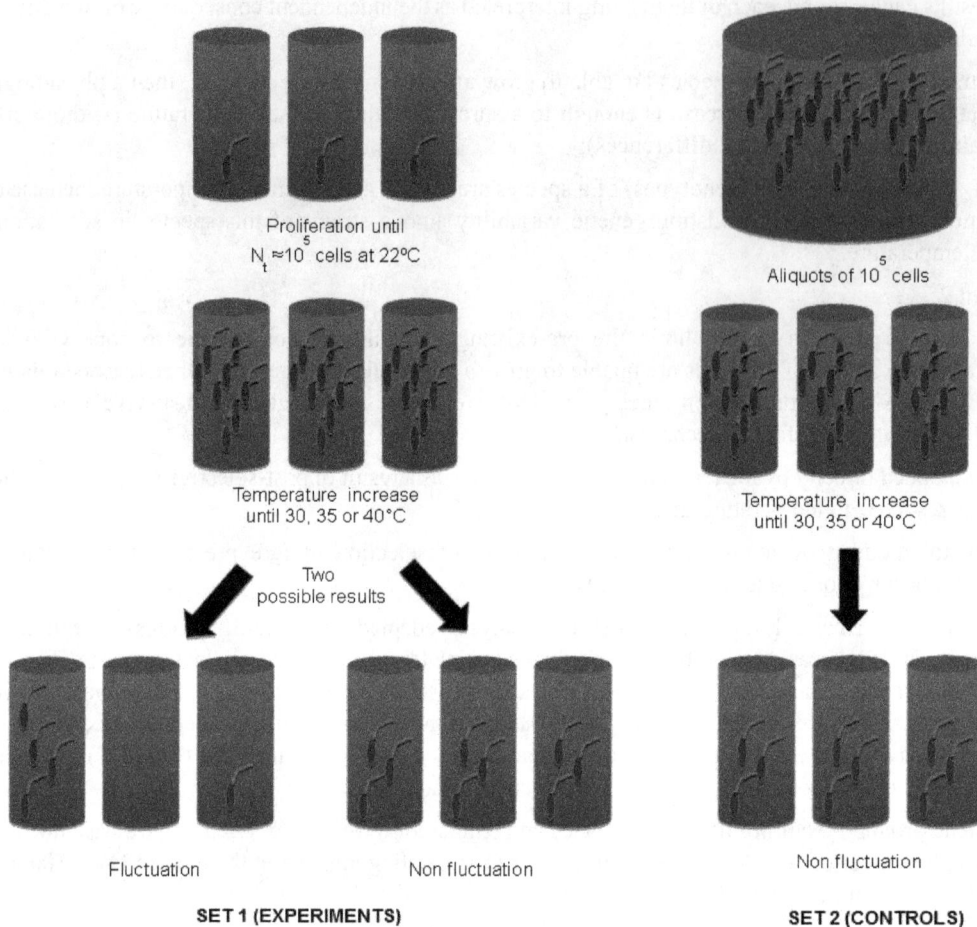

Figure 1. Possible results obtained in the fluctuation analysis

Set 1 experiments: different cultures (each started with $N_0 \approx 100$ cells) were grown under non-selective temperature (22 °C) until a cell density of $N_t = 10^5$ cells was reached, and then transferred to the selective temperatures. If fluctuation occurred, rare temperature-resistant mutants arose during the propagation of cultures under non-selective temperatures. The figure represents one mutational event early in the propagation of replicate 1 (therefore, the final density of temperature-resistant cells is high); one late in replicate 3 (therefore, the final density of temperature-resistant cells found is low); and replicate 2 without mutational events. In contrast, no fluctuation indicated that resistant cells arose as a response to selective temperature. Set 2 sampled the variance as an experimental control.

Obviously, another result (0 resistant cells in each culture) could also be found, indicating that non-adaptation to temperature increase took place.

In addition, fluctuation analysis allows estimation of the mutation rates of appearance of resistant cells. The proportion of cultures from set 1 showing non-resistant cells after temperature increase (P_0 estimator) was used to calculate the mutation rate (μ) as follows:

$$P_0 = e^{-\mu (N_t - N_0)} \tag{1}$$

where P_0 is the proportion of cultures with no resistant cells; N_0 and N_t represent the initial and the final cell densities, respectively; and μ stands for the mutation rate expressed in mutants per cell division (Luria & Delbrück, 1943).

If, as expected, the mutation from a wild-type temperature-sensitive allele to a temperature-resistant allele is recurrent (Spiess, 1989), new temperature-resistant mutants will arise in each generation. Usually, these new resistant mutants are detrimental in fitness under nonselective conditions (Spiess, 1989) (i.e. normal temperature). Consequently, new temperature-resistant mutants will arise in each generation, but most of them will be eliminated sooner or later by natural selection under normal temperatures. Eventually, at any given time there will be a certain number of resistant mutants (at frequency q) that will not be eliminated. The balance between recurrent occurrence of temperature-resistant mutants (at rate μ) and the rate of selective elimination of these mutants (at coefficient s) will determine the average number of such temperature-resistant mutants within the populations, according to the equation, (Kimura & Maruyama, 1966):

$$q = \mu / (\mu + s) \tag{2}$$

The selection coefficient was calculated according to the equation:

$$s = 1 - (m_r / m_s) \tag{3}$$

where m_r and m_s are the Malthusian fitness (growth rates) of temperature-resistant and wild-type sensitive genotypes, respectively, measured in non-selective temperature (i.e. 22 °C).

3. Results

The role played by the different mechanisms that allow adaptation of phytoplankton to temperature increase (physiological acclimatization, adaptation based on pre-existing genetic variability and occurrence of new mutations that confer resistance) of phytoplankton organisms is summarized in Table 2.

Table 2. Fluctuation analysis to study adaptation of different phytoplankton species to temperature increase.

	N° of replicates	
Microcystis aeruginosa (at 35 °C) (CWB)	Set 1	Set 2
No. of replicate cultures	90	35
No. of cultures containing the following no. of resistant cells:		
0	48	0
$< 10^{-4}$	5	0
from 10^{-4} to $5 \cdot 10^{-4}$	4	0
from $5 \cdot 10^{-4}$ to 10^{-5}	10	35
$> 10^{-5}$	23	0
Variance/mean (of the no. of resistant cells per replicate)	21	1.1
Fluctuation	yes	no
Prorocentrum triestinum (at 30 °C) (COA)		
No. of replicate cultures	50	30
No. of cultures containing the following no. of resistant cells:		
0	44	0
$< 10^{-4}$	3	30
from 10^{-4} to $5 \cdot 10^{-4}$	2	0
from $5 \cdot 10^{-4}$ to 10^{-5}	0	0
$> 10^{-5}$	1	0
Variance/mean (of the no. of resistant cells per replicate)	5.2	0.9
Fluctuation	yes	No
Tetraselmis suecica (at 35 °C) (COA)		
No. of replicate cultures	90	40
No. of cultures containing the following no. of resistant cells:		
0	78	0
$< 10^{-4}$	2	40
from 10^{-4} to $5 \cdot 10^{-4}$	2	0
from $5 \cdot 10{-4}$ to 10^{-5}	6	0
$> 10^{-5}$	2	0
Variance/mean (of the no. of resistant cells per replicate)	7.3	1.0
Fluctuation	yes	no

Table 2 (cont.)

	N° of Replicates	
Prorocentrum triestinum (at 30 °C) (COA)	Set 1	Set 2
No. of replicate cultures	70	30
No. of cultures containing the following no. of resistant cells:		
0	65	0
$< 10^{-4}$	3	0
from 10^{-4} to $5 \cdot 10^{-4}$	2	0
from $5 \cdot 10^{-4}$ to 10^{-5}	0	35
$> 10^{-5}$	0	0
Variance/mean (of the no. of resistant cells per replicate)	2.3	1.1
Fluctuation	yes	No
Isochrysis galbana (at 35 °C) (OPE)		
No. of replicate cultures	65	35
No. of cultures containing the following no. of resistant cells:		
0	58	0
$< 10^{-4}$	3	0
from 10^{-4} to $5 \cdot 10^{-4}$	3	35
from $5 \cdot 10^{-4}$ to 10^{-5}	1	0
$> 10^{-5}$	0	0
Variance/mean (of the no. of resistant cells per replicate)	9.1	0.9
Fluctuation	Yes	No

The phytoplankton organisms are able to cope with slight temperature increases by mean of physiological acclimation. So, physiological acclimatization allow adaptation of *P. triestium* and *N. closterium* up to 20 °C and of *M. aeruginosa, T. suecica* and *I. galbana* up to 25 °C (Table 2). Until these temperatures all the clones (genotypes) of each species were able to grow.

When the temperature increases, only some strains (genotypes) of each species are able to proliferate (e.g. from the 26 strains of *M. aeruginosa* only 14 were able to grow at 30 ° C, or from the 18 strains of *P. triestinum* only 5 were able to grow at 25 °C; Table 2). These differences existing among the different strains (genotypes) of each species denote the pre-existing genetic variability for temperature resistance.

If the temperature is further increased, the thresholds for physiological acclimatization and pre-existing genetic variability are exceeded and only the occurrence of new mutations could guarantee the survival of the species. When *M. aeruginosa T. suecica* and *I. galbana* were exposed to 35 °C or *P. triestium* and *N. closterium* were exposed to 30 °C during the fluctuation experiments, the cell density was drastically reduced. This result was evident in all experimental cultures of sets 1 and 2, and for all species, owing to destruction of sensitive cells by the lethal effect of the temperature. However, after further incubation for 75 days, some cultures increased in density again, apparently due to growth of temperature-resistant variants. In the case of set 1, only some cultures showed appreciable cell growth. By contrast, all set 2 cultures recovered and temperature-resistant cells were detected in all the replicated tubes (Table 3). Fluctuations in the number of temperature-resistant cells were observed in set 1 experiment (variance/mean ratio > 1; $P < 0.001$ using $\chi 2$ as a test of goodness of fit), in contrast with low fluctuation of set 2 controls (variance/mean ratio ≈ 1, consistent with Poisson distribution; $P < 0.05$). As the large fluctuation found in set 1 cultures is related to processes different from the sampling error, it could be inferred that temperature-resistant cells have arisen from rare, pre-selective spontaneous mutations occurring randomly during replication of organisms prior to exposure to the high temperatures, rather than by specific occurrence of resistant cells appearing in response to the temperature increase. The mutation rates from temperature sensitivity to temperature resistance ranged arround 10^{-6}.

Table 3. Mechanisms for adaptation (physiological acclimatization, adaptation based on pre-existing genetic variability and occurrence of new mutations that confer resistance) of phytoplankton organisms to temperature increase

M. aeruginosa (26 strains analysed)	15°C	20°C	25°C	30°C	35°C	40°C
Experiment 1:						
Number of strains that were able to grow	26	26	26	14	0	0
Adaptation based on physiological acclimatization	yes	yes	yes	no	no	no
Adaptation based on physiological acclimatization				yes	no	no
Experiment 2: (more details are given in Table 3)						
Fluctuation (F = CV set1 / CV set 2)					21.0	
Mutation rates from temperature sensitivity to temperature resistance					1.2×10^{-6}	
s (coefficient of selection against resistant mutant					0.51	
q (frequency of thermal-resistant allele in mutation-selection balance)					1.2×10^{-6}	
Adaptation based on the occurrence of new mutants that confer resistance					yes	no
P. triestinum (18 strains analysed)						
Experiment 1:						
Number of strains that were able to grow	18	18	5	0	0	0
Adaptation based on physiological acclimatization	yes	yes	no	no	no	no
Adaptation based on physiological acclimatization			yes	no	no	no
Experiment 2: (more details are given in Table 3)						
Fluctuation (F = CV set1 / CV set 2)				6.2		
Mutation rates from temperature sensitivity to temperature resistance				1.4×10^{-6}		
s (coefficient of selection against resistant mutant				0.27		
q (frequency of thermal-resistant allele in mutation-selection balance)				5.2×10^{-6}		
Adaptation based on the occurrence of new mutants that confer resistance					yes	no	no
T. suecica (20 strains analysed)						
Experiment 1:						
Number of strains that were able to grow	20	20	20	15	0	0
Adaptation based on physiological acclimatization	Yes	yes	yes	no	no	no
Adaptation based on physiological acclimatization				yes	no	no
Experiment 2: (more details are given in Table 3)						
Fluctuation (F = CV set1 / CV set 2)					7.1	
Mutation rates from temperature sensitivity to temperature resistance					1.5×10^{-6}	
s (coefficient of selection against resistant mutant					0.21	
q (frequency of thermal-resistant allele in mutation-selection balance)					6.6×10^{-6}	
Adaptation based on the occurrence of new mutants that confer resistance					yes	no

Table 3: (cont.)

N. closterium (24 strains analysed)	15°C	20°C	25°C	30°C	35°C	40°C
Experiment 1:						
Number of strains that were able to grow	24	24	9	0	0	0
Adaptation based on physiological acclimatization	yes	yes	no	no	no	no
Adaptation based on physiological acclimatization			yes	no	no	no
Experiment 2: (more details are given in Table 3)						
Fluctuation (F = CV set1 / CV set 2)				2.0		
Mutation rates from temperature sensitivity to temperature resistance				7.4×10^{-6}		
s (coefficient of selection against resistant mutant				0.36		
q (frequency of thermal-resistant allele in mutation-selection balance)				2.0×10^{-6}		
Adaptation based on the occurrence of new mutants that confer resistance				yes	no	no
I. galbana (12 strains analysed)						
Experiment 1:						
Number of strains that were able to grow	12	12	12	8	0	0
Adaptation based on physiological acclimatization	yes	yes	yes	no	no	no
Adaptation based on physiological acclimatization				yes	no	no
Experiment 2: (more details are given in Table 3)						
Fluctuation (F = CV set1 / CV set 2)					2.9	
Mutation rates from temperature sensitivity to temperature resistance					3.4×10^{-6}	
s (coefficient of selection against resistant mutant					0.15	
q (frequency of thermal-resistant allele in mutation-selection balance)					2.3×10^{-6}	
Adaptation based on the occurrence of new mutants that confer resistance					yes	no

Temperature-resistant mutants growing under normal temperatures showed lower fitness values than those found in the wild-type temperature sensitive strains. The relative values of fitness of resistant mutants and sensitive wild types were used to estimate the coefficient of selection (s) of temperature-resistant mutants under normal temperature (Table 2). By using the values of μ and s, the frequency (q) of resistant alleles in the wild-type population as a consequence of the balance between mutation and selection was calculated. As the result, the frequency of resistant alleles maintained in populations under normal temperature ranged between 1.2 temperature-resistant mutants per 10^6 sensitive cells in *M. aeruginosa* and 6.6 temperature resistant mutants per 10^6 sensitive cells in *T. suecica* (Table 3).

Obviously, temperature reaches a threshold at which organisms are unable to adapt (*M. aeruginosa*, *T. suecica* and *I. galbana* were unable to adapt to 40 °C, whereas *P. triestium* and *N. closterium* were unable to adapt to 35 °C; Table 2).

4. Discussion

Very little is currently known on phytoplankton capabilities in response to climatic selection pressures under an adaptive point of view. Experimental research mimicking temperature-increasing scenarios can be considered a novel way to explore the adaptation of phytoplankton to predicted future thermal scenarios. Such experiments show that phytoplankton organisms can survive under increasing temperatures as a result of physiological adaptation (i.e. acclimatization). Physiological acclimatization allowed all the strains from the five analyzed species survive under a low temperature increase. Consequently, it can be concluded that usually the phenotypic plasticity must be able to cope to the global temperature increase during the last century (around 0.6 °C, Houghton et al., 2001), and perhaps allow adaptation to temperature increase under a low CO_2 emission scenario (around 1.18 °C, Meehl et al., 2007).

However, when environmental stress exceeds the physiological limits, survival depends exclusively on adaptive evolution supported by the occurrence genetic changes (Sniegowski & Lenski, 1995). As the temperature rises above physiological acclimatization capacity, only the genetic mechanisms allows adaptation. It is commonly assumed that the preexisting genetic variability ensures the adaptation of the population (reviewed in Fisher, 1930 and Lewontin, 1975). All the species analyzed here presented genetic variability for thermal resistance. This preexisting genetic variability allows that some genotypes can be adapted to high temperature (i.e. some genotypes resistant until 5 °C more than others). About this, Costas, Baselga-Cervera, García-Balboa and López-Rodas, (2014) measured the heritability of fitness (i.e. proportion of variance in fitness that has genetic basis) under increasing temperatures, using an experimental quantitative genetic procedure suitable to phytoplankton populations. The results reveal that there is enough genetic variability in fitness within the population of phytoplankton to assure adaptation to temperature increase, even under a intermediate CO_2 emission scenario.

It is accepted that short-term unpredictable stress is best met by physiological acclimatization, whereas continuous stress can be met by genetic adaptation (Bradshaw & Hardwick, 1989). Conceivably, phytoplankton could face to low CO_2 emission scenario by mean of physiological acclimatization, but if a warming is maintained over time, the phytoplankton populations could change genetically.

It is usually assumed that genetic adaptation is achieved slowly. However, recent studies are changing many preconceived notions about the adaptation of microalgae to extreme environments. So, microalgae can adapt very quickly to powerful algaecides such as copper (Garcia-Villada et al., 2004), modern herbicides like glyphosate, simazine and diquat (López-Rodas et al., 2007, Marva et al., 2010; Huertas et al., 2010) pesticides as lindane (Gonzalez et al., 2012), crude oil spills (Carrera-Martinez, Mateos-Sanz, López-Rodas, & Costas, 2010; 2011; Romero-López et al., 2012), extremely contaminated mining waters (López-Rodas et al., 2008a; López-Rodas, Marva, Rouco, Costas, & Flores-Moya, 2008b), extreme geothermal waters (Costas, Flores-Moya, & López-Rodas, 2008, López-Rodas et al., 2009) and even uranium (Baselga-Cervera., López-Rodas, García-Balboa, Costas, 2013; Garcia-Balboa et al., 2013). In all these cases a single mutation is enough to achieve rapid adaptation. Adaptation to warming is much the same. When the temperature rises above a certain level, the pre-existing genetic variability is not able to ensure the adaptation, as in the case of *M. aeruginosa T. suecica* and *I. galbana* at 35 °C or *P. triestium* and *N. closterium* at 30 °C. However, the occurrences of new mutations that confer thermal resistance allow phytoplankton adaption to temperature increase. Fluctuation analysis shows that thermal-resistant cells arose by rare spontaneous mutations, and not through direct and specific adaptation in response to temperature increase.

The mutation rates in the five strains of microalgae were found to be in the middle of the range of the mutations rates for resistance to algaecides, herbicides, pesticides, heavy metals, extreme environments and others (Costas et al., 2013; Costas & López-Rodas., 2013; Fernandez-Arjona et al., 2013; Marvá et al., 2014; Romero-López, Costas, & López-Rodas, 2014). Mutation is recurrent and new resistant mutants arise each generation, but most of them disappear eventually due to natural selection or chance. At any one time, the balance between the continuous appearance of mutants and their selective elimination determines the number of remaining temperature-resistant mutants in microalgae populations growing under their normal temperature. Accordingly, the population would be predominantly a line of wild-type temperature-sensitive genotypes, accompanied by a very small fraction of temperature-resistant clone lines. Therefore, the frequency of mutants present in the population should be enough to assure survival of phytoplankton populations in this age of global warming.

Additionally, recurrent exposure to changing temperatures could enhance the frequency of temperature-resistant alleles as the consequence of mutation-selection equilibrium. This could be very relevant in the case of inland waters species such as *M. aeruginosa*, or some coastal species such as *T. suecica*. Taking into account that all these species live in ecosystems where temperature variation is very common, our result seems coherent with the fact that the frequency of the resistant allele under growth in the normal temperature range was 10 times higher in these species than in *P. triestinum* and *N. closterium*, which are not naturally exposed to sudden changes in water temperature. For example, accurate circannual rhythms controlling the formation and germination in dinoflagellates cysts ensure that the free-living stages only face favourable environmental conditions (Costas & Varela, 1989).

This capability of microalgae to adapt to temperature increase by means of mutation at one gene could have significant implications for the survival of microalgae under the temperature increase expected in a scenario of high CO_2 emissions. These heat-resistant mutants present a detriment in fitness. Although phytoplankton can adapt to global heating, its growth and primary production will decrease, not only for oceanographic reasons, but because the heat-resistant genotypes that will survive are less productive. Furthermore, no adaptation is possible for neither of the species of study at 40 °C, and only three out of five species (*M. aeruginosa, T. suecica* and *I.*

galbana) were able to adapt at 35 °C. This information give us an idea of the temperature limit that phytoplankton could stand.

Although there still remain many uncertainties concerning the impact of global warming on phytoplankton communities, rare spontaneous mutations conferring resistance seem able to ensure the survival of many phytoplankton species. Warming also could induce a progressive replacement of the most sensitive functional phytoplankton genotypes and species by those that are more resistant. The impact of this possible scenario of temperature resistant mutants arising and replacing the sensible ones is virtually inscrutable.

Acknowledgments

This work has by been supported financially by the Spanish Ministry of Sciences and Innovation through the Grant CTM2012-34757. Special thanks are given to L. de Miguel for technical support.

References

Baselga-Cervera, B., López-Rodas, V., García-Balboa, C., & Costas, E. (2013). Microalgae: first nuclear engineers? *Anales de la Real Academia Nacional de Farmacia RANF, 79*(4), 634-645.

Behrenfeld, M. J., O'Malley, R. T., Siegel, D. A., McClain, C. R., Sarmiento, J. L., Feldman, G. C., ... Boss, E. S. (2006). Climate-driven trends in contemporary ocean productivity. *Nature, 444*, 752-755. http://dx.doi.org/10.1038/nature05317

Behrenfeld, M. J., Randerson, J. T., McClain, C. R., Feldman, G. C., Los, S. O., Tucker, C. J., ... Pollack, N. H. (2001). Biospheric Primary Production during an ENSO Transition. *Science, 29*, 2594-2597. http://dx.doi.org/10.1126/science.1055071

Boyce, D., Lewis, M., & Worm, B. (2010). Global phytoplankton decline over the past century. *Nature, 466*, 591-596. http://dx.doi.org/10.1038/nature09268

Bradshaw, A. D., & Hardwick, K. (1989). Evolution and stress–genotype and phenotype components. *Bioogical Journal of Linnean Society, 37*, 137–155. http://dx.doi.org/10.1046/j.1095-8312.2002.00020.x

Carrera-Martinez, D., Mateos-Sanz, A., Lopez-Rodas, V., & Costas, E., (2011). Adaptation of microalgae to a gradient of continuous petroleum contamination. *Aquatic Toxicology, 101*, 342-350. http://dx.doi.org/10.1016/j.aquatox.2010.11.009

Carrera-Martínez, D., Mateos-Sanz, A., López-Rodas, V., & Costas, E. (2010). Microalgae response to petroleum spill: An experimental model analyzing physiological and genetic response of *Dunaliella tertiolecta* (Chlorophyceae) to oil samples from the tanker Prestige. *Aquatic Toxicology, 97*, 151-159. http://dx.doi.org/10.1016/j.aquatox.2009.12.016

Costas, E., & Lopez-Rodas., V., (2013). The role played for spontaneous mutation. *Stress Biology of Cyanobacteria: Molecular Mechanisms to Cellular Responses.* (pp. 307). CRC Press.

Costas, E., & Varela, M. A. (1989). A Circannual rhythm in cysts formation and growth-rates in the dinoflagellate *Scripsiella-Trochoidea* Stein. *Chronobiologia, 16*, 265-270. Retrieved from http://europepmc.org/abstract/MED/2805945

Costas, E., Baselga-Cervera, B., García-Balboa, C., & López-Rodas, V. (2014). Estimating the genetic capability of different phytoplankton organisms to adapt to climate warning. *Environmental Science group, Oceanography Open Access.* (In press) http://dx.doi.org/10.4172/2332-2632.1000123

Costas, E., Carrillo, E., Ferrero, L.M., Agrelo, M., Garcia-Villada, L., Juste, J., & López-Rodas, V. (2001). Mutation of algae from sensitivity to resistance against environmental selective agents: the ecological genetics of *Dictyosphaerium chlorelloides* (Chlorophyceae) under lethal doses of 3-(3,4-dichlorophenyl)-1,1 dimethylurea herbicide. *Phycologia, 40*, 391-398. http://dx.doi.org/10.2216/i0031-8884-40-5-391.1

Costas, E., Flores-Moya, A., & López-Rodas, V. (2008). Rapid adaptation of algae to extreme environments (geothermal waters) by single mutation allows "Noah's Arks" for photosynthesizers during the Neoproterozoic "snowball Earth"? *New Phytology, 189*, 922-932. http://dx.doi.org/10.1111/j.1469-8137.2008.02620.x

Costas, E., Gonzalez, R., López-Rodas, V., & Huertas, E. (2013). Mutation of microalgae from antifouling sensitivity to antifouling resistance allows phytoplankton dispersal through ship's biofouling. *Biological Invasions, 15*(8), 1739-1750. http://dx.doi.org/ 10.1007/s10530-012-0405-8

Falkowski, P. G., & Raven, J. A. (1997). *Aquatic Photosynthesis* (pp. 375). Blackwell, Oxford. http://dx.doi.org/ 10.1016/S0176-1617(99)80078-9

Falkowski, P. G., Barber, R. T., & Smetacek, V. (1998). Biogeochemical controls and feedbacks on ocean primary production. *Science, 281*, 200–206. http://dx.doi.org/ 10.1126/science.281.5374.200

Fernández-Arjona, M del Mar., Bañares-España, E., García-Sánchez, M. J., Hernández-López, M., López-Rodas, V., Costas, E., & Flores-Moya, A. (2013). Disentangling mechanisms involved in the adaptation of photosynthetic microorganisms to the extreme sulphureous water from Los Baños de Vilo (S Spain). *Microbial Ecology, 66*(4),742-51. http://dx.doi.org/ 10.1007/s00248-013-0268-2

Fisher, R. A. (1930). *The Genetical Theory of Natural Selection*, Clarendon Press, Oxford.

Flores-Moya, A., Costas, E., Bañares-España, E., García-Villada, L., Altamirano, M., & López-Rodas, V. (2005). Adaptation of *Spirogyra insignis* (Chlorophyta) to an extreme natural environment (sulphureous waters) through pre-selective mutations. *New Phytology, 166*, 655–661. http://dx.doi.org/ 10.1111/j.1469-8137.2005.01325.x

García-Balboa, C., Baselga-Cervera, B., García-Sanchez, A., Mariano-Igual, J., López-Rodas, V., & Costas, E. (2013). Rapid adaptation of microalgae to extremely polluted waterbodies from uranium mining: an explanation of how the mesophilic organisms can rapidly colonize extremely toxic environments. *Aquatic Toxicology, 144-145*, 166-123. http://dx.doi.org/10.1016/j.aquatox.2013.10.003

Garcia-Villada, L., Rico, M., Altamirano, M., Sánchez, L., López-Rodas, V., & Costas, E. (2004). Occurrence of copper resistant mutants in the toxic cyanobacteria *Microcystis aeruginosa*: characterization and future implications in the use of copper sulphate as algaecide. *Water Research, 38*, 2207–2213. http://dx.doi.org/10.1016/j.watres.2004.01.036

González, R., García-Balboa, C., Rouco, M., López-Rodas V., & Costas, E. (2012). Adaptation of microalgae to lindane: a new approach for bioremediation. *Aquatic Toxicology, 109*, 25-32. http://dx.doi.org/10.1016/j.aquatox.2011.11.015

Houghton, J. T., Ding, Y., Griggs, D. J., Noguer, M., van der Linden, P. J., Dai, X., … Johnson, C. A. (2001). *IPCC, 2001: Climate Change 2001: The Scientific Basis. Contribution of Working Group I to the Third Assessment Report of the Intergovernmental Panel on Climate Change.* Cambridge University Press.

Huertas, I. E., Rouco, M., López-Rodas, V., & Costas E. (2010). Estimating the capability of different phytoplankton groups to adapt to contamination: herbicides will affect phytoplankton species differently. *New Phytologist, 188*, 478-487. http://dx.doi.org/ 10.1111/j.1469-8137.2010.03370.x

Huertas, I. E., Rouco, M., López-Rodas, V., & Costas, E. (2011). Warming will affect phytoplankton differently: Evidence through a mechanistic approach. *Proceedings of the Royal Society B, 278*, 3534-3543. http://dx.doi.org/10.1098/rspb.2011.0160

Kimura, M., & Maruyama, T. (1966). The mutational load with epistatic gene interactions in fitness. *Genetics, 54*, 1337–1351.

Kirk, J. T. O. (1994). Optics of UV-B radiation in natural waters. *Ergebnisse der Limnology, 43*, 1-16.

Lewontin, R. C. (1974). *The Genetic Basis of Evolutionary Change*. New York and London: Columbia University Press.

Lewontin, R. C. (1975). The problem of genetic diversity. *Harvey Lecture Series, 70*, 1-20.

López-Rodas, V., Agrelo, M., Carrillo, E., Ferrero, L. M., Larrauri, A., Martín-Otero, L., & Costas, E. (2001). Resistance of microalgae to modern water contaminants as the result of rare spontaneous mutations. *European Journal of Phycology, 36*, 179–190. http://dx.doi.org/10.1080/09670260110001735328

López-Rodas, V., Carrera-Martinez, D., Salgado, E., Mateos-Sanz, A., Baez, J. C., & Costas, E. (2009). A fascinating example of microalgae adaptation to extreme crude oil contamination in a natural spill in Arroyo Minero, Río Negro, Argentina. *Anales de la Real Academia Nacional de Farmacia RABF, 75*, 883–899.

López-Rodas, V., Flores-Moya, A., Maneiro, E., Perdigones, N., Marva, F., Garcia, M. E., & Costas, E. (2007). Resistance to glyphosate in the cyanobacterium *Microcystis aeruginosa* as a result of pre-selective mutations. *Evolutionary Ecology, 21*, 535–547. http://dx.doi.org/10.1007/s10682-006-9134-8

López-Rodas, V., Maneiro, E., Lanzarot, M. P., Perdigones, N., & Costas, E. (2008a) Mass wildlife mortality due to cyanobacteria in the Doñana National Park, Spain. *Veterinary Record, 162*, 317-318.

López-Rodas, V., Marva, F., Rouco, M., Costas, E., & Flores-Moya, A. (2008b). Adaptation of the chlorophycean *Dictyosphaerium chlorelloides* to the stressful acidic, mine metal-rich waters from Aguas Agrias Stream (SWSpain) as result of pre-selective mutations. *Chemosphere, 72*, 703–707. http://dx.doi.org/10.1016/j.chemosphere.2008.04.009.

Luria, S., & Delbrück, M., (1943). Mutations of bacteria from virus sensitivity to virus resistance. *Genetics, 28*, 491–511. http://www.genetics.org/content/28/6/491.full.pdf+html

Magnuson, J. J., Webster, K. E., Assel, R. A., Bowser, C. J., Dillon, P. J., Eaton, J. G., ... Quinn, F. H. (1997). Potential effects of climate changes on aquatic systems: Laurentian Great Lakes and Precambrian Shield Region. *Hydrological Processes, 11*, 825-871. http://dx.doi.org/10.1002/(SICI)1099-1085(19970630)11:8<825::AID-HYP509>3.0.CO;2-G

Marvá, F., García-Balboa, C., Baselga-Cervera, B., & Costas, E. (2014). Rapid adaptation of some phytoplankton species to osmium as a result of spontaneous mutations. *Ecotoxicology, 23*, 213-220. http://dx.doi.org/101007/810646-013-1164-8.

Marvá, F., López-Rodas, V., Rouco, M., Navarro, M., Toro, F. J., Costas, E., & Flores-Moya, A. (2010). Adaptation of green microalgae to the herbicides simazine and diquat as result of pre-selective mutations. *Aquatic Toxicology, 96*, 130-134. http://dx.doi.org/10.1016/j.aquatox.2009.10.009

Meehl, G. A., Cove, C., Delworth, T., Latif, M., McAvaney, B., Mitchell, J. B., & Taylor, K. E. (2007). The WCRP CMIP3 Multimodel Dataset: A New Era in Climate Change Research. *Bulletin of the American Meteorological Society, 88*, 1383-1394. http://dx.doi.org/10.1175/BAMS-88-9-1383

Parmesan, C., & Yohe, G. (2003). A globally coherent fingerprint of climate change impacts across natural systems. *Nature, 421*, 37-42. http://dx.doi.org/ 10.1038/nature01286

Romero, J., López-Rodas, V., & Costas, E. (2012). Estimating the capability of microalgae to physiological acclimatization and genetic adaptation to petroleum and diesel oil contamination. *Aquatic Toxicology, 124*, 227- 237. http://dx.doi.org/ 10.1016/j.aquatox.2012.08.001

Romero-López, J., Costas, E., & Lopez-Rodas, V. (2014). Selected microalgae for petroleum bioremediation: towards a bio-depuration based on von Neumann -like machines. In J. B. Velázquez-Fernández & S. Muñiz-Hernández (Eds), *Bioremediation, Process, Challenges and Future Prospects* (pp. 211-221). New York: NOVA Publishers.

Rouco, M., López-Rodas, V., Flores-Moya, A., & Costas, E. (2011). Evolutionary changes in growth rate and toxin production in the cyanobacterium *Microcystis aeruginosa* under a scenario of eutrophication and temperature increase. *Microbial Ecology, 62*, 265-273. http://dx.doi.org/ 10.1007/s00248-011-9804-0

Sniegowski, P. D., & Lenski, R. E. (1995). Mutation and adaptation: the directed mutation controversy in evolutionary perspective. *Annual Review of Ecology, Evolutions, and Systematics, 26*, 553–578. http://dx.doi.org/10.1146/annurev.es.26.110195.003005

Solomon, S. (Ed.). (2007). *Climate change 2007-the physical science basis: Working group I contribution to the fourth assessment report of the IPCC* (Vol. 4). Cambridge University Press.

Spiess, E. B. (1989). *Genes in Populations* (2th Ed.). New York: John Wiley.

Stenseth, N. C., Mysterud, A., Ottersen, G., Hurrell, J. W., Chan, K. S., & Mauricio, L. (2002). Ecological Effects of Climate fluctuactions. *Science, 297* (5585), 1292-1296. http://dx.doi.org/10.1126/science.1071281

Walther, G. R., Post, E., Convey, P., Menzel, A., Parmesan, C., Beebee, T. J., ... Bairlein, F. (2002). Ecological responses to recent climate change. *Nature, 416*, 389-395. http://dx.doi.org/ 10.1038/416389a

Wrona, F. J., Prowse, T. D., Reist, J. D., Hobbie, J. E., Lévesque, L. M. J., & Vincent, W. F. (2006). Key Findings, Science Gaps and Policy Recommendations. *Ambio, 35*(7), 411-415. http://dx.doi.org/10.1579/0044-7447 (2006)35[411:KFSGAP]2.0.CO;2

16

Morphological Variation and Species Distribution of *Baccaurea dulcis* (Jack) Müll. Arg. in West Java, Indonesia

Reni Lestari[1]

[1] Center for Plant Conservation Bogor Botanical Gardens, Indonesian Institute of Sciences, Indonesia

Correspondence: Reni Lestari, Center for Plant Conservation Bogor Botanical Gardens, Indonesian Institute of Sciences, Jl. Ir. H. Djuanda 13, Bogor, West Java, Indonesia. E-mail: reni_naa@yahoo.com

Abstract

Baccaurea dulcis is an underutilized plant, primarily grown for its fruit, distributed and cultivated only in Sumatra, Borneo and western part of Java Island, and its population is under threat. On the other hand, very few studies have been carried out of this species. The objective of this study was to estimate population distribution and ecology of *Baccaurea dulcis* in West Java, characterize plant morphological characters and correlate plant habitat of *B. dulcis* and the plant and fruit. In West Java *B. dulcis* is only distributed in the sub district Taman Sari of Bogor and sub district Cijeruk of Sukabumi. Even though the species in West Java has a restricted distribution, its morphological characters is quite varied, including size and nature of tree; color and texture of the bark; size and shape of leaves and fruits; color of fruit peel and pulp; and the size, shape and color of seeds. The species grow well in the low land (250 m - 610 m above sea level) of tropical region at neutral soil pH of regosol or latosol soil type and smooth rather coarse soil texture at land slope from 0 % until 45 %. Using 32 variables of trunk, leaf, fruit and seed, all samples could be clustered into 6 groups with the proportion correct 0.903 and with specific fruit characters in each group. There were some significant positive and negative correlations found between habitat and fruit characters and among tree and fruit variables.

Keywords: *Baccaurea dulcis*, underutilized tropical fruit, West Java, characterization, variation, clustering, correlation

1. Introduction

Baccaurea dulcis (Jack) Müll.Arg. is a dioecious plant species belonging to the family Phyllantaceae (previously under Euphorbiaceae) (Haegens, 2000; Wurdack et al., 2004). Its common names are ketupa (English), cupa, tupa, kapul, menteng negri, menteng besar (Indonesian), tjupa, tupa (Malaysian). This species is distributed and cultivated only in Sumatra and western part of Java Island (Uji, 1992).

This species is primarily grown for its fruit. It is propagated mainly by seeds, but rarely by vegetative means. The tree produces fruits in high quantities and the fruits are usually taste sour or fairly sweet. The fruits can be pickled and used in stew or fermented to make wine (Uji, 1992; Munawaroh, 2001). The leaf of *B. dulcis* is boiled and the resulted decoction is used to treat stomachache during menstruation (Munawaroh, 2001). The tree trunk is used in construction (Heyne, 1987). The fruits of *B. dulcis* are rounded, with a diameter of 3.5 - 4 cm and brownish yellow in color. The edible arils are cream, white or reddish in color. Nutrition of fresh pulp of *B. dulcis* per 100 g were 82.3 g water, 0.4 g protein, 7.5 g saccharosa, 0.2 g fibers, 0.5 g ash, 5 mg vitamin C, 0 g vitamin B1 and B2 (Uji, 1992).

In general, the species is underutilized and used only locally. The species is usually grown in the home-yard and the fruits are usually self-consumed and rarely sold in local markets in West Java. Even though, the species is relatively known by the local people, but until now the distribution in West Java is only in very small areas. There is no information about the cultivation of the plant, reproductive period, growth pattern, pest and diseases and production and harvest of the species (Uji, 1992). Vegetative propagation of the species by shoot tip grafting could be applied, which almost 100 % success rate, whereas the airlayering propagation technique resulted only 27.5 % - 40 % of rooted shoot (Lestari, 2009, 2010). The risk of extinction of a fruit species is higher because there are not many people interested in the fruits and the fruits availability in the market is rare (Subekti et al., 2005). Field trips were conducted in West Java to study the species.

The aims of the study were (1) to find out the distribution of the species *Baccaurea dulcis* in West Java and to characterize the plants and fruits based on the morphology and qualitative parameter such as color and texture, (2) to find out the variation and clustering *B. dulcis* plants and fruits in West Java, and (3) to find out the correlation between the characters of the plants and fruits, and the plant's habitat.

A   B   C

Figure 1. A. *Baccaurea dulcis* plant; B. and C. *B. dulcis* fruits

2. Materials and Methods

2.1 Distribution, Characterization and Variation of Plant and Fruit

Field trips, and ecological observations of *B. dulcis* focused on trees, fruits and their habitat and interviews with the farmers were conducted two days a week during the harvest period, starting from February until April 2008 mostly in Bogor, Cianjur, Sukabumi, Tangerang, Depok, Bekasi and Purwakarta of West Java, Indonesia (Figure 2).

Figure 2. Study site in West Java of Indonesia

The distribution of a total of 103 fruiting plants was recorded. The habitat of the plants and the variables of the tree and fruit were documented. The observation of the plant habitat included location of plant growth, altitude, longitude and altitude, air temperature, relative humidity, slope, light intensity, soil pH, soil relative humidity, soil type and soil texture/drainage. The distribution and habitat of the plant was also examined from the herbaria at the Bogor Herbarium (BO). Detail character variables of tree and fruit observed could be seen in Table 1.

As many as 30 fruits randomly were harvested from every plant observed, the minimum and maximum length and width of fruit and seed were measured. The fruits were grouped into 3 and each group of fruits was weighed to calculate the average fruit weight. Using kitchen knife, all of those fruits were peeled and then observed for the easiness to peel, the minimum and maximum amount of the pulp segment and the seed weight. The average of each of those parameters was calculated. The thickness of fruit peel and pulp, and soluble solid content of pulp were measured from 3 samples of fruits randomly. Some qualitative parameters were observed during the field trips and the result of the observation was ranked for the data analysis (Table 2).

Table 1. The Variables observed of trees and fruits of *B. dulcis* in West Java

No	Variables	No	Variables
1	Tree height	17	Fruit peel color
2	Trunk diameter	18	Easiness to peel the fruit
3	Canopy width	19	Thickness of fruit peel
4	Canopy condition (sparse/dense)	20	Peel weight per fruit
5	Lowest branch height	21	Pulp color
6	Bark color	22	The number of fruit segment
7	Bark texture	23	Soluble solid content of pulp
8	Leaf shape	24	Seed shape
9	Leaf length	25	Seed length
10	Leaf width	26	Seed width
11	The Ratio of maximum leaf length/leaf width	27	Ratio of the longest seed length/ seed width
12	Fruit shape	28	Seed weight
13	Fruit length	29	Percentage of peel weight per fruit
14	Fruit width	30	Percentage of pulp weight per fruit
15	The ratio of maximum fruit length/ fruit width	31	Percentage of seed weight per fruit
16	Fruit weight	32	Seed color

Table 2. Qualitative parameters and rank of qualitative number of the observation of *B. dulcis* trees, fruits and seeds in West Java

No	Qualitative parameter	Rank of Qualitative Number and Information
1	Canopy condition	(1) Dense, (2) Medium dense, (3) Sparse
2	Bark color	(1) Light brown/cream, (2) Medium brown, (3) Grey - medium brown, (4) Grey, (5) Browned - grey, (6) Grey, yellowed - medium brown, (7) Grey - dark brown, (8) Dark brown, (9) Browned - black
3	Bark texture	(1) Smooth, (2) Flaky - smooth, (3) Lenticelate, (4) Flaky - lenticelate, (5) Lump - flaky, (6) Flaky, (7) Flaky - fissured, (8) Flaky - rectangular
4	Fruit peel color	(1) Yellow, (2) Light orange/Yellowish orange, (3) Orange/ dark orange, (4) Reddish yellowish - orange
5	Easiness to peel the fruit	(1) Easy, (2) Medium, (3) Difficult
6	Fruit pulp color	(1) White, (2) Transparent - white, (3) Cream, (4) Pink, transparent - cream, (5) Pink, white - transparent, (6) Cream - transparent, (7) Pink - transparent, (8) Pinked/purpled lines - cream/white
7	Seed color	(1) Light brown, (2) Pink - light brown, (3) Brown, (4) Pink - brown, (5) Purplish - brown, (6) Light brown - Pink, (7) Pink

2.2 Measurement Equipments

The equipment used during the observation were GPS Garmin (latitude), Termohygrometer Haar-Synth-Hygro, Germany (air temperature and relative humidity), Soil tester TEW Type 36, Demetra, Japan (soil pH and relative humidity), Clinometer, Suunto PM-5/360, Finland (slope), Light meter, LX-101 A, Lutron, Taiwan (light intensity), Altimeter (altitude). Other equipment used are to measure the tree height (BL 6, Carl Leiss, Berlin, Germany), the diameter of trunk, fruit and seed (diameter tape 20 m x 5 m, Tool No. D-5M, YAMAYO, Japan), canopy width (Tape measure, 50 m), the width of fruit and seed (Digital caliper, 200 mm, Mitutoyo CO., Japan), the fruit and seed weight (Balance, capacity 2 kg), Soluble solid content (Digital refractometer, Palette Series PR 101 α, ATAGO CO., LTD, Japan)

2.3 Statistical Analysis of the Data

As many as 32 tree and fruit variables observed from 103 numbers of *B. dulcis* in West Java were clustered to find out the groups based on the similarity characters using the MINITAB program version 14. The variables chosen were those that are not influenced by the age of the tree observed. Those variables were bark color and texture, maximum and minimum of leaf length and width, ratio of maximum leaf length and width, maximum and minimum of fruit length and width, ratio of maximum fruit length and width, fruit weight, peel color, easiness to peel the fruit, maximum and minimum of thickness of peel, peel weight, pulp color, maximum and minimum number of fruit segment, soluble solid content of pulp, maximum and minimum of seed length and width, ratio maximum seed length and width, seed weight, percentage of peel, pulp and seed weight per fruit and seed color. The qualitative parameters and detailed rank of qualitative number and information of the observation of *B. dulcis* trees, fruits and seeds for clustering are shown in Table 2.

The same variables were also tested by Linier Discriminant Analysis using the MINITAB program version 14 to find distinctive characteristic of each group. The correlations between the tree and fruit variables and its habitat as well as among the trees and fruits variables were tested using the statistical MINITAB program version 14.

3. Results and Discussion

3.1 Distribution, Characterization and Variation of Baccaurea dulcis in West Java

From the study of herbaria, it was known that *Baccaurea dulcis* is distributed in Sumatra, Borneo and Western part of Java. In Sumatra, the species is distributed in Palembang, Lampung, Riau, Payakumbu, Bangka, Siberut and Jambi, whereas in Borneo, the species was spread in Ketapang, Gunung Palung National Park, Sarawak, and West Samarinda. In West Java, the species was found in Batutulis and Kotabatu of Bogor. Most of the herbaria observed were collected by Dutch explorers long time ago, before or in the early to mid 1900's. Therefore, the distribution of the species at present including in Sumatra and Borneo may have changed. Indeed, increasing human population and land-use intensification resulted in the loss habitats and increasing species extinction rates. Several strong climate oscillation and disaster could affect vegetation shape and species distribution (Ounsavi & Sokpon, 2010). Whereas in West Java at the moment, there is no *B. dulcis* plant found in Batutulis and Kotabatu anymore.

From the field trips and observation in West Java, it is known that besides as collection in the botanical garden, the distribution of *B. dulcis* in West Java was only in sub districts Taman Sari of Bogor district and Cijeruk of Sukabumi district. The result showed that the occurrence of the species in West Java is only in very restricted areas. According to people in the local areas, the species used to abundant in the past, including in other villages of Bogor district. However, people usually cut the trees and used the trunk for many purposes such as material for building/house and equipments. On the other hand, the species is rarely planted. The plants usually grew from seeds that drop at the surrounding of the mother plants. The risk of extinction of the plant species is high because there were not many people interested in the fruits. Therefore, conservation and development of species to become more commercialized are needed.

The results of the characterization of trees and fruits of *B. dulcis* and plant habitat can be seen in Table 3. The plants were only found in the home-yard, small garden or botanical garden at an altitude range of 250 m - 610 m above sea level, this means that the species could grow well and be planted in relatively low lying areas. From Table 3, it is also known that the trees grow well in the tropical region at neutral soil pH of regosol or latosol soil type and smooth until rather coarse soil texture at land slope from 0% until 45%. From the study of herbaria, it is also known that the species grew well in gully river bank, in hillside of primary forest, swampy places, riverside and also cultivated at "kampoeng" or remote village. The plants could grow at the location until 1100 m above sea level at sandyloam soil and swampy places.

From the observation and measurement results, the morphological characters of *B. dulcis* were various (Table 3).

Those included nature of tree, color and texture of the bark, size and shape of leaves and fruits, color of fruit peel and pulp and the size, shape and color of seeds. It is clear from the study that there is considerable phenotypic variation in almost every parameter observed and measured. Similar to another study on the variation of fruit of *Irvingia gabonensis*, an indigenous fruit tree of west and central Africa, there were significant variation in fruit, nut and kernel size and weight (Leakey et al., 2000). Differences were also identified in shell weight and brittleness, fruit taste, fibrousity and pulp color (Leakey et al., 2000). Salisbury (1942) mentioned that seed size varies tremendously among plant species and was investigated early as a life-history trait of obvious importance. Variation in seed size and weight of *Desmodium paniculatum* (Leguminosae) was also reported in a population in two locations in North Carolina, USA, which caused by environmental conditions and nutrient supply (Wulff, 1986). The result of other study on population of a single seeded fruit *Ocota tenera* (Lauraceae) from Monteverde, Costa Rica showed that the fruits that vary from 1.4 to 2.4 cm and much variation occurred within individual trees (Wheelwright, 1993). The relative size of fruits produced by different trees remained generally constant over an 11-year period despite slight differences between years in the average size of fruits produced by a given tree (Wheelwright, 1993)

Table 3. Character and variation of *B.dulcis* trees and fruits in West Java and the plant habitat

No	Data Recorded	Measurement/Description
Habitat		
1	Location of the plant	Homeyard, small people's garden, Botanical garden
2	Altitude	320 - 610 m above sea level
3	Longitude and	06° 34'49.7"- 06° 40' 06.9" and
	Latitude	106° 43' 55,9"- 106° 49' 25,0"
4	Air temperature	18 °C - 33 °C
5	Relative humidity	59 % - 100 %
6	Slope	0 % - 45 %
7	Light intensity	341 - 299,000 lux
8	Soil pH	5.8 - 7
9	Soil relative humidity	17 % - 90 %
10	Soil type	Brown regosol and red-brown latosol
11	Soil texture	Smooth-rather coarse
Morphology of Tree and Fruit Measurement/Description		
12	Tree height	4.5 - 21 m
13	Trunk diameter	10.5 cm - 80.5 cm
14	Canopy width	3.2 m - 13.2 m
15	Canopy condition	Sparse - Medium - Dense
16	Lowest branch height	0.55 m - 7.0 m
17	Bark color	Light brown/cream, Medium brown, Grey-medium brown, Grey, Browned-grey, Grey, yellowed-medium brown, Grey-dark brown, Dark brown, Browned-black
18	Bark texture	Smooth, Flaky-smooth, Lenticelate, Flaky-lenticelate, Lump-flaky, Flaky, Flaky-fissured, Flaky-rectangular
19	Leaves shape	Obovate, lanceolate
20	leaf length	7.7 cm - 30 cm
21	Leaf width	3.1 cm - 14.9 cm
22	Ratio maximum leaf length/maximum leaf width	2.01

23	Fruit shape	Rounded, slightly oval, or truncate at one end
24	Fruit length	2.4 cm - 4.6 cm
25	Maximum fruit width	4.8 cm
26	Ratio of fruit length/ fruit width	2.2 cm
27	Length of main fruits stalk	0.65 cm - 10.67 cm
28	Length of branch fruit stalk	2.02 mm - 14.7 mm
29	Fruit weight	7.0 - 38.33 g
30	Fruit peel color	Yellow, Light orange/Yellowish orange, Orange/dark orange, Reddish yellowish-orange
31	Easiness to peel the fruit	Easy-difficult
32	Fruit peel thickness	1.65 mm - 9.72 mm
33	Peel weight per fruit	4.17 gr - 23.67 gr
34	Pulp weight	1.56 mg - 15.26 mg
35	Pulp color	White, Transparent-white, Cream, Pink, transparent-cream, Pink, white-transparent, Cream-transparent, Pink-transparent, Pinked/purpled lines-cream/white
36	Fruit pulp segment	1 - 6
37	Soluble solid content of pulp	11 Brix - 20 Brix
38	Seed shape	Ovate, thin, 1 - 6 curves
39	Seed length	0.75 cm - 2.31 cm
40	Seed width	0.52 - 1.95 cm
41	Seed thickness	1 - 4 mm
42	Ratio of longest seed length/ seed width	1.18
43	Seed weight	0.12 mg - 0.74 mg
44	Percentage of peel weight per fruit	36.64 % - 78.31 %
45	Percentage of pulp weight per fruit	26 % - 63 %
46	Percentage of seed weight per fruit	1.17 % - 2.96 %
47	Seed color	Light brown, Pink-light brown, Brown, Pink-brown, Purplish-brown, Light brown-Pink, Pink
48	Fruit production per tree	5 kg - 200 kg

3.2 Clustering of Baccaurea dulcis Variation in West Java

Using 32 variables of trunk, leaf, fruit and seed, all samples could be clustered into 6 groups with the proportion correct 0.903 (Figure 4, Table 5). As can be seen at Table 4, the characters that belong to group 1 were small and light fruit, light fruit peel, small seed and sweeter fruit taste, whereas those of group 2 were fruit pulp color white/transparent, small seed, low portion of fruit pulp and high portion of fruit peel. Group 3 belongs to the tree and fruit of *B. dulcis* with the characters of big and heavier fruit, high portion of fruit pulp, more heavy seed, more segment of fruit pulp (Table 4). The characters of group 4 were thin and light fruit peel, more segment of fruit pulp, high portion of fruit pulp, low portion of fruit peel, color of fruit pulp reddish/purple. Group 5 were characterized by big and more heavy fruit, thick and more heavy fruit peel, fruit pulp color white/transparent, big seed and high portion of fruit pulp, whereas group 6 were characterized by light fruit, sour taste, small seed, color of fruit pulp reddish or purple (Table 4). Another study on the variation of pomelo (*Citrus grandis*) in Nepal found that from the multivariate analysis of the data produced five discrete groups, which differed significantly in fruit shape and size, pulp, juice, total soluble solids and acid content, seed number, leaf shape and size (Paudyal & Haq, 2008).

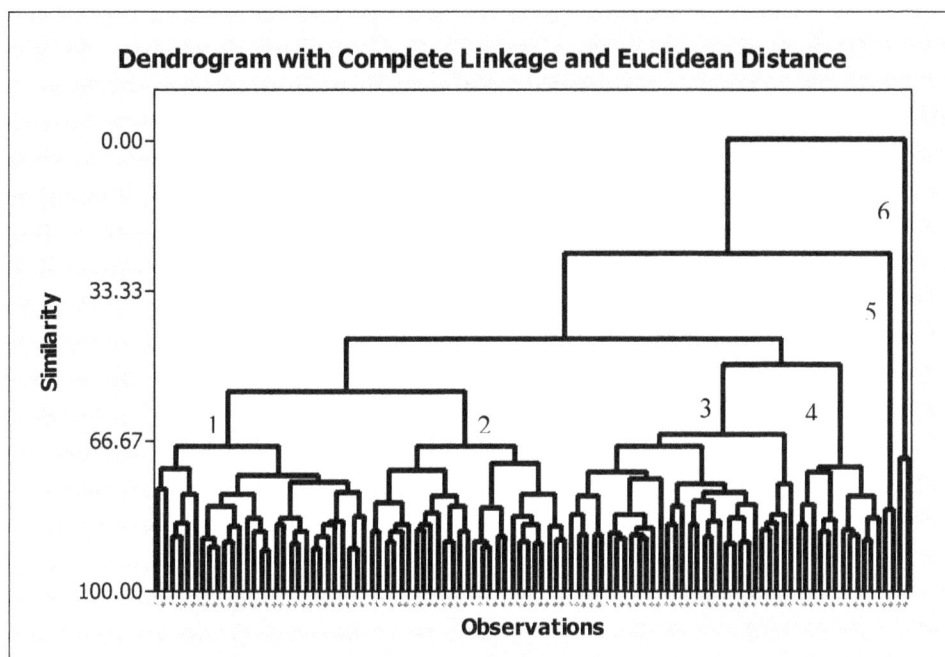

Figure 3. Dendrogram of *B. dulcis* in West Java based on the similarity variables

Table 4. Average value of characteristic component of 6 groups *B. dulcis* in West Java

Characteristic variable	Group 1	Group 2	Group 3	Group 4	Group 5	Group 6	Population mean
Bark color	2.90	3.09	3.73	1.60	4.50	4.00	3.32
Bark texture	4.79	4.74	5.31	5.60	5.25	5.00	5.01
Leaf length	21.75	23.09	22.78	21.58	23.46	20.23	22.49
Leaf width	10.68	10.68	10.65	9.80	10.90	9.65	10.62
Fruit size	35.94	37.12	39.60	38.26	41.87	36.80	38.01
Fruit weight	18.44	21.18	26.44	22.95	31.96	20.01	23.03
Peel color	2.39	2.82	2.55	2.40	2.75	2.75	2.60
Peel thickness	6.50	6.94	6.27	5.29	7.87	6.39	6.65
Peel percentage	58.25	65.92	54.39	46.81	62.23	59.9	59.26
Pulp percentage	39.84	32.33	43.85	51.45	36.27	38.20	38.97
Minimum pulp segment	1.7	1.7	2.0	2.4	1.8	1.8	1.8
Maximum pulp segment	3.5	3.4	3.7	3.6	3.3	3.3	3.5
Pulp color	3.03	2.96	3.86	3.80	2.75	4.25	3.27
SSC*	16.51	15.55	15.82	15.72	15.69	14.69	15.88
Seed length	16.9	17.0	18.0	18.1	18.6	17.0	17.5
Seed width	12.6	13.0	14.1	13.9	14.1	13.6	13.4

*SCC = Soluble solid content of fruit pulp.

Table 5. Results of test of correctness based on discriminant analysis for summary of classification

Put into Group	True Group					
	1	2	3	4	5	6
1	28	2	2	0	0	0
2	0	23	0	0	0	1
3	0	0	23	0	1	0
4	0	0	1	5	0	0
5	0	0	0	0	11	0
6	1	2	0	0	0	3
Total N	29	27	26	5	12	4
N correct	28	23	23	5	11	3
Proportion	0.966	0.852	0.885	1.000	0.917	0.750

Note: N = 103, N Correct = 93 and Proportion Correct = 0.903.

3.3 Correlations Among Habitat, Plant and Fruit Variables

The result of the statistical data analysis for significant correlation between plant habitat and variables of tree and fruit could be seen in Table 6. There was one significant positive correlation and nine significant negative correlations known. The positive correlation was between slope and fruit pulp color, whereas significant negative correlation were between altitude and pulp weight; light intensity and pulp soluble solid content (SSC); soil pH and number of pulp segment; soil texture and trunk diameter, canopy width, pulp color, pulp weight, peel weight portion; drainage and number of pulp segment (Table 6).

This findings could indicate that higher slope the plant grow, the pulp color of the fruit tend to be pink or purple. On the other hand, higher the altitude of the plant position could correlate with less weight of the fruit pulp; more light intensity could correlate with less pulp SSC. Light intensity could have an effect on the pulp color and pulp SSC, while air temperature could affect the weight of fruit pulp.

The other significant negative correlation indicated that higher soil pH correlates with less pulp segment; more coarse the soil texture correlates with less trunk diameter, less canopy width, more white/transparent pulp color, less pulp weight, less peel portion; higher soil drainage correlates with less number of pulp segment. The soil texture and soil drainage could correlate with the soil fertility condition, which then could affect the size of the plant. Other study on pomelo (*Citrus grandis*) in Nepal found that yield related characters, such as fruit weight had positively correlated with tree size and soil fertility level, but none of these factors were correlated with fruit quality, such as percent of pulp and pulp SSC (Paudyal & Haq, 2008)

Table 6. Significant correlation between plant habitat and variables of *Baccaurea dulcis*

Plant/Habitat	Altitude	Slope	Light intensity	Soil pH	Soil Texture	Drainage
Trunk Diameter	-0.094	0.077	0.073	0.123	-0.199*	-0.098
Canopy width	0.011	0.029	-0.070	-0.081	-0.200*	-0.024
Pulp color	-0.213	0.226*	-0.056	-0.166	-0.205*	-0.137
Minimum number of pulp segment	-0.047	0.070	-0.133	-0.223*	0.180	0.006
Maximum number of pulp segment	0.120	-0.168	0.079	0.110	-0.004	-0.194*
Pulp soluble solid content	0.106	0.135	-0.208*	0.047	0.166	0.086
Pulp weight	-0.246*	0.007	0.058	-0.006	-0.261*	-0.154
Peel weight portion	-0.155	0.016	-0.071	0.093	-0.294*	-0.063

* Significant correlation.

It could be seen at Table 7 that there was significant correlation among the variable of *B. dulcis* trees and fruits in West Java. Significant positive correlations were found between (1) Canopy diameter and trunk width; (2) Maximum fruit width and fruit weight, pulp thickness, peel color, peel thickness, pulp soluble solid content (SSC), seed length, seed width, seed weight; (3) Maximum fruit length and seed weight per fruit; (4) Fruit weight and peel color, peel thickness, pulp SSC, seed length, seed width, seed weight; (5) Pulp thickness and peel weight, pulp SSC, seed width, seed weight; (6) Easiness to peel and peel color; (7) Peel color and peel thickness; (8) Peel thickness and pulp SSC, seed length, seed width, seed weight, pulp portion; (9) Pulp color and seed weight, pulp weight; (10) Pulp SSC and seed width; (11) Maximum seed length and maximum seed width, seed weight; (12) Maximum seed width and seed weight.

The results of positive significant correlation among the variable of *B. dulcis* trees and fruits in West Java (Table 7) indicate that the trunk width is in accordance with canopy diameter; the fruit size is in accordance with size of seed, peel and pulp; more difficult to peel the fruit correlates with more red/orange peel color and more thick the peel.

On the other hand, the significant negative correlation among the variable of *B. dulcis* trees and fruits in West Java (Table 7) were between (1) Peel weight and tree height, trunk diameter, peel color, pulp color; (2) Pulp thickness and easiness to peel, peel portion; (3) Easiness to peel and pulp color, maximum pulp segment, pulp SSC, maximum seed length, seed weight per fruit, pulp weight per fruit; (4) Peel color and peel weight, pulp weight per fruit, peel portion; (5) Peel thickness and pulp weight, peel portion; (6) Pulp color and pulp SSC; (7) Maximum pulp segment and easiness to peel, pulp SSC, maximum seed width; (8) Seed weight and pulp weight per fruit; (9) Pulp weight per fruit and leaf length, leaf width, easiness to peel, peel color, peel thickness, pulp portion; (10) Peel portion and leaf length, leaf width, fruit length, fruit width, fruit weight, pulp thickness, peel color, peel thickness, seed weight per fruit.

The result of significant negative correlation among the variable of *B. dulcis* trees and fruits in West Java (Table 7) indicates the composition of fruit parts is in accordance; more weight of fruit peel correlate with smaller tree and more dull peel and pulp color; more thick the pulp, more difficult to peel the fruit; more red/purple pulp color less sweet of the pulp; more weight of pulp and peel portion correlates with less size of leaf. Other study regarding fruit variation of *Irvingia gabonensis*, an indigenous fruit tree of West and Central Africa found that there were very weak relationship between fruit size and weight with nut and kernel size and weight (Leakey et al., 2000).

Table 7. Significant correlation among the variables of *Baccaurea dulcis* trees and fruits from West Java

CODE	X1	X2	X4	X5	X6	X7	X8	X9	X10	X11	X12	X13	X14	X15	X16	X17	X18	X19	X20	X21
X3	0.54	**0.67***	0.135	0.149	0.082	-0.146	-0.042	-0.077	0.091	-0.082	0.049	**-0.24***	-0.094	-0.135	0.089	0.108	-0.106	0.009	-0.032	0.067
X9	0.01	-0.03	-0.18	0.18	0.08	0.72	**0.75***	1	0.243	**-0.19***	0.079	**0.22***	**0.90***	0.068	0.103	0.046	-0.109	**0.41***	**0.52***	**0.60***
X10	0.05	0.14	0.13	0.04	-0.07	0.24	**0.22***	0.24	1	**-0.30***	0.08	**0.23***	0.01	-0.01	-0.10	**0.32***	0.11	**0.11***	**0.18***	0.01
X11	-0.18	-0.21	-0.08	0.08	0.10	-0.13	-0.16	-0.19	**-0.30***	1	**0.20***	-0.09	0.11	**-0.21***	**-0.28***	**-0.19***	**-0.23***	-0.06	**-0.27***	**-0.20***
X12	-0.08	-0.14	-0.13	0.08	0.18	0.32	**0.31***	**0.22***	0.08	**0.20***	1	**0.39***	**-0.18***	-0.11	-0.03	0.07	0.07	0.06	-0.07	**-0.39***
X13	-0.01	-0.02	**-0.17***	0.25	0.18	0.69	**0.71***	**0.90***	**0.23***	-0.09	**0.39***	1	0.01	-0.03	-0.11	**0.33***	**0.41***	**0.48***	**0.48***	**-0.36***
X14	**-0.19***	**-0.22***	-0.09	-0.07	-0.10	0.08	-0.06	0.07	0.01	0.11	**-0.18***	0.01	1	**-0.23***	-0.15	-0.06	0.08	0.10	0.12	0.14
X15	0.06	0.02	0.06	0.11	0.10	-0.02	0.12	0.10	-0.01	**-0.21***	-0.11	-0.03	**-0.23***	1	0.02	0.02	0.15	-0.02	**0.25***	**0.28***
X16	-0.12	0.07	0.09	-0.09	-0.04	-0.04	-0.02	-0.11	-0.10	**-0.28***	-0.03	-0.11	-0.15	0.02	1	**-0.22***	-0.09	**-0.18***	-0.07	0.03
X17	0.18	0.13	-0.10	-0.03	-0.09	0.47	**0.33***	**0.41***	**0.32***	**-0.19***	0.07	**0.33***	-0.06	0.02	**-0.22***	1	0.11	**0.29***	0.38	0.10
X18	0.00	-0.09	-0.06	0.08	0.10	0.43	**0.61***	**0.52***	0.11	**-0.23***	0.07	**0.41***	0.08	0.15	-0.09	0.11	1	**0.53***	**0.49***	0.13
X19	0.07	-0.06	-0.07	0.00	-0.07	0.40	**0.55***	**0.60***	**0.11***	-0.06	0.06	**0.48***	0.10	-0.02	**-0.18***	**0.29***	**0.53***	1	0.13	-0.14
X20	0.03	-0.02	-0.13	0.04	-0.07	**0.54***	**0.55***	**0.82***	**0.18***	**-0.27***	-0.07	**0.48***	0.12	**0.25***	-0.07	0.38	**0.49***	**0.54***	1	0.62
X21	0.02	-0.01	0.01	**-0.18***	**-0.24***	-0.02	-0.04	0.08	0.01	**-0.20***	**-0.39***	**-0.36***	0.14	**0.28***	0.03	0.10	0.13	0.13	0.62	1
X22	-0.02	0.01	-0.02	0.19	0.24	0.04	0.06	-0.05	-0.01	0.19	0.40	**0.39***	-0.14	**-0.27***	-0.03	-0.09	-0.13	-0.14	-0.60	**-1.00***
X23	0.10	0.00	0.24	**-0.21***	**-0.13***	**-0.46***	**-0.33***	**-0.56***	**-0.12***	0.16	**-0.20***	**-0.55***	0.00	-0.14	-0.06	-0.19	-0.09	0.31	**-0.41***	0.03

Note: * Significant correlation.

X1 = Tree height, X2 = Trunk diameter, X3 = Canopy width, X4 = Lowest branch height, X5 = Maximum leaf length, X6 = Maximum leaf width, X7 = Maximum fruit length, X8 = Maximum fruit width, X9 = Fruit weight, X10 = Pulp thickness, X11 = Easiness to peel, X12 = Peel color, X13 = Peel thickness, X14 = Peel weight, X15 = Pulp color, X16 = Maximum pulp segment, X17 = Pulp SSC, X18 = Maximum seed length, X19 = Maximum seed width, X20 = Seed weight per fruit, X21 = Pulp weight per fruit, X22 = Pulp portion, X23 = Peel portion

4. Conclusion

This paper explored the distribution and variation of *Baccaurea dulcis* (Jack) Müll. Arg. in West Java, Indonesia. The population of the species is under threath. The species has restricted distribution, however the morphological characters were various. Therefore, its conservation is needed. Further study on the influence of genetic and environment aspects to the variation of the species is also important. Moreover, other interesting topic should be gaining high quality of fruits to become more commercialized one.

Acknowledgements

The presented results are a product of the scientific project carried out with the support of the competitive grant of the Indonesian Institutes of Sciences number 11.9/SK/KPPI/DKP/2008. The author would like to thank Junaedi and Ngatari for technical support.

References

Haegens, R. (2000). Taxonomy, phylogeny, and biogeography of *Baccaurea, Distichirhops*, and *Nothobaccaurea* (Euphorbiaceae). *Blumea (Supplement), 12*, 1-217.

Heyne, K. (1987). Tumbuhan berguna Indonesia, Jilid II (Translation Edition). *Yayasan Sarana Wana Jaya* (p. 1247). Jakarta, Indonesia.

Leakey, R. R. B., Fondoun, J. M., Atangana, A., & Tchoundjeu, Z. (2000). Quantitative descriptors of variation in the fruits and seeds of *Irvingia gabonensis*. *Agrobiodiversity Systems, 50*, 47-58. http://dx.doi.org/10.1023/A:1006434902691

Lestari, R. (2009). Propagation of *Baccaurea dulcis* by airlayering with the addition of plant regulator substance. *Proceeding of National Seminar on Flora Conservation to Overcome the Effect of Global Warming* (pp. 227-231). "Eka Karya" Bali Botanical Gardens, Indonesian Institute of Sciences, Bali, Indonesia.

Lestari, R., & Aprilianti, P. (2010). Propagation of "big menteng" (*Baccaurea dulcis* (Jack) Müll Arg.) by grafting as the efforts of species conservation and development. *Proceeding of National Seminar Horticulture on "Research Reorientation for Optimalize Horticultural Product and Value Chain"* (pp. 68-75). PERHORTI, Indonesia.

Munawaroh, E. (2001). The potency of *Baccaurea* spp. as fresh fruit source and its conservation in the Bogor Botanical Gardens. *A Day Seminar Proceeding at the National Flora Fauna Day* (pp. 81-88). Center for Plant Conservation Bogor Botanical Gardens, Indonesian Institute of Sciences, Bogor.

Ouinsavi, C., & Sokpon, N. (2010). Morphological variation and ecological structure of Iroko (*Milicia excelsa* Welw. C. C. Berg) population accross different biogeographical zones in Benin. *International Journal of Forestry Research, 1*, 1-11. http://dx.doi.org/10.1155/2010/658396

Paudyal, K. P., & Haq, N. (2008). Variation of pomelo (*Citrus grandis* (L.) Osbeck) in Nepal and participatory selection of strains for further improvement. *Agroforest Syst., 72*, 195-204. http://dx.doi.org/10.1007/s10457-007-9088-z

Salisbury, E. J. (1942). The reproductive capacity of plants. *G. Bell and Sons*. London.

Subekti, A., Yeni, S., Sumaryadi, T., Anggraito, B., & Ibrahim, T. M. (2005). The examine of domestication and commercialization of fruit species and variety in West Kalimantan. *Workshop Proceeding I, Domestication and Commercialization of Horticultural Plant* (pp. 23-34). Center for Research and Development of Horticultura, Ministry of Agricultura, Jakarta.

Uji, T. (1992). *Baccaurea* Lour. In E. W. M. Verheij, & R. E. Cornel (Eds.), *Plant Resources of South East Asia No.2. Edible fruits and nuts* (pp. 98-100). Leiden: Backhuys Publishers.

Wheelwright, N. T. (1993). Fruit size in a tropical tree species: variation, preference by birds and heritability. *Vegetatio, 107-108*, 163-174

Wulff, R. D. (1986). Seed size variation in *Desmodium paniculatum*, I. Factors affecting seed size. *Journal of

Ecology, 74, 87-97. http://dx.doi.org/10.2307/2260351

Wurdack, K. J., Hoffman, P., Samue, R., de Bruijn, A., van der Bank, M., & Chase, M. W. (2004). Molecular phylogenetic analysis of Phyllanthaceae (Phyllanthoideae pro parte, Euphorbiaceae sensu lato) using plastid rbcl DNA sequences. *American Journal of Botany, 91*(11), 1882-1900. http://dx.doi.org/10.3732/ajb.91.11.1882

Supplemental Tables

Supplemental Table 1. Linear Discriminant Function for Groups

Variables	Groups					
	1	2	3	4	5	6
Constant	-1.919	-2.066	-3.039	-8.559	-7.571	-6.580
C42	-0.701	-0.243	0.497	-0.123	1.011	0.612
C43	0.138	-0.279	0.046	1.101	-0.265	0.003
C44	-0.261	1.509	-0.645	-3.309	-0.209	0.669
C45	0.716	-3.269	1.334	**10.999**	1.320	**-9.505**
C46	-0.278	-0.887	0.650	1.976	0.651	-0.648
C47	-0.952	3.241	-0.800	**-11.732**	-1.707	**10.016**
C48	-1.293	2.705	-0.626	**-8.412**	-0.576	**7.426**
C49	-1.914	0.208	-0.190	2.227	3.198	1.330
C50	**-4.671**	**-3.609**	**6.154**	**8.508**	**4.450**	**-5.760**
C51	0.394	-1.632	2.082	0.206	-1.029	-2.545
C52	3.520	3.502	-5.073	-7.086	-4.258	5.448
C53	3.893	2.359	-3.890	-7.533	-4.531	4.152
C54	**-7.480**	**-4.515**	**18.165**	**23.060**	**-10.383**	**-31.038**
C55	-0.480	0.501	-0.218	-0.889	0.629	0.742
C56	0.151	0.421	-0.730	0.411	0.044	0.167
C57	0.271	-0.621	0.065	0.541	0.443	-0.203
C58	-0.446	0.468	-0.049	-1.910	0.835	0.274
C59	**5.645**	**6.563**	**-21.309**	**-26.441**	**16.976**	**35.403**
C60	-0.337	-0.352	0.742	1.158	-0.446	-0.112
C61	-0.100	0.609	-0.255	-0.398	-0.581	0.513
C62	0.223	-0.062	0.091	-0.813	0.007	-0.792
C63	0.362	-0.779	0.318	0.320	0.405	-1.047
C64	-0.398	-0.417	0.057	0.212	1.121	1.701
C65	**-0.587**	**1.870**	**-3.977**	**-3.173**	**5.852**	**3.893**
C66	-0.317	0.176	-0.681	-2.298	2.935	-0.393
C67	0.442	-1.600	2.808	3.369	-4.489	-1.398
C68	0.899	-1.967	4.179	3.417	-6.527	-5.095
C69	1.965	-1.866	1.869	-0.756	-3.015	-3.816
C70	40.719	-21.741	12.235	1.751	-60.747	-47.938
C71	38.601	-21.998	18.918	8.233	-67.798	-61.238
C73	-0.658	0.215	0.307	0.604	0.063	0.383

17

Optimum Fermentation Condition of Soybean Curd Residue and Rice Bran by *Preussia aemulans* using Solid-State Fermentation Method

Yiting Li[1], Shili Meng[2], Linbo Wang[2] & Zhenya Zhang[2]

[1] Key Laboratory of Food Nutrition and Safety, Tianjin University of Science and Technology, Ministry of Education, Tianjin 300457, China

[2] Graduate School of Life and Environmental Sciences, University of Tsukuba, Ibaraki 305-8577, Japan

Correspondence: Yiting Li, Key Laboratory of Food Nutrition and Safety, Tianjin University of Science and Technology, Ministry of Education, Tianjin 300457,China.E-mail: liyitingyoyo@hotmail.com

Abstract

An environmental method for using soybean curd residue (SCR) and rice bran (RB) was developed in this study. SCR and RB were utilized as growth medium for *Preussia aemulans,* a new fungus isolated from *Cordyceps sinensis* fruiting body. According to Orthogonal test and Duncan's multiple range test, the optimum fermentation condition of fermented SCR and RB for producing polysaccharide, adenosine and ergosterol were summarized. Under the optimum fermentation condition of SCR, the polysaccharide, adenosine and ergosterol contents were reached to 39.18 ± 1.06 mg/g dry matter, 127.94 ± 1.82 mg/100g dry matter and 37.53 ± 0.11 mg/100g dry matter, respectively. And under the optimum fermentation condition of RB, the content of polysaccharide, adenosine and ergosterol were also enhanced 3-fold, 10-fold and 10-fold, respectively. Therefore, the fermented SCR and RB could be utilized as nutritious functional food or food additives in the future.

Keywoeds: *Cordyceps sinensis*, *Preussia aemulans*, soybean curd residue, rice bran, solid-state fermentation

1. Introduction

In recent years, due to the serious economic and environmental concerns, the utilization of food by-products is unprecedentedly expected to increase and become more efficient. Thus, reduce, reuse, and recycle (3R) of by-products is getting more and more important in food industries (Wang & Nishino, 2008).

Soybean curd residue (SCR), is produced from the tofu industry in China and Japan. It was once consumed as a traditional food, but modernization and urbanization in the lifestyle has reduced its status to that of a mere industrial waste, which is now mainly incinerated like other industrial wastes (Ohno, Ano, & Shoda, 1996; O'Toole, 1999). The main disadvantage of SCR is natural spoilage when storage is not under refrigeration. In Japan, 0.7 million tons of SCR is disposed annually, mostly by incineration which has caused severe environmental pollution (Mizumoto, Hirai, & Shoda, 2006). In fact, SCR is rich in carbohydrate, protein and many other nutrients, suggesting that it is a potential source of low cost medium for the growth of mycelia. Many researchers have investigated the possibility of bioconversion of the residues by submerged and solid-state cultivation (Yokoi, Maki, Hirose, & Hayashi, 2002; Shi, Yang, Li, Wang, & Zhang, 2011).

Rice bran (RB), which includes the pericarp, the aleurone and sub-aleurone layers, parts of the germ and the embryo as well as small portions of the starchy endosperm (Jiamyangyuen, Srijesdaruk, & Harper, 2005), is a valuable milling by-product. However, RB contains enzyme lipase, which rapidly degrades the oil making the bran rancid and inedible, therefore 63 to 76 million tons of rice bran is produced in the world and more than 90% of RB is sold cheaply as animal feed each year. Actually, RB contains 12-22% oil, 11-17% protein, 6-14% fiber, 10-15% moisture and 8-17% ash and also rich in vitamins, minerals, amino acids and essential fatty acids (Hernandez, Rodriguez, Gonzalez, & Lopez, 2000; Jiang & Wang, 2005; Piironen, Lindsay, Miettinen, Toivo, & Lampi, 2000). RB is highly nutritious and hence used as a food additive (Nagendra, Sanjay, Shravya, Vismaya, & Nanjunda, 2011).

Cordyceps sinensis (Berk.) Sacc. is a parasitic fungus and has long been used to treat multitude of ailments, promote longevity, increase athletic power and improve quality of life. The physiological activators of *Cordyceps sinensis* have been detected, including adenosine, cordycepin, cordycepic acid, d-mannitol, polysaccharides, vitamins and trace elements, etc. (Kumara et al., 2011). Polysaccharide exhibited antioxidative and antitumor activities, and regulating immune functions (Paterson, 2008). Adenosine is an endogenous purine nucleoside that modulates many physiological processes, and it has been used as a marker for the quality control of *C. sinensis* in Chinese Pharmacopoeia (Li, Yang, & Tsim, 2006). Ergosterol have multiple pharmacological activities, such as cytotoxic activity (Bok, Lermer, Chilton, Klingeman, & Towers, 1999) and antiviral activity (Lindequist, Lesnau, Teuscher, & Pilgrim, 1989). Furthermore, according to previous researches, 572 species fungi (*Preussia intermedia, Penicillium boreae* etc.) were isolated from different parts (stromata, sclerotia, and external mycelial cortices) of natural *Cordyceps sinensis* fruiting body, and all of the isolated fungus had the similar metabolites and exhibited the similar pharmacological activities as *Cordyceps sinensis* (Zhang et al., 2010).

For reuse of SCR and RB, reduce of waste and recycle of organic material, according to the nutrition profile of SCR and RB, it could be considered as a growth medium for fungi. In this study, the SCR and RB were used as a culture medium for *Preussia aemulans* (*P. aemulans*) which was isolated from *Cordyceps sinensis* fruiting body. The objective of this research was to find out the optimum fermentation condition, which maximize quantity of polysaccharide, ergosterol and adenosine using SCR and RB, respectively. Fermented SCR and RB were detected to produce potential functional animal feed to substitute the antibiotic added to the feed, and improve the safety of food.

2. Materials and Methods

2.1 Chemicals and Reagents

D-glucose, sucrose, peptone, KH2PO4, MgSO4, Na2CO3, NaOH, potato extract, yeast extract, agar, ethanol, sulfuric acid, phenol were obtained from Wako Pure Chemical Industries, Ltd, Osaka, Japan. adenosine and ergosterol were purchased from Sigma Aldrich, Inc. (Saint Louis, MO, USA). All other chemical reagents were of analytical grade.

2.2 Isolation and Cultivation of P. aemulans

The fruiting body of *Cordyceps sinensis* was purchased from Qinghai, China, and the isolated *P. aemulans* mycelium (SIID11759-01) was identified by Techno Suruga laboratory co., ltd, Japan. The stroma of *Cordyceps sinensis* fruiting body was sterilized with ethanol three times, air-dried, cut into small segment and transferred to slant tube fermentor to incubate for 7 days, at room temperature. The white mycelium appeared on the surface during slant fermentation. Then, mycelium was transferred to agar medium, which contained (per liter): 20 g of sucrose, 10 g of peptone, 20 g of agar powder, 1.5 g of $MgSO_4$, 3 g of KH_2PO_4. After 7 days of the culture, when white mycelium appeared on the surface of the medium, the mycelium was transferred into the liquid medium, which was containing (per liter): 20 g of sucrose, 10 g of peptone, 4 g of potato powder, 1.5 g of $MgSO_4$, 3 g of KH_2PO_4. The *Cordyceps sinensis* mycelium was incubated in a 200 mL of flask with 100 mL of PDA liquid medium, and the mixture was stationary cultured for 7 days. After the stationary culture, the *P. aemulans* mycelium was inoculated to SCR and RB followed by the orthogonal test design.

2.3 Orthogonal Test Design

SCR was obtained from the inamoto toufu factory in tsukuba, Japan. The carbon nitrogen ratio, moisture content, and pH value of SCR were 10.8, 80% and 5.5, respectively. According to the initial conditions, the fermentation conditions were designed to investigate the optimum condition for yield of polysaccharide, ergosterol and adenosine by solid-state fermentation. The carbon resources, nitrogen resources (3% w/w), adding dosage of carbon sources and the fermentation time, were regarded as correlated factors of culture condition. The optimum fermentation condition was obtained by an orthogonal layout $L_9(3^4)$ in a 200 mL flask with 20 g of SCR. The level of factor is shown in Table 1. The inoculum size of *P. aemulans* mycelium (liquid medium) was 20 % (v/w). After fermentation, the fermented SCR mycelia mixture was dried and grounded into powder for further experiment.

RB was collected from Automatic rice-polishing machine in tsukuba, Japan. The carbon nitrogen ratio and moisture content of RB were 12 and 10%, respectively. According to the initial conditions, the fermentation conditions were designed to investigate the optimum condition for yield of polysaccharide, ergosterol and adenosine by solid-state fermentation. The carbon resources, nitrogen resources (3% w/w), adding dosage of carbon and nitrogen sources, moisture content and the fermentation time, were regarded as correlated factors of culture condition. The optimum fermentation condition was obtained by an orthogonal layout $L_{16}(4^5)$ in a 500 mL flask with 20 g various moisture content of RB. The level of factor is shown in Table 3. The inoculum size of *P. aemulans* mycelium (liquid medium) was 20 % (v/w). After fermentation, the fermented RB mycelia mixture was dried and grounded into powder for further experiment.

Table 1. L$_9$ (3^4) orthogonal layout and results of fermented SCR by *P. aemulans*

Experimental group	CS	NS (3% w/w)	ADCS (% w/w)	FT (day)	Polysaccharide content (mg/g dry matter)	Ergosterol content (mg/100g dry matter)	Adenosine Content (mg/100g dry matter)
1	Glucose	Peptone	5%	10	22.74 ± 0.66	11.90 ± 1.22	40.95 ± 1.62
2	Glucose	Beef extract	10%	15	27.25 ± 1.77	35.65 ± 2.76	99.62 ± 1.89
3	Glucose	Yeast extract	15%	20	21.43 ± 1.50	11.99 ± 1.89	101.99 ± 1.74
4	Sucrose	Peptone	10%	20	19.21 ± 1.08	10.60 ± 0.31	45.32 ± 1.88
5	Sucrose	Beef extract	15%	10	31.43 ± 0.37	26.19 ± 1.39	69.92 ± 1.43
6	Sucrose	Yeast extract	5%	15	25.06 ± 1.60	13.64 ± 1.08	117.96 ± 1.24
7	Fructose	Peptone	15%	15	21.07 ± 2.41	14.79 ± 1.92	41.66 ± 1.58
8	Fructose	Beef extract	5%	20	22.39 ± 1.86	20.11 ± 1.87	82.74 ± 1.63
9	Fructose	Yeast extract	10%	10	20.32 ± 1.67	9.81 ± 1.03	100.63 ± 1.33

*Note: CS, carbon source; NS, nitrogen source; ADCS, adding dosage of carbon source; FT, fermentation time. Mean values were mean of three determinations with standard deviation (±).

2.4 Determination of Polysaccharide Content

The fermented SCR and RB dried powder was extracted with boiling water for two hours. The water-soluble polysaccharide was precipitated by adding eight volumes of 99.5% ethanol and stored at 4℃ overnight. The precipitated polysaccharide was collected by centrifuging at 7000 rpm for 30 min. Then the precipitate was dissolved in 10 mL of distilled water. The total polysaccharide was determined by the phenol-sulfuric acid method with some modifications (Li, Ding & Ding, 2007). The color reaction was initiated by mixing 1 mL of the polysaccharide solution with 0.5 mL of 5% phenol solution and 2.5 mL of concentrated sulfuric acid, and the reaction mixture was incubated in a boiling water bath for 15 min. After cooling it to room temperature, the optical density (OD) of the mixture was determined at 490 nm and the polysaccharide content was calculated with D-glucose as the standard. The results were expressed as milligram of glucose equivalent per gram of the fermented SCR and RB.

2.5 Determination of Ergosterol Content

The fermented SCR and RB dried powder were extracted with a mixture of methanol and dichloromethane in the ratio of 75/25 (v/v) and the solid-to-liquid ratio was 1/10 (w/v) using ultrasonic-assisted extract method for 1 h (50 W) at ambient temperature. Then, the supernatant was collected and filtered by filter (0.45 μm) for HPLC determination. The samples were analyzed by the HPLC (JASCO International Co., Ltd) with a reverse-phase Capcell-Pak C$_{18}$ column (4.6 mm I.D. × 150 mm, particle size of 5 μm Nacalai Tesque, Inc. Japan) in a flow rate of 1.0 mL/min, the column temperature was set at 30°C and the UV detection was operated at 254 nm. The mobile phase was methanol (99.5%), and the concentration of ergosterol was calculated by comparing peak areas with appropriate standards.

2.6 Determination of Adenosine Content

The fermented SCR and RB dried powder were extracted with deionized water (1/10 w/v) by using ultrasonic-assisted extract method for 1 h (50 W) at ambient temperature. Then, the supernatant was collected and filtered by filter (0.45 μm) for HPLC determination. The samples were analyzed by the HPLC (JASCO International Co., Ltd) with a reverse-phase Capcell-Pak C$_{18}$ column (4.6 mm I.D. × 150 mm, particle size of 5 μm Nacalai Tesque, Inc. Japan) in a flow rate of 1.0 mL/min, the column temperature was set at 30°C and the UV detection was operated at 260 nm. The mobile phase was a mixture of acetonitrile and water (5:95, v/v). And the concentration of adenosine was calculated by comparing peak areas with appropriate standards.

2.7 Statistical Analysis

Experimental results were means ± standard deviation (SD) of triple determinations. The data were analyzed by one-way analysis of variance (ANOVA). Tests of significant differences were determined by Student's t-test analysis at $P = 0.05$ or independent sample t-test ($P = 0.05$).

3. Results and Discussion

3.1 Orthogonal Test Results of Fermented SCR

The yield of polysaccharide, ergosterol and adenosine from fermented SCR were shown in Table 1, the optimum fermentation conditions and the significant levels were shown in Table 2.

The highest mean yield of polysaccharide was 31.43 ± 0.37 mg/g dry matter. The optimum levels of factors were sucrose as carbon source, beef extract as the nitrogen source, 15% of adding dosage of carbon source, and 10 days of fermentation time, respectively. The R value of various factors indicated that the nitrogen source was the highest among these factors. And the significant levels indicated that all of the factors significantly related with the yield of polysaccharide.

About ergosterol, the highest mean yield of ergosterol was 35.65 ± 2.76 mg/100g dry matter. The optimum levels of factors were glucose, beef extract, 10% of adding dosage of carbon source, and 15 days of fermentation time, respectively. The R value of various factors indicated that the nitrogen source was the highest among these factors. And the significant levels showed that the ergosterol yield of the fermented SCR was significantly related to all of the factors.

The highest mean yield of adenosine was 117.96 ± 1.24 mg/100 g dry matter. The optimum levels of factors were glucose, yeast extract, 10% of adding dosage of carbon source, and 15 days of fermentation time, respectively. The R value of various factors indicated that the nitrogen source was the highest among these factors. And the significant levels revealed that the adenosine content of the fermented SCR was significantly related to all of the factors.

Further, in order to evaluate the fermented SCR, the solid-state fermentation was enlarged by using 500 mL flask with 50 g SCR, under optimum conditions. The polysaccharide yield of the fermented SCR was reached to 43.49 ± 2.48 mg/g dry matter. Compared with the unfermented SCR (12.91 ± 0.39 mg/g dry matter), the polysaccharide content was 4-fold improvement during the fermentation by *P. aemulans* under the optimum fermentation conditions of polysaccharide yield (OPCPS-SCR). The ergosterol yield of the fermented SCR was reached to 37.53 ± 1.34 mg/100 g dry matter. In contrast with the unfermented SCR (3.13 ± 0.26 mg/100 g dry matter), the ergosterol content was enhanced about 10-fold during the fermentation by *P. aemulans* under the optimum fermentation conditions of ergosterol yield (OFCER-SCR). According to previous reports, the ergosterol content of the fermented SCR as much as cultured *C. sinensis* of Wanfong (38 mg/100 g dry matter) (Li, Yang & Tsim, 2006). And the adenosine yield of the fermented SCR was reached to 148.32 ± 4.21 mg/g dry matter. Compared with the unfermented SCR (12.68 ± 1.36 mg/100g dry matter), the adenosine content was increased by 10-fold under the optimum fermentation conditions of adenosine yield (OFCAD-SCR). And on basis of the previous reports, the adenosine content of fermented SCR was 5-fold higher than that of nature *C. sinensis* (Tibet and Qinghai) (Li, Yang & Tsim, 2006).

Table 2. Range and variance analysis of L_9 (3^4) orthogonal experiment results

	Polysaccharide content				Ergosterol content				Adenosine content			
	CS	NS	ADCS	FT	CS	NS	ADCS	FT	CS	NS	ADCS	FT
I_j	214.24	189.05	210.54	223.43	178.61	111.89	136.96	143.69	727.67	383.76	724.95	634.49
II_j	227.04	243.16	200.34	220.14	151.30	245.84	168.17	192.23	699.60	756.87	736.70	777.72
III_j	191.34	200.41	221.75	189.05	134.12	106.29	158.89	128.10	675.08	961.72	640.70	690.14
R	35.70	54.11	21.41	34.38	44.49	139.55	31.22	64.13	52.59	577.96	96.00	143.23
Optimum level	2	2	3	1	1	2	2	2	1	3	2	2

		Polysaccharide content				Ergosterol content				Adenosine content			
Factor	Y	SS	MS	F ratio	SL	SS	MS	F ratio	SL	SS	MS	F ratio	SL
CS	2	72.69	36.35	15.01	***	111.87	55.93	14.94	***	153.90	76.95	29.76	***
NS	2	180.94	90.47	37.37	***	1386.85	693.42	185.19	***	19081.91	9540.95	3689.41	***
ADCS	2	25.50	12.75	5.27	**	57.10	28.55	7.62	***	609.33	304.66	117.81	***
FT	2	79.96	39.98	16.51	***	248.61	124.30	14.94	***	1158.53	579.26	224.00	***
e	18	43.58	2.42			111.87	55.93			46.55	2.59		

*Note: I_j, II_j, III_j, were sum of the polysaccharide, ergosterol and adenosine contents from fermented SCR of level 1, level 2 and level 3; R means the maximum of I_j, II_j and III_j minus the minimum of I_j, II_j and III_j; CS, carbon source; NS, nitrogen source; ADCS, adding dosage of carbon source; FT, fermentation time. F 0.10 (2, 18) = 2.78; F 0.05 (2, 18) = 3.55; F 0.01 (2, 18) = 6.01; * F ratio > F0.1; ** 0.01 > F ratio > F0.05; *** F ratio > F0.01. υ: Degree of freedom; e: error., SS: Sum of square deviation., MS: Mean square., SL: Significance level.

3.2 The Optimum Fermentation Condition of Fermented SCR

According to the results of orthogonal test, the optimum fermentation condition of polysaccharide, ergosterol and adenosine were different. Therefore, it was necessary to discuss the integrated optimum fermentation condition. The polysaccharide contents of OFCPS-SCR (43.49 ± 1.48 mg/g dry matter), OFCER-SCR (39.18 ± 1.06 mg/g dry matter) and OFCAD-SCR (35.83 ± 1.24 mg/g dry matter) were shown in Figure 1 a. According to Duncan's multiple range test, the polysaccharide content of OFCPS-SCR was significantly higher than that of OFCER-SCR and OFCAD-SCR. The ergosterol contents of OPCPS-SCR, OPCER-SCR and OPCAD-SCR were 22.06 ± 0.16, 37.53 ± 0.11 and 16.62 ± 0.62 mg/100 g dry matter, respectively (Figure 1 b). The adenosine contents of OPCPS-SCR, OPCER-SCR and OPCAD-SCR were 124.59 ± 1.53, 127.94 ± 1.82 and 148.32 ± 1.61 mg/100 g dry matter, respectively (Figure 1 c). As the results, the significant levels of OPCER-SCR ([AB, A, B]) were higher than those of OPCPS-SCR ([A, B, B]) and OPCAD-SCR ([BC, C, A]), thus OPCER-SCR was the optimum fermentation condition for producing polysaccharide, ergosterol and adenosine.

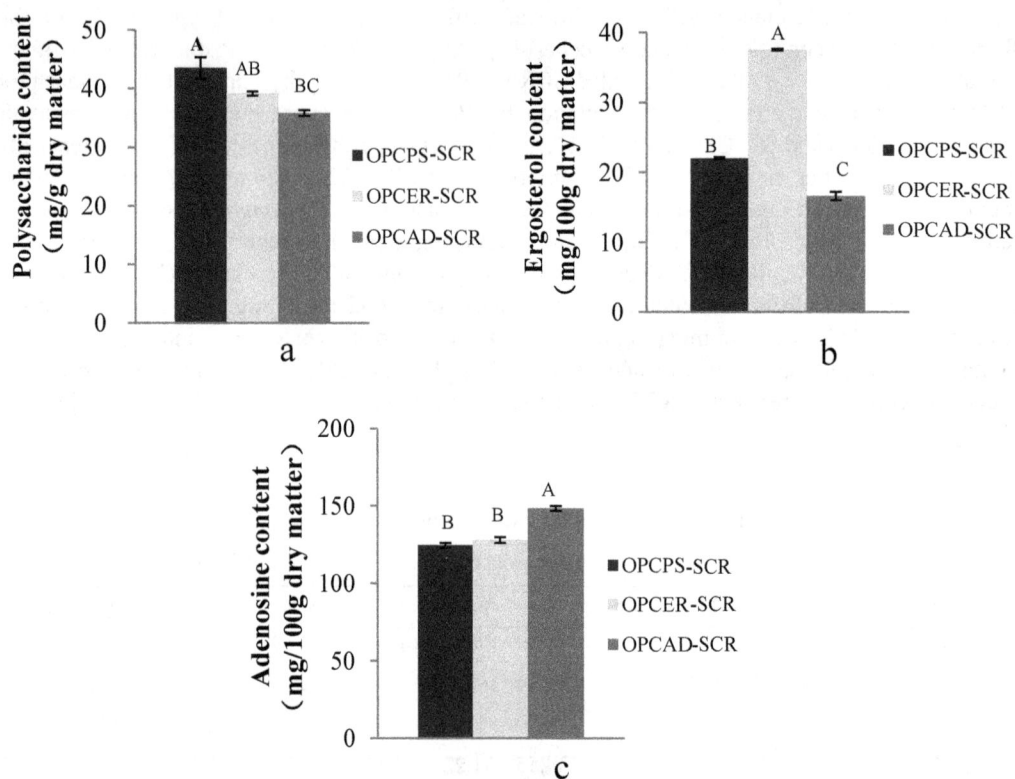

Figure. 1 The optimum fermentation condition of fermented SCR

(a), (b) and (c) were the polysaccharide, ergosterol, adenosine content of three optimum fermentation conditions for SCR (OFCPS-SCR, OPCER-SCR, OFCAD-SCR), respectively ([A. B. C.] $p < 0.01$, Data were expressed as means $\pm$ S.D. n=3 using Duncan's multiple range test).

3.3 Orthogonal Test Results of Fermented RB

The polysaccharide, ergosterol and adenosine yield of fermented RB were shown in Table 3, the optimum fermentation conditions and the significant levels were shown in Table 4.

The highest mean yield of polysaccharide in the orthogonal experiment was 70.02 ± 1.94 mg/g dry matter. The optimum levels of factors were maltose of carbon source, yeast extract of the nitrogen source, 10% of adding dosage of carbon source, 60% of moisture content and 15 days of fermentation time, respectively. The *R* value of various factors indicated that the nitrogen source was the highest among these factors. And the significant levels indicated that all of the factors significantly related with the yield of polysaccharide.

Table 3. L_{16} (4^5) orthogonal layout and results of fermented RB by *P. Aemulans*

Experimental group	CS	NS (3% w/w)	ADCS (% w/w)	MC (%)	FT (day)	Polysaccharide content (mg/g dry matter)	Adenosine Content (mg/100g dry matter)	Ergosterol content (mg/100g dry matter)
1	Glucose	Peptone	5%	60%	5	57.60 ± 1.35	55.65 ± 0.68	57.69 ± 2.65
2	Glucose	Beef extract	10%	70%	10	50.23 ± 2.24	81.61 ± 1.32	41.96 ± 2.99
3	Glucose	Yeast extract	15%	80%	15	46.79 ± 1.06	213.86 ± 2.06	ND
4	Glucose	Ammonium sulfate	20%	90%	20	22.96 ± 1.39	273.48 ± 1.89	10.20 ± 0.32
5	Sucrose	Peptone	10%	80%	20	41.74 ± 1.58	104.30 ± 1.22	76.79 ± 3.36
6	Sucrose	Beef extract	5%	90%	15	56.85 ± 1.96	281.31 ± 2.12	86.47 ± 1.76
7	Sucrose	Yeast extract	20%	60%	10	38.12 ± 1.33	108.72 ± 1.61	42.33 ± 1.88
8	Sucrose	Ammonium sulfate	15%	70%	5	46.58 ± 2.64	38.63 ± 0.33	ND
9	Fructose	Peptone	15%	90%	10	29.03 ± 0.85	30.57 ± 0.58	40.88 ± 0.97
10	Fructose	Beef extract	20%	80%	5	31.49 ± 0.97	50.84 ± 0.63	ND
11	Fructose	Yeast extract	5%	70%	20	50.13 ± 1.81	83.92 ± 0.51	66.03 ± 2.05
12	Fructose	Ammonium sulfate	10%	60%	15	70.02 ± 1.94	104.87 ± 1.06	ND
13	Maltose	Peptone	20%	70%	15	43.47 ± 2.44	29.04 ± 0.22	45.55 ± 1.38
14	Maltose	Beef extract	15%	60%	20	55.88 ± 2.13	81.04 ± 0.72	ND
15	Maltose	Yeast extract	10%	90%	5	66.50 ± 2.06	228.00 ± 1.35	47.19 ± 0.78
16	Maltose	Ammonium sulfate	5%	80%	10	27.60 ± 0.43	59.24 ± 0.68	68.22 ± 1.53

*Note: CS, carbon source; NS, nitrogen source; ADCS, adding dosage of carbon source; MC, moisture content; FT, fermentation time. Mean values were mean of three determinations with standard deviation (±). ND means not detected.

Table 4. Range and variance analysis of l_{16} (4^5) orthogonal experiment results

	Polysaccharide content					Adenosine content				
	CS	NS	ADCS	MC	FT	CS	NS	ADCS	MC	FT
I_j	532.71	515.50	576.55	664.87	606.50	1873.81	658.70	1427.24	1050.83	1119.35
II_j	549.91	583.35	685.48	571.24	434.95	1585.74	1471.27	1556.30	699.60	840.40
III_j	542.02	604.64	534.85	442.84	651.40	810.59	1903.47	1092.30	1284.70	1874.11
IVj	580.35	501.50	408.11	526.03	512.14	1191.95	1428.64	1386.26	2426.96	1628.23
R	30.44	103.14	277.37	222.03	216.45	1063.22	1244.77	464.00	1727.36	1033.71
Optimum level	4	3	2	1	3	1	3	2	4	3

		Polysaccharide content				Adenosine content			
Factor	Y	SS	MS	*F* ratio	SL	SS	MS	*F* ratio	SL
CS	3	106.48	35.49	3.89	**	53744.30	17914.77	1460.34	**
NS	3	636.19	212.06	23.26	***	67012.91	22337.64	1820.88	***
ADCS	3	3284.68	1094.89	120.07	***	9607.12	3202.37	261.05	***
MC	3	2141.59	713.86	78.29	***	139639.20	46546.40	3794.28	***
FT	3	2344.85	781.62	85.72	***	55336.07	18445.36	1503.59	***
e	32	291.80	9.12			392.56	12.27		

*Note: I_j, II_j, III_j and IVj were sum of the polysaccharide and adenosine contents from fermented RB of level 1, level 2, level 3 and level 4; R means the maximum of I_j, II_j, III_j and IVj minus the minimum of I_j, II_j, III_j and IVj; CS, carbon source; NS, nitrogen source; ADCS, adding dosage of carbon source; MC, moisture content; FT, fermentation time. $F0$ .10(3, 32) = 2 .28; $F0$ .05(3, 32) = 2 .90; $F0$ .01(3, 32) = 4 .46; *F ratio > F0 .1; ** F 0.01 > F ratio > F0.05; ***F ratio > F0 .01; CS: carbon Source; NS: nitrogen source; ADCS: adding dosage of carbon source; FT: fermentation time; υ: Degree of freedom; e: error., SS: Sum of square deviation., MS: Mean square., SL: Significance level.

The highest mean yield of ergosterol was 86.47 ± 1.76 mg/100g dry matter (Table 3). Because the contents of several samples were extremely low, the range and variance analysis could not be used in this experiment. Therefore, the optimum levels of factors were sucrose, beef extract, 5% of adding dosage of carbon source, 90% moisture content and 15 days of fermentation time, respectively.

The highest mean yield of adenosine was 281.31 ± 2.12 mg/100 g dry matter. The optimum levels of factors were glucose, yeast extract, 10% of adding dosage of carbon source, 90% moisture content and 15 days of fermentation time, respectively. The R value of various factors indicated that the moisture content was the highest among these factors. And the significant levels were indicated that the adenosine content of the fermented RB was significantly related to all of the factors.

Further, in order to evaluate the fermented RB, the solid-state fermentation was demonstrated by using 500 mL flask with 50 g RB, under optimum conditions. The mean polysaccharide content of the fermented RB was reached to 71.16 ± 2.63 mg/g dry matter. Compared with the unfermented RB (19.80 ± 1.23 mg/g dry matter), the polysaccharide content was increased to almost 4-fold during the fermentation by *P. aemulans* under the optimum fermentation conditions of polysaccharide content (OPCPS-RB). The ergosterol content was reached to 88.04 ± 0.36 mg/100 g dry matter, enhanced about 10-fold during the fermentation by *P. aemulans* under the optimum fermentation conditions of ergosterol content (OFCER-RB). The mean adenosine content of the fermented RB was also enhanced to 282.25 ± 1.83 mg/g dry matter. Contrast with the unfermented RB (30.13 ± 1.53 mg/100g dry matter), the adenosine content was increased by 10-fold during the fermentation by *P. aemulans* under the optimum fermentation conditions of adenosine content (OFCAD-RB).

3.4 The Optimum Fermentation Condition of Fermented RB

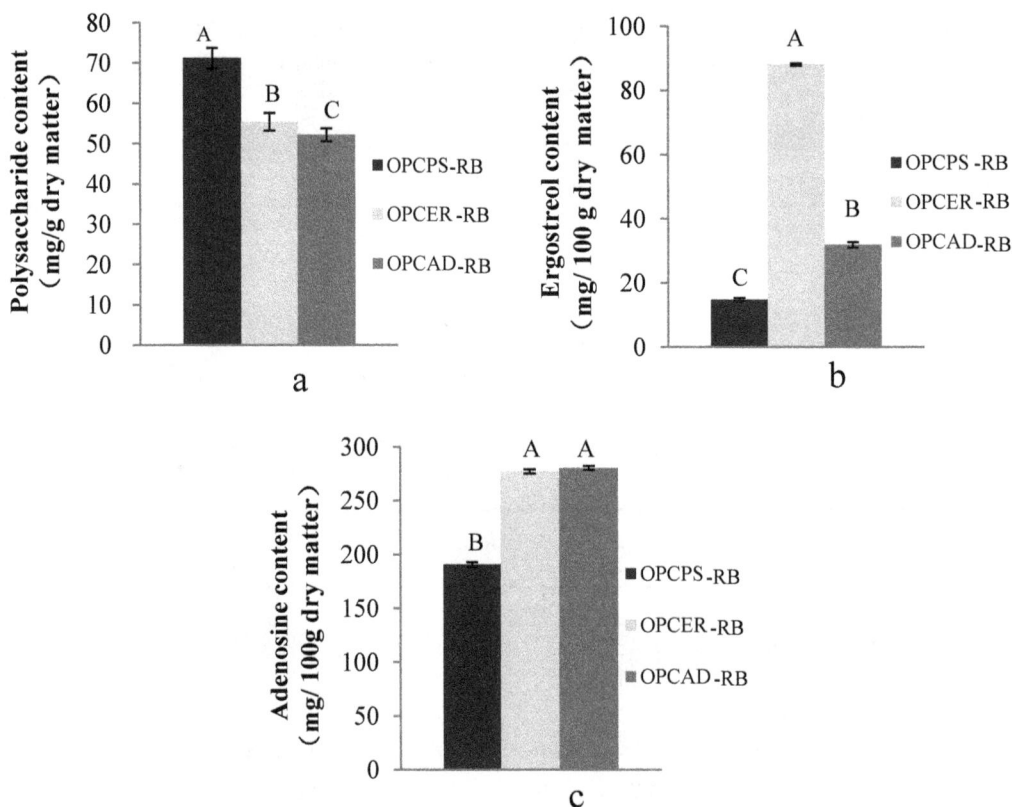

Figure 2. The optimum fermentation condition of fermented RB

(a), (b) and (c) were the polysaccharide, ergosterol, adenosine content of three optimum fermentation conditions for RB (OFCPS-RB, OPCER-RB, OFCAD-RB), respectively ([A. B. C.] p < 0.01, Data were expressed as means ± S.D. n=3 using Duncan's multiple range test).

According to the results of orthogonal test, the optimum fermentation condition of polysaccharide, ergosterol and adenosine were different. Therefore, it was necessary to discuss the integrated optimum fermentation condition.

The polysaccharide contents of OFCPS-RB (71.16 ± 2.63 mg/g dry matter), OFCER-RB (55.40 ± 2.29 mg/g dry matter) and OFCAD-RB (52.16 ± 1.58 mg/g dry matter) were shown in Figure 2 a. According to Duncan's multiple range test, the polysaccharide content of OFCPS-RB was significantly higher than that of OFCER-RB and OFCAD-RB. The ergosterol contents of OPCPS-RB, OPCER-RB and OPCAD-RB were 14.79 ± 0.48, 88.04 ± 0.36 and 31.85 ± 0.87 mg/100 g dry matter, respectively (Figure 2 b). The adenosine contents of OPCPS-RB, OPCER-RB and OPCAD-RB were 190.57 ± 2.11, 276.94 ± 1.96 and 282.25 ± 1.83 mg/100 g dry matter, respectively (Figure 2 c). As the results, the significant levels of OPCER-RB ([B, A, A]) were higher than those of OPCPS-RB ([A, C, B]) and OPCAD-RB ([C, B, A]), thus OPCER-RB was the optimum fermentation condition for producing polysaccharide, ergosterol and adenosine.

4. Conclusions

The optimum fermentation conditions of SCR were: glucose, beef extract, 10% of adding dosage of carbon source, and 15 days of fermentation time, respectively. Under the optimum fermentation conditions, the polysaccharide, ergosterol and adenosine content were 39.18 ± 1.06 mg/g, 37.53 ± 0.11 mg/100 g dry matter and 127.94 ± 1.82 mg/100 g dry matter, respectively.

The optimum fermentation conditions of RB were: sucrose, beef extract, 5% of adding dosage of carbon source, 90% moisture content and 15 days of fermentation time, respectively. Under the optimum fermentation conditions, the polysaccharide, ergosterol and adenosine content were 55.40 ± 2.29 mg/g dry matter, 88.04 ± 0.36 and 276.94 ± 1.96 mg/100 g dry matter, respectively.

The results indicated that the polysaccharide, ergosterol and adenosine content of SCR and RB were improved by solid-state fermentation using *P. aemulans*. The effective utilization of such agricultural waste not only solves environmental problems, but also promotes the economic value of the agricultural products. The fermented SCR and RB were rich in physiological active substances, low in cost, could be explored as ecological feed or functional food material in the further.

Reference

Bok, J. W., Lermer, L., Chilton, J., Klingeman, H. G., & Towers, G. H. (1999). Antitumor sterols from the mycelia of *Cordyceps sinensis*. *Phytochemistry, 51*, 891-898. http://dx.doi:10.1016/S0031-9422(99)00128-4

Hernandez, N., Rodriguez, A. M. E., Gonzalez, F., & Lopez, M. A. (2000). Enzymatic treatment of rice bran to improve processing. *Journal of the American Oil Chemists' Society, 77*, 177-180. http://dx.doi.org/10.1007/s 11746-000-0028-2

Jiamyangyuen, S., Srijesdaruk, V., & Harper, W. J. (2005). Extraction of rice bran protein concentrate and its application in bread. *Songklanakarin Journal of Science and Technology, 27*, 55-64.

Jiang, Y., & Wang, T. (2005). Phytosterols in cereal by-products. *Journal of the American Oil Chemists' Society, 82*, 439-444. http://dx.doi.org/10.1007/s11746-005-1090-5.

Kumara, R., Negib, P. S., Singhc, B., Ilavazhagana, G., Bhargavaa, K., & Sethya, N. K. (2011). *Cordyceps sinensis* promotes exercise endurance capacity of rats by activating skeletal muscle metabolic regulators. *Journal of Ethnopharmacology, 136*(1), 260-266. http://dx.doi:10.1016/j.jep.2011.04.040

Li, J. W., Ding, S. D., & Ding, X. L. (2007). Optimization of the ultrasonically assisted extraction of polysaccharides from *Zizyphus jujuba* cv. *jinsixiaozao*. *Journal of Food Engineering, 80*, 176-183. http://dx.doi:10.1016/j.jfoodeng.2006.05.006.

Li, S. P., Yang, F. Q., & Tsim, K. W. K. (2006). Quality control of *Cordyceps sinensis*, a valued traditional Chinese medicine. *Journal of Pharmaceutical and Biomedical Analysis, 41*, 1571-1584. http://dx.doi.org/10. 1016/j.jpba.2006.01.046

Lindequist, U., Lesnau, A., Teuscher, E., & Pilgrim, H. (1989). Antiviral activity of ergosterol peroxide. *Pharmazie, 44*, 579-580

Mizumoto, S., Hirai, M., & Shoda, M. (2006). Production of lipopeptide antibiotic iturin A using soybean curd residue cultivated with *Bacillus subtilis* in solid-state fermentation. *Biotechnological Products and Process Engineering, 72*, 869-875. http://dx. doi:10.1007/s00253-006-0389-3.

Nagendra, P. M. N., Sanjay, K. R., Shravya, K. M., Vismaya, M. N., & Nanjunda, S. S. (2011). Health Benefits of Rice Bran - A Review. *Journal of Nutrition and Food Sciences.* http://dx.doi.org/10.4172/2155-9600. 1000108.

Ohno, A., Ano, T., & Shoda, M. (1996). Use of soybean curd residue, Okara, for the solid state substrate in the production of lioipeptid antibiotic, iturin A, by *Bacillus subtilis* NB22. *Process Biochemistry, 31*, 801-806. http://dx.doi:10.1016/S0032-9592(96)00034-9.

O'Toole, D. K. (1999). Characteristics and use of okara, the soybean residue from soy milk production. *Journal of Agricultural and Food Chemistry, 47*, 363-371. http://dx. doi:10.1021/jf980754lCCC.

Paterson, R. R. (2008). *Cordyceps*-A traditional Chinese medicine and another fungal therapeutic biofactory. *Phytochemistry, 69*, 1469-1495. http://dx.doi: 10.1016/j.phytochem.2008.01.027

Piironen, V., Lindsay, D., Miettinen, T., Toivo, J., & Lampi, A. (2000) Plant Sterols: biosynthesis, biological function and their importance to human nutrition. *Journal of the Science of Food and Agriculture, 80*, 939-966. http://dx.doi.og/10.1002/(SICI)1097-0010(20000515)80:7<939::AID-JSFA644>3.0.CO;2-C

Shi, M., Yang, Y. N., Li, Y. T., Wang, Y. P., & Zhang, Z. Y. (2011). Optimum Condition of Ecologic Feed Fermentation by *Pleurotus Ostreatus* Using Soybean Curd Residue as Raw Materials. *International Journal of Biology, 3*(4), 2-12. http://dx.doi:10.5539/ijb.v3n4p2

Wang, F., & Nishino, N. (2008). Ensiling of soybean curd residue and wet brewers grains with or without other feeds as a total mixed ration. *Journal of Dairy Science, 91*, 2380-2387. http://dx.doi: 10.3168/jds.2007-0821

Yokoi, H., Maki, R., Hirose, J., & Hayashi, S. (2002). Microbial production of hydrogen from starch-manufacturing wastes. *Biomass and Bioenergy, 22*, 389-395. http://dx.doi:10.1016/S0961-9534(02)0 0014-4

Zhang, Y. J., Sun, B. D., Zhang, S., Wang, M., Liu, X. Z., & Gong, W. F. (2010). Mycobiotal investigation of natural Ophiocordyceps sinensisbased on culture-dependent investigation. *Mycosystema, 29*(4), 518-527. Retrieved from http://journals.im.ac.cn/jwxtcn/ch/reader/view_abstract.aspx?file_no=10040518

18

Population Diversity of *Leptosphaeria maculans* in Australia

Dhwani A. Patel[1], Manuel Zander[1], Angela P. Van de Wouw[2], Annaliese S. Mason[3], David Edwards[4] & Jacqueline Batley[4]

[1] School of Agriculture and Food Sciences and Centre for Integrative Legume Research, University of Queensland, Brisbane, Australia

[2] School of BioSciences, University of Melbourne, Parkville, Australia

[3] Department of Plant Breeding, Land Use and Nutrition, Justus Liebig University, Giessen, Germany

[4] School of Plant Biology, University of Western Australia, Perth, Australia

Correspondance: Jacqueline Batley, School of Plant Biology, University of Western Australia, Crawley, WA 6009, Australia.Email: jacqueline.batley@uwa.edu.au

Abstract

The fungal pathogen *Leptosphaeria maculans*, causal agent of blackleg disease, is a primary cause of canola (*Brassica napus*) crop loss in Australia. Expanding our knowledge of the occurrence of this pathogen in Australia will provide valuable insights into developing methods of resistance against it. In this study, we examine the population diversity of *L. maculans* in Australia using single nucleotide polymorphisms (SNPs). An Illumina GoldenGate 384 SNP assay was developed and used to genotype 59 blackleg isolates collected from across Australia, in different years and from different stubble sources. Limited linkage disequilibrium, absence of significant clustering in the principal component analysis and a mixed dendrogram suggest that the Australian *L. maculans* population as a whole is panmictic. Some evidence of clonality concentrated in each state was also observed. There was a lack of correlation between SNP haplotypes, stubble cultivar and year of collection. These results suggest a high rate of sexual reproduction and evolutionary diversification in the pathogen. These features could enable the pathogen to overcome resistance and continue to cause disease in *Brassica* crops. Analysis of these fungal population isolates will help shed some light on evolution and pathogenicity questions in this important crop pathogen.

Keywords: *Leptosphaeria maculans*, blackleg, Single Nucleotide Polymorphisms, goldengate, genetic diversity

1. Introduction

The ubiquitous fungal pathogen *L. maculans* is the causal agent of phoma stem canker (blackleg) in *Brassica napus*, *B. juncea*, *B. rapa* and *B. oleracea*: canola, vegetable and mustard crops (West, Kharbanda, Barbetti, & Fitt, 2001). This ascomycete was first described in 1791 by Tode and 1849 by Desmaziéres (Gout, Eckert, Rouxel, & Balesdent, 2006). Severe epidemics of blackleg disease occurred in Australia during the 1970s, wiping out the nascent canola industry (Rouxel & Balesdent, 2005). The 1972 epidemic in Australia caused almost 90% crop losses, highlighting the susceptibility of canola to *L. maculans*. Annually, this pathogen causes an average loss of AUD $100 million to the Australian economy (Zander et al., 2013). Furthermore, *L. maculans* isolates present in Australia are classified as highly virulent, able to cause disease even in the more resistant *Brassica* species: *B. juncea*, *B. nigra* and *B. carinata* (Purwantara, Salisbury, Burton, & Howlett, 1998).

Daverdin et al. (2012) describe that rapid evolution in pathogens gives rise to new strains to combat crop defences. Maintenance of crop resistance directly depends on the field population size of the pathogen, its evolutionary potential and cropping practices that directly affect its reproductive system (Daverdin et al., 2012). *L. maculans* can survive as a saprobe in the stubble of infected plants for many years and this is usually favoured by dry hot summers and cold winters (West, Kharbanda, Barbetti, & Fitt, 2001). During this period, it produces sexual inoculum (ascospores), which can travel from several hundred metres to several hundred kilometres (Travadon et al., 2011) and infect plants followed by asexual spore (conidia) production at the site of infection (Rouxel & Balesdent, 2005).

Recent genome sequencing revealed that the *L. maculans* genome has an isochore-like structure (Rouxel et al., 2011), where the genome is divided into AT and GC-rich blocks, probably caused by the amplification of transposable elements and repeat-induced point (RIP) mutations. The RIP mechanism causes nucleotide substitutions from C to T and G to A and is a premeiotic repeat-inactivation mechanism specific to fungi that creates genetic diversification in the fungal genome (Rouxel et al., 2011). Fudal et al. (2009) reported that RIP affects the *AvrLm6* locus, causing gene inactivation and leading to virulence. Furthermore, isolates have been found to undergo continuous deletions and mutations apart from RIP that lead to further genome diversity. Current approaches to establish blackleg resistance in canola have not been successful in fully controlling this pathogen (Hayward, McLanders, Campbell, Edwards, & Batley, 2012). Therefore, understanding how this fungus has evolved, diversified and spread in Australia is important in providing information for the breeding and sowing of improved resistant varieties of *Brassica*.

Population diversity in *L. maculans* has been examined using a variety of markers. Genetic differences between eastern and western Australian isolates have previously been found using microsatellites and minisatellites, which were attributed to the presence of arid desert between the coasts (Hayden, Cozijnsen, & Howlett, 2007). Minisatellite markers used to analyse four field populations in France found high levels of gene and genotypic diversity within populations and high gene flow between populations, consistent with randomly mating populations (Gout et al., 2006) Dilmaghani et al. (2012) also used minisatellite markers to show that the *L. maculans* population in Western Canada comprises two genetically distinct populations. A further study implementing fourteen minisatellite markers also found clonal sub populations of this pathogen on *B. oleracea* in Mexico (Dilmaghani et al., 2013). However, Travadon et al. (2011) found the French *L. maculans* population to be panmictic. This study also employed minisatellite and microsatellite markers. Other investigations into blackleg population structure have also been conducted using amplified fragment length polymorphisms (AFLPs) (Purwantara, Barrins, Cozijnsen, Ades, & Howlett, 2000) and restricted fragment length polymorphisms (RFLPs) (Barrins, Ades, Salisbury, & Howlett, 2004). Single Nucleotide Polymorphisms (SNPs) have recently become a popular choice of molecular marker for population diversity studies, and offer significant benefits in terms of abundance in genomes and ease of high-throughput assessment. SNPs are single base-pair differences between two individuals at a particular locus (Appleby, Edwards, & Batley, 2009). SNPs can be classified as transitions (C to T, G to A), transversions (C to G, A to T, T to G or C to A) and insertions/deletions (indels) of a single base pair. Such molecular markers are good tools to analyse the various processes encompassing the population genetics and evolutionary processes of an organism. These include mating systems, patterns of speciation, dispersal, mutation, migration and selection etc. (Giraud, Enjalbert, Fournier, Delmotte, & Dutech, 2008; Gout et al., 2006).

The Illumina GoldenGate genotyping assay can be used to simultaneously analyse 384-3072 SNP loci across multiple individuals (Tindall et al., 2010). Previous studies using the Illumina GoldenGate assay have shown that it can be used to reliably score SNPs for genetic analysis (Durstewitz et al., 2010). Furthermore, it is cost-effective and flexible for analysing large numbers of SNPs (Appleby et al., 2009). We applied 384 previously developed *L. maculans* SNPs (Zander et al., 2013) in a GoldenGate assay to analyse 59 Australian *L. maculans* population isolates collected from different years, regions and cultivars, assessing the diversity of this pathogen across Australia.

2. Materials and Methods

2.1 Fungal Samples

A total of 59 fungal isolates were analysed using the Illumina GoldenGate assay, this comprised of 96 samples including replicates and controls (Table 1). Isolates were carefully selected to cover a wide range of parameters including region of collection, cultivar grown at collection site, year isolated and *Avr* gene complement (not shown) (Table 1, Appendix A). The isolates received were either stored in liquid form (agar piece in water) or filter form (filter discs in silica beads). The isolates were grown and genomic DNA extracted as in Zander et al. (2013). The extracted DNA was quantified using a Qubit Fluorometer (Life Technologies, 2013). The reference isolate v23.1.3, (for which the genome sequence is available) (Table 1) (Rouxel et al., 2011) was used for data analysis.

Table 1. List of *L. maculans* population isolates used in this study

As referred to in text	Isolate	Year Cultured	Species isolated from	Stubble cultivar	Stubble collection site	Country/State	Replicates	Reference
Ref	Reference (v23.1.3)	Mid-1990		-	-	Europe	N/A	(Rouxel et al., 2011)
Lm-1	04MGPS021 (21)	2004	*B. napus*	AG-Emblem	Eyre Peninsula	SA	2	
Lm-2	06MGPP041 (41)	2006	*B. napus*	Skipton	Lake Bolac	Vic	2	
Lm-3	04MGPP003	2004	*B. napus*	TI1 Pinnacle	Geelong	VIC	N/A	
Lm-4	04MGPP008	2004	*B. napus*	Unknown	Wonwondah	VIC	N/A	
Lm-5	04MGPP016	2004	*B. napus*	AG-Emblem	Bordertown	SA	N/A	
Lm-6	04MGPP022	2004	*B. napus*	Grace	Moyhall	SA	N/A	
Lm-7	04MGPP026	2004	*B. napus*	Grace	Moyhall	SA	N/A	
Lm-8	04MGPP035	2004	*B. napus*	TI1 Pinnacle	Geelong	VIC	N/A	
Lm-9	04MGPP041	2004	*B. napus*	Grace	Wonwondah - Pymers	VIC	N/A	
Lm-10	04MGPP043	2004	*B. napus*	Grace	Wonwondah - Pymers	VIC	N/A	
Lm-11	04MGPP045	2004	*B. napus*	Grace	Wonwondah - Pymers	VIC	N/A	
Lm-12	04MGPP046	2004	*B. napus*	TI1 Pinnacle	Laharum	VIC	N/A	
Lm-13	04MGPP049	2004	*B. napus*	TI1 Pinnacle	Laharum	VIC	N/A	
Lm-14	04MGPS006	2004	*B. napus*	Surpass 400	Eyre Peninsula	SA	N/A	
Lm-15	04MGPS016	2004	*B. napus*	Surpass 603CL	Bordertown -Ballinger	SA	N/A	
Lm-16	04MGPS024	2004	*B. napus*	ATR-Beacon	Bordertown - Ivan	SA	N/A	
Lm-17	05MGPP002	2005	*B. napus*	ATR-Beacon	Woseley	SA	N/A	
Lm-18	05MGPP033	2005	*B. napus*	Skipton	Yeelana	SA	N/A	
Lm-19	06MGPP019	2006	*B. napus*	ATR-Beacon	Wagga Wagga	NSW	N/A	
Lm-20	06MGPP025	2006	*B. napus*	ATR-Beacon	Wagga Wagga	NSW	N/A	
Lm-21	06MGPS032	2006	*B. napus*	Surpass 501TT	Keith	SA	N/A	
Lm-22	07VTJH002	2007	*B. juncea*	JC05002	Horsham	Vic	N/A	
Lm-23	07VTJH020	2007	*B. juncea*	JC05007	Horsham	Vic	N/A	
Lm-24	D13	2009	*B. napus*	Hyola50	Cummins	SA	N/A	(Marcroft et al., 2012)
Lm-25	09SMJ087	2009	*B. juncea*	EXCEED OasisCL	Kaniva	VIC	N/A	
Lm-26	10SMJ041	2010	*B. juncea*	EXCEED OasisCL	Tamworth	NSW	N/A	
Lm-27	LM300	2002	*B. napus*	TI1 Pinnacle	Mt Barker	WA	1	
Lm-28	LM580	2003	*B. napus*	ATR-Beacon	Wonwondah	Vic	N/A	
Lm-29	LM592	2003	*B. napus*	TI1 Pinnacle	Mt Barker	WA	N/A	
Lm-30	LM659	2003	*B. napus*	Hyden	Wongan Hills	WA	N/A	
Lm-31	LM661	2003	*B. napus*	Hyden	Wongan Hills	WA	N/A	
Lm-32	IBCN13	1991	B. napus	Unknown	Mt Barker	WA	1	(Balesdent et al., 2005)
Lm-33	IBCN15	1988	*B. napus*	Unknown	Streatham	Vic	2	(Purwantara et al., 2000)
Lm-34	IBCN16	1988	*B. napus*	Unknown	Mt Barker	WA	2	(Purwantara et al., 2000)
Lm-35	IBCN17	1988	*B. napus*	Unknown	Millicent	SA	1	(Balesdent et al., 2005)
Lm-36	IBCN18	1988	*B. napus*	Unknown	Penshurst	Vic	2	(Purwantara et al., 2000)
Lm-37	IBCN75	1987	*B. napus*	Unknown	Mt Barker	WA	2	(Purwantara et al., 2000)
Lm-38	IBCN76	1987	*B. napus*	Unknown	Mt Barker	WA	2	(Purwantara et al., 2000)

Lm-39	D8 (M)	2005	*B. napus*	Surpass 501TT	Mt Barker	WA	2	(Marcroft et al., 2012)
Lm-40	D9 (M)	2005	*B. napus*	ATR-Beacon	Mt Barker	WA	2	(Marcroft et al., 2012)
Lm-41	PHW1223	1987	*B. napus*	Unknown	Mt Barker	WA	2	(Purwantara et al., 2000)
Lm-42	V4	1988	*B. napus*	Unknown	Numurkah	Vic	N/A	(Van de Wouw et al., 2010)
Lm-43	35	1988	*B. napus*	Unknown	Penshurst	Vic	N/A	(Van de Wouw et al., 2010)
Lm-44	80	1988	*B. napus*	Unknown	Millicent	SA	N/A	(Van de Wouw et al., 2010)
Lm-45	89	1988	*B. napus*	Unknown	Millicent	SA	N/A	(Van de Wouw et al., 2010)
Lm-46	535	2003	*B. napus*	TI1 Pinnacle	Lake Bolac	Vic	1	
Lm-47	1245	1988	*B. napus*	Unknown	Galong	NSW	N/A	(Van de Wouw et al., 2010)
Lm-48	04S012	2004	*B. napus*	Surpass603CL	Bordertown	SA	1	
Lm-49	04S005	2004	*B. napus*	Surpass400	Eyre Peninsula	SA	1	
Lm-50	04P042	2004	*B. napus*	Grace	Wonwondah	Vic	N/A	
Lm-51	05P032	2005	*B. napus*	Skipton	Yeelanna	SA	N/A	
Lm-52	06P039	2006	*B. napus*	Skipton	Lake Bolac	Vic	1	(Van de Wouw et al., 2010)
Lm-53	06S014	2006	*B. napus*	Surpass 501TT	Bordertown	SA	N/A	
Lm-54	06S012	2006	*B. napus*	ATR-Beacon	Bordertown	SA	1	
Lm-55	06S039	2006	*B. napus*	Hyola60	Lake Bolac	Vic	N/A	(Van de Wouw et al., 2010)
Lm-56	06J085	2006	*B. juncea*	Unknown	Horsham	VIC	1	
Lm-57	06J095	2006	*B. juncea*	Unknown	Horsham	Vic	N/A	
Lm-58	06J112	2006	*B. juncea*	Unknown	Horsham	Vic	N/A	
Lm-59	04MGPP029	2004	B. napus	TI1 Pinnacle	Geelong	VIC	N/A	

Note: VIC-Victoria; NSW-New South Wales; WA-Western Australia; SA-South Australia; Isolate v23.1.3 is the result of a series of in vitro crosses between European field isolates (Balesdent et al., 2001); Not all data on these isolates was available, "-" denotes an unknown variable. IBCN numbers represent the IDs of "International Blackleg Collection Network" isolates (Marcroft et al., 2012).

2.2 Illumina GoldenGate assay

A total of 384 SNPs were selected for the Illumina GoldenGate assay. The SNPs were chosen to cover a range of the 76 supercontigs on which SNPs were predicted, from the list of 21,814 SNPs described in Zander et al. (2013). A designability assessment conducted using the Illumina Assay Design Tool (ADT) scored the 384 SNPs at 0.4 or above, which is deemed a good score for the Illumina GoldenGate assay (Durstewitz et al., 2010). Sample preparation for the Illumina GoldenGate assay was performed according to the Illumina GoldenGate Genotyping Assay guide (According to manufacturer's instructions). The software "Genome Studio" (Illumina Inc., 2013) was used to manually cluster the SNPs into one of the two possible genotype clusters (A and B) for this haploid organism. SNPs that clustered confidently were selected for future data analyses and monomorphic and non-clustering SNPs (did not clearly separate into either the 'A' group or the 'B' group) were eliminated from further analyses, resulting in 214 high-quality SNPs. A sub- set of 193 SNPs was used for linkage disequilibrium (LD) analysis (SNPs monomorphic in all isolates except Lm-1 and Lm-2, the isolates which were used to identify polymorphic SNPs for the assay, were omitted).

2.3 Data Analysis

The data set of 214 SNPs was sorted according to predicted positions on the *L. maculans* supercontigs, as outlined in Supplementary Figure S1 of Rouxel et al. (2011). In order to look for potential SNP blocks relating to a parameter, the isolates were sorted individually, based on each parameter eg. State, stubble species or stubble cultivar (Table 1). Manhattan plots generated using the R package 'Gapit' (Lipka et al., 2012), were used to visualise any possible association between SNPs and these parameters.

All isolates were considered to be part of one population for the statistical analyses. The SNP positions were given 1 and 0 values (for the dendrogram and PCA analyses) or 'A/A' and 'B/B' (for LD analysis) for each genotype call and 'NA' was assigned to missing values. A binary distance matrix was generated and used to create a phylogenetic dendrogram. The R package "pvclust" (Suzuki & Shimodaira, 2006) was used to generate a dendrogram with 1000 bootstrap iterations, binary distance and complete clustering. Population LD was calculated using the R package 'genetics' (Warnes, Gorjanc, Leisch, & Man, 2012). R^2 LD values were used to generate the heatmap using the R package 'LDHeatmap' (Shin, Blay, McNeney, & Graham, 2006) to visualise LD. PCA was performed using the R packages 'ade4' (Dray & Dufour, 2007) and 'maptools' (Lewin-Koh et al., 2012). SNPs with possible null and private alleles were checked against the reference (Rouxel et al., 2011) using the alignment tool in Geneious Pro version 5.6 (Biomatters Ltd., 2015; Kearse et al., 2012).

3. Results

3.1 Illumina GoldenGate results

The results from the GoldenGate assay supported the SNP prediction of Zander et al. (2013). The data generated was sorted for SNPs that had high confidence clusters. 2.6% of SNPs had missing values (NA) for all isolates, 29.9% were monomorphic and 11.7% were non-clustering SNPs. SNPs belonging to these three categories were eliminated from further analyses. No correlation between eliminated SNPs and SNP score or supercontig on which they were positioned could be observed. Filtering for quality polymorphic data resulted in a dataset of 214 SNPs. A subset of 193 SNPs was used for linkage disequilibrium (LD) after elimination of a further 21 SNPs. Reproducibility less than 100% was due only to missing data in one or other of the replicates.

Table 2. Percent reproducibility of replicates used in the assay

Isolate name	Replicates	Reproducibility (%)
Lm-1	2	99.22
Lm-2	2	98.96
Lm-27	1	98.96
Lm-32	1	99.74
Lm-33	3	97.92
Lm-34	3	92.52
Lm-35	1	99.22
Lm-36	3	97.20
Lm-37	3	96.26
Lm-38	3	96.26
Lm-39	3	93.93
Lm-40	3	92.06
Lm-41	3	94.39
Lm-46	1	95.57
Lm-48	1	98.44
Lm-49	1	98.96
Lm-52	1	97.40
Lm-54	1	98.96
Lm-56	1	99.22

3.2 General Marker and Population Statistics

The SNP data was mined for possible private alleles (Table 3). Private alleles are unique alleles in isolates that denote genetic distinctiveness. Of the 21 private allele SNPs, eight occurred in the isolate Lm-1 and seven occurred in the isolate Lm-2, both of which were used for SNP discovery. Of the other four alleles occurring at low frequency, four were in Lm-2 and Lm-55 only, and two were present in only Lm-2 and Lm-24 and Lm-2 and Lm-30. SNPs private in Lm-2 and Lm-55 were the only ones that consistently occurred in intergenic regions of the genome. The SNP polymorphic information content (PIC) scores (See Appendix B) ranged from 0.3-0.5, indicative of relatively high polymorphism. General statistics can be found in Appendix B and association analysis of SNPs to the state the isolates were collected from can be found in Figures C1-C4 of Appendix C.

Table 3. Possible private alleles identified and their location in the *L. maculans* genome

SNP name	SuperContig	Location (bp)	Polymorphic in	Location in genome
SNP 55	SuperContig_2	1120505	Lm-1	Intergenic
SNP 81	SuperContig_13	1254623	Lm-1	Intergenic
SNP 85	SuperContig_13	1424634	Lm-1	Intergenic
SNP 115	SuperContig_8	774901	Lm-2 and Lm-55	Intergenic
SNP 125	SuperContig_10	992433	Lm-2 and Lm-55	Intergenic
SNP 131	SuperContig_6	638539	Lm-2 and Lm-30	Intergenic
SNP 135	SuperContig_11	578821	Lm-2	Intergenic
SNP 137	SuperContig_11	970997	Lm-2 and Lm-55	Intergenic
SNP 140	SuperContig_11	1524610	Lm-1	End of SC
SNP 141	SuperContig_3	95443	Lm-2	In gene similar to peroxisomal membrane protein CBX93615.1
SNP 151	SuperContig_4	346457	Lm-2 and Lm-55	Intergenic
SNP 153	SuperContig_4	873687	Lm-1	Downstream of gene product similar to lipolytic protein G-D-S-L family CBX93216.1
SNP 160	SuperContig_4	1358095	Lm-2	Intergenic
SNP 161	SuperContig_4	1362010	Lm-2	Intergenic
SNP 181	SuperContig_9	796976	Lm-1	Intergenic
SNP 188	SuperContig_14	264857	Lm-2	Hypothetical protein CDS CBX98019.1
SNP 194	SuperContig_14	1268037	Lm-1	Hypothetical protein CDS CBX98389.1
SNP 198	SuperContig_16	447349	Lm-1	Exon of gene whose product is similar to epoxide hydrolase CBX97267.1
SNP 203	SuperContig_17	350439	Lm-2	Hypothetical protein CDS CBX96839.1
SNP 204	SuperContig_17	523022	Lm-2 and Lm-24	Hypothetical protein CDS CBX96888.1
SNP 205	SuperContig_17	757759	Lm-2	Intergenic

3.3 Population Analysis

No significant correlation was observed in the data between the SNPs and the isolate collection site, the stubble cultivar, resistance complement of the cultivar or the year the isolates were collected (Table 1). A low level of association was observed between SNP73 on SC 12 and *B. napus* (as the stubble species) compared to *B. juncea* cultivars as the stubble species (Figure 1). No association to state, stubble cultivar, year collected or place collected could be found (Appendix C Figures C1-C4; other data not shown).

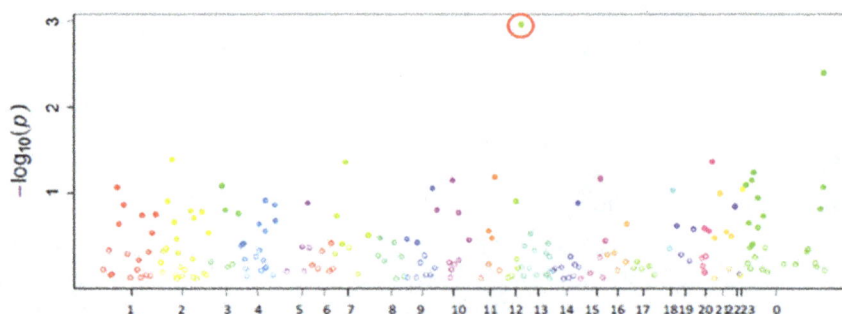

Figure 1. Manhattan plot of all isolates showing association between SNPs and *B. napus* species

Note: SNP 73 circled in red; x-axis supercontigs; y-xis –log10 values of association between stubble species and SNPs.

3.4 Phylogenetic Tree

In order to visually analyse the relationships between the isolates, a dendrogram was generated. The isolates used for SNP prediction, Lm-1 and Lm-2 are on separate clades as can be discerned from the resulting phylogenetic tree (Figure 2). Based on this analysis, a number of isolates appeared to be genetically identical. DNA replicates used in the assay were noted to be the same (Table 2).

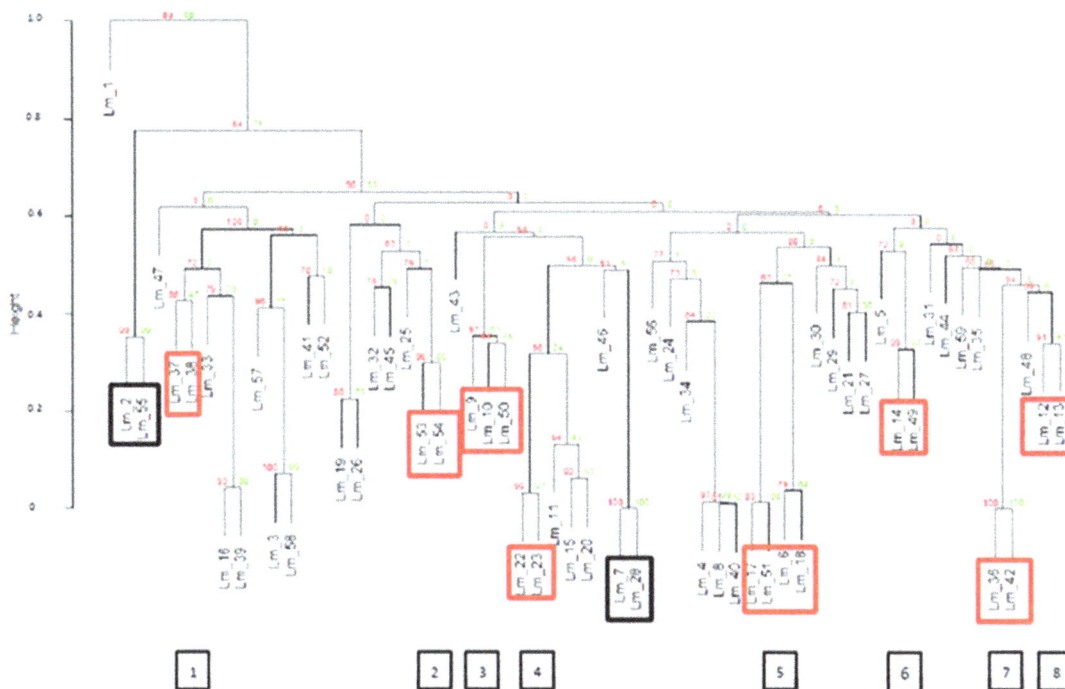

Figure 2. Phylogenetic tree based on all isolates

Note: Bootstrap values: red-Approximately Unbiased (AU) p-values calculated by multiscale bootstrap resampling; green-Bootstrap Probability (BP) p-values calculated by normal bootstrap resampling; AU values >95 % strongly supported by data (Suzuki & Shimodaira, 2006); Replicates not shown; Left to right red boxes 1-8; Black boxes indicate similar isolates.

From the dendrogram it can be seen that the isolate Lm-55 was most similar to Lm-2 (Figure 2). Both isolates were collected in the same year (2006) and the same region of the state (Victoria) but from different stubble cultivars of *B. napus*. Isolates Lm-7 and Lm-28 were also seen to be similar. However, Lm-7 was collected in 2004 and Lm-28 in 2003 from different stubble cultivars (Grace and ATR-Beacon respectively) and from different states (Wonwondah in Victoria and Moyhall in South Australia respectively, both close to a common border about 120 km apart). Boxes 1-8 denote isolates that group locally based on year collected, site of collection, state, stubble species and/or stubble cultivar (Figure 2), the details for which are seen in Table 1. The majority of clusters were from Victoria and South Australia. The most common groupings were based on state and stubble cultivar.

Some differences between certain isolates grouping together were also noticeable. Lm-16 and Lm-39 were isolated in different years (2004, 2005), from different stubble cultivars (ATR-Beacon, Surpass501TT) and in different places (Bordertown, SA and Mt. Barker, WA). The same differences could be seen for Lm-41 and Lm-52, Lm-32 and Lm-45, Lm15 and Lm-20, Lm-4, Lm-8 and Lm-40, Lm-6 and Lm-18 and Lm-21 and Lm-27. Therefore, small groupings based on the parameters listed in Table 1 were seen throughout the tree along with larger differences. No other patterns could be elicited from the positioning of sub-clades in this tree.

3.5 Principal Component Analysis (PCA)

The Principal Component Analysis (PCA) primarily showed a random distribution of isolates along the two principal component axes (Figure 3). Isolates Lm-1 and Lm-2 were completely different to each other, as expected based on their use for identification of polymorphic SNPs for designing the assay. Genetically identical isolates were plotted at the same point or grouped together in the same area, validating the results of the dendrogram (Figure 2). Isolates Lm-51 and Lm-17 were both collected in 2005 from South Australia. Isolates Lm-9 and Lm-10 were collected in 2004 isolated from "Grace" in Wonwondah. No other conclusive correlations could be elucidated.

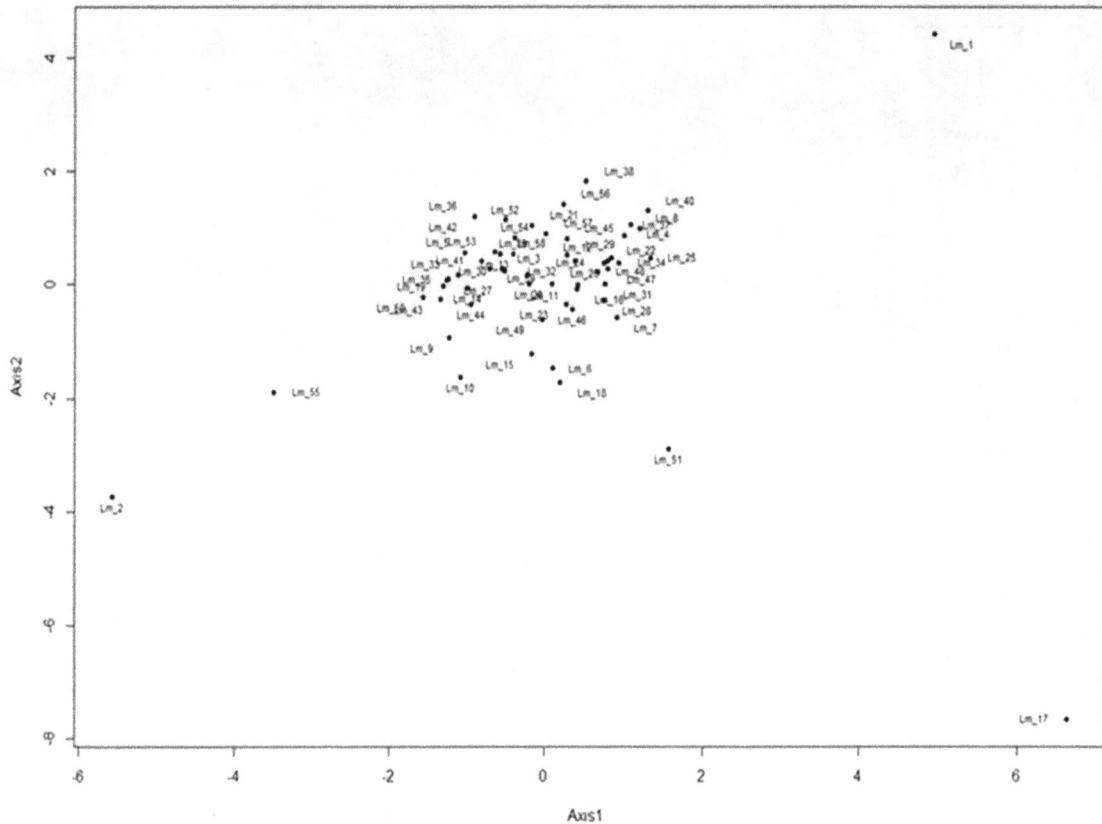

Figure 3. PCA displaying correlation between Australian *L. maculans* population isolates across 214 SNP loci isolates

Note: Axis1-Principal Component 1, Axis2-Principal Component 2; SNP prediction based on Lm-1 and Lm-2; replicates not shown.

3.6 Linkage Disequilibrium

The sub-data set of 193 SNPs, after elimination of 21 SNPs considered private alleles, was analysed to measure pairwise linkage disequilibrium (LD) between SNPs. The heatmap in Figure 4 displays results of the LD calculation. Based on the R^2 values displayed in the heatmap, little significant LD was observed in the *L. maculans* population. $R^2 = 1$ indicates no recombination, and thus high LD, and $R^2 = 0$ indicates considerable recombination thus no LD. Only 0.83% of p-values associated with pairwise comparisons were significant. Furthermore, the heatmap failed to show any noticeable patterns or blocks of LD.

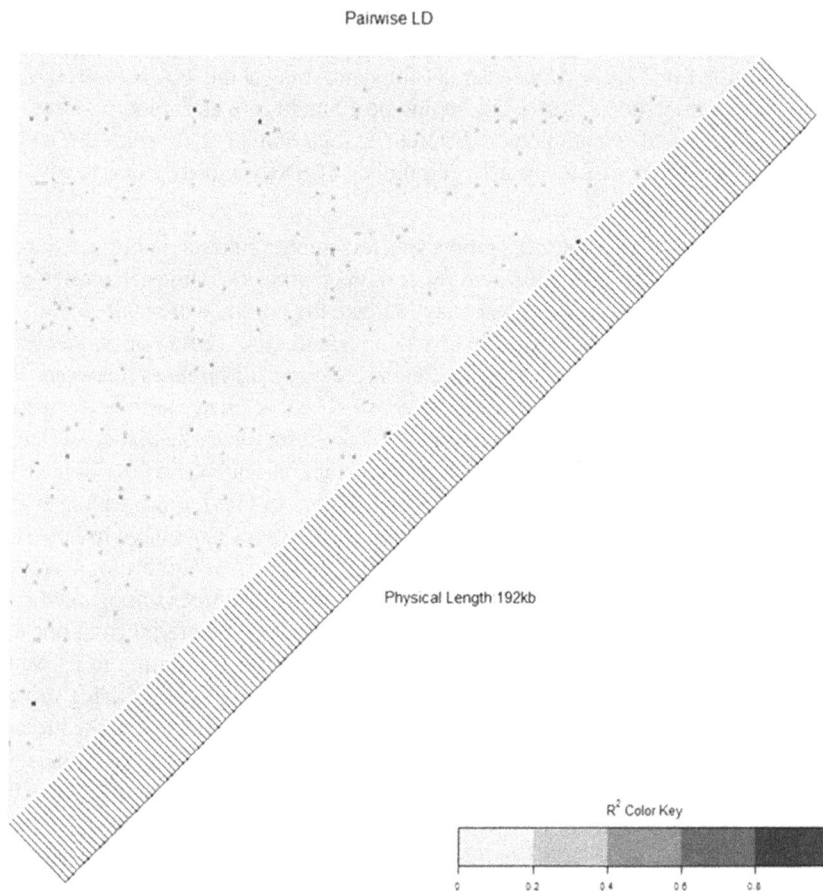

Figure 4. Heatmap displaying Linkage disequilibrium in Australian *L. maculans* population isolates across 193 SNP loci

Note: $R^2=0$ considerable recombination; $R^2=1$ no recombination.

4. Discussion

The isolates used in this study were collected from all around Australia. The absence of any obvious genetic variance specific to a certain parameter suggests that the Australian *L. maculans* isolates comprise a single population, but that some possible subpopulations localised to the state or site of collection do occur. Cultivar stubble may assist in maintaining the large population size (Daverdin et al., 2012; Travadon et al., 2011). No noticeable patterns or haplotypes were detected in the data, suggesting that this fungus evolves rapidly under selection pressure from the host. Sexual reproduction in this pathogen facilitates the production of ascospores which is its primary inoculum (Rouxel & Balesdent, 2005). Human transport may aid in transporting infected material to different regions (Travadon et al., 2011). This in turn leads to random mating between isolates from different regions, creating genetic variance at the avirulence loci and assisting the pathogen to overcome host resistance (Dilmaghani et al., 2012).

It is expected that the population of a sexual reproductively active pathogen will be panmictic. In a panmictic population, members may interact with one another at random, which creates extensive recombination and genetic diversity (Polk & Peek, 2010). Previous *L. maculans* population studies have reached the same conclusion of panmixia (Barrins et al., 2004; Travadon et al., 2011). We assumed the null hypothesis of a panmictic blackleg population across Australia while conducting population analyses, which was supported by the results of the manhattan plots, dendrogram, principal component analysis and linkage disequilibrium analysis.

Overall, the phylogenetic tree and PCA analysis suggested that *L. maculans* possesses a high evolutionary potential, indicative of populations able to overcome genetic resistance (McDonald & Linde, 2002). Large population size, high rate of mutation, high genotype flow and mixed reproduction all confer high evolutionary potential to the pathogen, putatively enabling it to overcome host genetic resistance (McDonald & Linde, 2002). The length of the tree branches indicates genetic similarity between isolates. Based on this, isolates Lm-28 and

Lm-7 appear to be genetically identical with 78% shared alleles (22% missing values). These isolates have been collected from Moyhall, South Australia in 2003 and Wonwondah, Victoria in 2004 respectively. Van de Wouw et al. (2010) also used isolates Lm-6, Lm-17, Lm-18 and Lm-51 for genotyping at the *AvrLm1* and *AvrLm6* loci. They classified Lm-18 and Lm-51 as haplotype 24, Lm-17 as haplotype 10 and Lm-6 as haplotype 4 based on their *Avr* genotype. These haplotypes displayed a completely different association in their study as compared to the dendrogram. This highlights the efficacy of using a large number of SNPs chosen from across the genome to classify the relationships between isolates.

Some local groupings based on state, year collected, stubble species, stubble cultivar and/or site of collection were also observed. Isolates in Box 1-8 (Figure 2) all clustered for certain parameters, with each cluster group sharing a single state of collection. Local groupings such as these may indicate the presence of small clonal subpopulations of this pathogen within Australia, such as were observed by Dilmaghani et al. (2013) on *B. oleracea* in Mexico. Clusters in our study were spread across the tree, indicating genetic differences between these possible subpopulations. Different conditions particular to each state such as weather, cultivars grown and stubble resistance all cumulatively affect the evolution of this pathogen. Therefore, it may be that conditions particular to each state promote asexual reproduction rather than sexual reproduction leading to less diversity within each subpopulation. Dilmaghani et al. (2013) attributed clonality such as this to moving the pathogen from its native biogeographic range, loss of a mating-type by mutation and culture conditions conducive to large-scale dispersal of conidia. This conclusion was made based on the presence of high linkage disequilibrium. Certain isolates like Lm-16 and Lm-39 also clustered together but were vastly different in the parameters associated with them. It is known that human movement transports and introduces infected seed and plant material from one area to another (Dilmaghani et al., 2012) thereby mixing and changing the population, further attributing to its panmictic nature. The bootstrapping of the phylogenetic tree also supports the theory of a randomly interacting mixed population. Bootstrapping values for the main branches were <95%, which indicates low confidence in the hierarchical cluster analysis. On the other hand, most bootstrapping values within each box were >95%, indicating that they were strongly supported by the data (Suzuki & Shimodaira, 2006). An overall analysis of the tree yielded no particular association to any other parameter. The PCA results also displayed a random positioning of isolates. Certain isolates clustered together, such as Lm-36 and Lm-42 isolates in Box 7, validating the dendrogram and also supporting panmixia. These findings could be attributed to the high evolutionary potential of *L. maculans*. Spore dispersal also plays a role in increasing gene flow and generating a random mix of isolates across the population (Travadon et al., 2011). More samples from each region and year will need to be collected and examined to investigate the possibility of clonal sub-populations of *L. maculans* within Australia.

We hypothesised that a possible cause of LD in this population could be selection of *AvrLm* genes, due to their impact on host plant infection. This would lead to loci in the selected region segregating with each other more often than expected by chance and can be visualised as blocks on the LD heatmap. However, we failed to notice any such patterns. As the rate of recombination between loci increases, there is a greater chance of linkage equilibrium in the population, decreasing LD. Populations that are constantly recombining and have a high cross-over rate will show little LD (McVean, 2008). Xu (2006) stated that only 5% of locus-pairs have significant observed association to those expected in a completely panmictic population. Our LD analysis showed 0.83% of p-values associated with pairwise comparisons, to be significant. SNP73, which was seen to be significantly associated with *B. napus* cultivars and was located near *Avr4-7*, did not display significant LD. The SNP and the gene on SC12 of the *L. maculans* genome are 253.6 kb apart and the GC content of the traversing region is 45.2%. Parlange et al. (2009) reported two PCR markers on the border of the *AvrLm4-7* locus; the GC content between those markers was 35.2%. It has been concluded by Rouxel et al. (2011) that recombination in the *L. maculans* genome occurs more frequently within GC-rich regions than between GC-rich regions. Therefore we believe that recombination events in the GC-rich region between the SNP and the gene may have impacted the association between them in *B. napus* cultivars. The number of samples isolated from *B. napus* cultivars may also be too small to clearly display this association on the heatmap in the form of LD. Furthermore, the majority of Australian cultivars that have been genotyped contain *Rlm4* and therefore there has been strong selection pressure at the *AvrLm4-7* locus for a number of years in Australia (Marcroft et al., 2012). However, detailed studies comprising more isolates derived from *B. juncea* isolates will need to be conducted to confirm the validity of this association to *B. napus* cultivars. In the future, examining associations between SNPs and *Avr* genes in AT-rich regions such as these may prove fruitful in analysing the evolution of avirulence genes.

The results from this SNP genotyping assay successfully validated the work conducted by Zander et al. (2013) using the same SNP resource. The SNP prediction supported transferability of SNPs for use in the GoldenGate assay for the chosen SNPs. Stringent clustering criteria (ensuring that all SNPs visually separated into either the 'A' group or the 'B' group) yielded 214 SNPs, which were used for subsequent data analysis. Poor clustering could be a result of additional SNPs in the flanking regions of the predicted SNPs which can be resolved in the future by using more isolates for SNP discovery. Private alleles relating to five particular isolates (Lm-1, Lm-2, Lm-24, Lm-30 and Lm-55) were found. Lm-2 and Lm-55 were isolated from the same year and site of collection in Victoria. Alleles present only in the isolates used for SNP prediction (Lm-1 and Lm-2) are due to ascertainment bias of using these for the SNP prediction. Ascertainment bias is introduced because of the method used for SNP discovery (Albrechtsen, Nielsen, & Nielsen, 2010), which in this case, used two isolates (Lm-1 and Lm-2) to predict SNPs. Ascertainment bias can be corrected in the future by using more isolates for SNP prediction. Being closely related and hence sharing sequence similarity might explain the four alleles that were found only in Lm-2 and Lm-55. The two main clades on the phylogenetic tree separated the isolates Lm-1 and Lm-2. Replicates used in this assay were seen to be genetically identical except for missing values. This was validated by the phylogenetic tree and PCA output, which confirmed the reproducibility of this assay.

The purpose behind conducting a large-scale genotyping assay was to understand the genetic diversity in the Australian *L. maculans* population. Our cumulative analysis of these results supports our null hypothesis of a panmictic Australian *L. maculans* population with possible regional clonality. This contrasts with the conclusions of Hayden et al. (2007) who found two genetically distinct eastern and western blackleg populations in Australia using 6 microsatellite and 2 minisatellite markers in 513 isolates collected over two years. The study also found 85% difference within the 13 subpopulations identified and 10% difference between the coasts. However, our results concur with the panmictic conclusion of Travadon et al. (2011). This study analysed 29 field populations of French *L. maculans* isolates using minisatellite markers and also found low genetic differentiation within populations. Further study concentrated on sampling from each region will help provide insights into this theory.

Overall, the high rate of sexual reproduction, ability of the pathogen to survive on stubble for long periods of time and random mating between isolates likely assist in maintaining a large blackleg population in Australia. This combined with its high evolutionary potential enables it to overcome host resistance quickly and cause infection leading to wide-spread crop losses. It is therefore imperative to attempt to restrict the population size of this pathogen using existing methods such as stubble management and introgression of resistance genes in *Brassica* species. Future work involves identifying new disease-associated genes, which will help in developing novel strategies to further control this devastating pathogen.

Acknowledgments

The authors would like to acknowledge funding support from the Australian Research Council (Projects LP0882095, LP0883462, LP0989200, LP110100200, DE120100668 and DP0985953). Support from the Australian Genome Research Facility (AGRF), the Queensland Cyber Infrastructure Foundation (QCIF) and the Australian Partnership for Advanced Computing (APAC) is gratefully acknowledged.

References

Albrechtsen, A., Nielsen, F. C., & Nielsen, R. (2010). Ascertainment Biases in SNP chips affect measures of Population Divergence. *Molecular biology and Evolution, 27*(11), 2534-2547. http://dx.doi.org/10.1093/molbev/msq148

Appleby, N., Edwards, D., & Batley, J. (2009). New Technologies for Ultra-High Throughput Genotyping in Plants. *Plant Genomics Methods and Protocols*. UK: Humana Press. http://dx.doi.org/10.1007/978-1-59745-427-8_2

Balesdent, M.H., Attard, A., Ansan-Melayah, D., Delourme, R., Renard, M., & Rouxel, T. (2001) Genetic Control and Host Range of Avirulence toward *Brassica napus* Cultivars Quinta and Jet Neuf in *Leptosphaeria maculans*. *Phytopathology, 91*(1), 70–76. http://dx.doi.org/10.1094/PHYTO.2001.91.1.70

Balesdent, M. H., Barbetti, M. J., Li, H., Sivasithamparam, K., Gout, L., & Rouxel, T. (2005). Analysis of *Leptosphaeria maculans* Race Structure in a Worldwide Collection of Isolates. *Phytopathology, 95*(9), 1061-1071. http://dx.doi.org/10.1094/PHYTO-95-1061

Barrins, J. M., Ades, P. K., Salisbury, P. A., & Howlett, B. J. (2004). Genetic diversity of Australian Isolates of *Leptosphaeria maculans*, the Fungus that causes Blackleg of Canola (*Brassica napus*). *Australasian Plant Pathology, 33*(4), 529-536. http://dx.doi.org/10.1071/AP04061

Biomatters Ltd. (2015). Geneious Pro v5.6. Retrieved from http://www.geneious.com

Daverdin, G., Rouxel, T., Gout, L., Aubertot, J. N., Fudal, I., Meyer, M., . . . Balesdent, M. H. (2012). Genome Structure and Reproductive Behaviour influence the Evolutionary Potential of a Fungal Phytopathogen. *Plos Pathogens, 8*(11), e1003020. http://dx.doi.org/10.1371/journal.ppat.1003020

Dilmaghani, A., Gladieux, P., Gout, L., Giraud, T., Brunner, P. C., Stachowiak, A., . . . Rouxel, T. (2012). Migration Patterns and changes in Population Biology associated with the Worldwide Spread of the Oilseed Rape Pathogen *Leptosphaeria maculans*. *Molecular Ecology, 21*(10), 2519-2533. http://dx.doi.org/10.1111/j.1365-294X.2012.05535.x

Dilmaghani, A., Gout, L., Moreno-Rico, O., Dias, J. S., Coudard, L., Castillo-Torres, N., . . . Rouxel, T. (2013). Clonal Populations of *Leptosphaeria maculans* Contaminating Cabbage in Mexico. *Plant Pathology, 62*(3), 520-532. http://dx.doi.org/10.1111/j.1365-3059.2012.02668.x

Dray, S., & Dufour, A. B. (2007). The ade4 package: Implementing the Duality Diagram for Ecologists. *Journal of Statistical Software, 22*(4), 1-20.

Durstewitz, G., Polley, A., Plieske, J., Luerssen, H., Graner, E. M., Wieseke, R., & Ganal, M. W. (2010). SNP discovery by amplicon sequencing and multiplex SNP genotyping in the allopolyploid species *Brassica napus*. *Genome, 53*(11). http://dx.doi.org/10.1139/G10-079

Fudal, I., Ross, S., Brun, H., Besnard, A.-L., Ermel, M., Kuhn, M.-L., . . . Rouxel, T. (2009). Repeat-Induced Point Mutation (RIP) as an Alternative Mechanism of Evolution Toward Virulence in *Leptosphaeria maculans*. *Molecular Plant-Microbe Interactions, 22*(8), 932-941. http://dx.doi.org/10.1094/MPMI-22-8-0932

Giraud, T., Enjalbert, J., Fournier, E., Delmotte, F., & Dutech, C. (2008). Population Genetics of Fungal Diseases of Plants. *Parasite, 15*(3), 449-454. http://dx.doi.org/10.1051/parasite/2008153449

Gout, L., Eckert, M., Rouxel, T., & Balesdent, M. H. (2006). Genetic Variability and Distribution of Mating Type Alleles in Field Populations of *Leptosphaeria maculans* from France. *Applied and Environmental Microbiology, 72*(1), 185-191. http://dx.doi.org/10.1128/AEM.72.1.185-191.2006

Hayden, H. L., Cozijnsen, A. J., & Howlett, B. J. (2007). Microsatellite and Minisatellite analysis of *Leptosphaeria maculans* in Australia reveals Regional Genetic Differentiation. *Phytopathology, 97*(7), 879-887. http://dx.doi.org/10.1094/PHYTO-97-7-0879

Hayward, A., McLanders, J., Campbell, E., Edwards, D., & Batley, J. (2012). Genomic Advances will herald New Insights into the *Brassica*: *Leptosphaeria maculans* Pathosystem. *Plant Biology, 14*(1), 1-10. http://dx.doi.org/10.1111/j.1438-8677.2011.00481.x

Illumina Inc. (2013). GenomeStudio Software. Retrieved from http://www.illumina.com/applications/microarrays/microarray-software/genomestudio.html

Kearse, M., Moir, R., Wilson, A., Stones-Havas, S., Cheung, M., Sturrock, S., Buxton, S., Cooper, A., Markowitz, S., Duran, C., Thierer, T., Ashton, B., Mentjies, P., & Drummond, A. (2012). Geneious Basic: an Integrated and Extendable Desktop Software Platform for the Organization and Analysis of Sequence Data. *Bioinformatics, 28*(12), 1647-1649. http://dx.doi.org/10.1093/bioinformatics/bts199

Lewin-Koh, N. J., Bivand, R., Pebesma, E. J., Archer, E., Baddeley, A., Bibiko, H., . . . Turner, R. (2012). maptools: Tools for Reading and Handling Spatial Objects (Version R package version 0.8-20). Retrieved from http://CRAN.R-project.org/package=maptools

Life Technologies (2013). Qubit® 2.0 Fluorometer. Retrieved from http://www.invitrogen.com/site/us/en/home/brands/Product-Brand/Qubit/qubit-fluorometer.html

Lipka, A. E., Tian, F., Wang, Q., Peiffer, J., Li, M., Bradbury, P. J., . . . Zhang, Z. (2012). GAPIT: Genome Association and Prediction Integrated tool. *Bioinformatics, 28*(18), 2397-2399. http://dx.doi.org/10.1093/bioinformatics/bts444

Marcroft, S. J., Elliott, V. L., Cozijnsen, A. J., Salisbury, P. A., Howlett, B. J., & Van de Wouw, A. P. (2012). Identifying Resistance Genes to *Leptosphaeria maculans* in Australian *Brassica napus* Cultivars based on Reactions to Isolates with known Avirulence Genotypes. *Crop & Pasture Science, 63*(4), 338-350. http://dx.doi.org/10.1071/CP11341

McDonald, B. A., & Linde, C. (2002). Pathogen Population Genetics, Evolutionary Potential, and Durable Resistance. *Annual Review of Phytopathology, 40*, 349-379. http://dx.doi.org/10.1146/annurev.phyto.40.120501.101443

McVean, G. (2007) Linkage Disequilibrium, Recombination and Selection, in *Handbook of Statistical Genetics*, Third Edition, John Wiley & Sons, Ltd, Chichester, UK. http://dx.doi.org/10.1002/9780470061619.ch27

Parlange, F., Daverdin, G., Fudal, I., Kuhn, M. L., Balesdent, M. H., Blaise, F., . . . Rouxel, T. (2009). *Leptosphaeria maculans* Avirulence Gene *AvrLm4-7* confers a Dual Recognition Specificity by the *Rlm4* and *Rlm7* Resistance Genes of Oilseed Rape, and circumvents *Rlm4*-mediated Recognition through a Single Amino Acid Change. *Molecular Microbiology, 71*(4), 851-863. http://dx.doi.org/10.1111/j.1365-2958.2008. 06547.x

Polk, D. B., & Peek, R. M., Jr. (2010). *Helicobacter pylori*: Gastric Cancer and beyond. *Nature Reviews Cancer, 10*(6), 403-414. http://dx.doi.org/10.1038/nrc2857

Purwantara, A., Barrins, J. M., Cozijnsen, A. J., Ades, P. K., & Howle, B. J. (2000). Genetic Diversity of Isolates of the *Leptosphaeria maculans* species complex from Australia, Europe and North America using Amplified Fragment Length Polymorphism analysis. *Mycological Research, 104*(7), 772-781. http://dx.doi.org/10. 1017/S095375629900235X

Purwantara, A., Salisbury, P. A., Burton, W. A., & Howlett, B. J. (1998). Reaction of *Brassica juncea* (Indian mustard) lines to Australian isolates of *Leptosphaeria maculans* under Glasshouse and Field Conditions. *European Journal of Plant Pathology, 104*(9), 895-902. http://dx.doi.org/10.1023/A:1008609131695

Rouxel, T., & Balesdent, M. H. (2005). The Stem Canker (blackleg) Fungus, *Leptosphaeria maculans*, enters the Genomic Era. *Molecular Plant Pathology, 6*(3), 225-241. http://dx.doi.org/10.1111/j.1364-3703.2005.002 82.x

Rouxel, T., Grandaubert, J., Hane, J. K., Hoede, C., van de Wouw, A. P., Couloux, A., . . . Howlett, B. J. (2011). Effector Diversification within Compartments of the *Leptosphaeria maculans* Genome affected by Repeat-Induced Point Mutations. *Nature Communications, 2*. http://dx.doi.org/10.1038/ncomms1189

Shin, J. H., Blay, S., McNeney, B., & Graham, J. (2006). LDheatmap: An R Function for Graphical Display of Pairwise Linkage Disequilibria between Single Nucleotide Polymorphisms. *Journal of Statistical Software, 16*, code snippet 03.

Suzuki, R., & Shimodaira, H. (2006). Pvclust: an R Package for Assessing the Uncertainty in Hierarchical Clustering. *Bioinformatics, 22*(12), 1540-1542. http://dx.doi.org/10.1093/bioinformatics/btl117

Tindall, E. A., Petersen, D. C., Nikolaysen, S., Miller, W., Schuster, S. C., & Hayes, V. M. (2010). Interpretation of Custom Designed Illumina Genotype Cluster Plots for Targeted Association Studies and Next-Generation Sequence Validation. *BMC research notes, 3*, 39. http://dx.doi.org/10.1186/1756-0500-3-39

Travadon, R., Sache, I., Dutech, C., Stachowiak, A., Marquer, B., & Bousset, L. (2011). Absence of Isolation by Distance Patterns at the Regional Scale in the Fungal Plant Pathogen *Leptosphaeria maculans*. *Fungal Biology, 115*(7), 649-659. http://dx.doi.org/10.1016/j.funbio.2011.03.009

Van de Wouw, A. P., Cozijnsen, A. J., Hane, J. K., Brunner, P. C., McDonald, B. A., Oliver, R. P., & Howlett, B. J. (2010). Evolution of Linked Avirulence Effectors in *Leptosphaeria maculans* is affected by Genomic Environment and Exposure to Resistance Genes in Host Plants. *Plos Pathogens, 6*(11). http://dx.doi.org/10.1371/journal.ppat.1001180

Warnes, G., Gorjanc, G., Leisch, F., & Man, M. (2012). genetics: Population Genetics (Version R package version 1.3.7.). Retrieved from http://CRAN.R-project.org/package=genetics

West, J. S., Kharbanda, P. D., Barbetti, M. J., & Fitt, B. D. L. (2001). Epidemiology and Management of *Leptosphaeria maculans* (phoma stem canker) on Oilseed Rape in Australia, Canada and Europe. *Plant Pathology, 50*(1), 10-27. http://dx.doi.org/10.1046/j.1365-3059.2001.00546.x

Xu, J. R. (2006). Fundamentals of Fungal Molecular Population Genetic Analysis. *Current Issues in Molecular Biology, 8*, 75-89.

Zander, M., Patel, D. A., Van de Wouw, A., Lai, K., Lorenc, M. T., Campbell, E., . . . Batley, J. (2013). Identifying Genetic Diversity of Avirulence Genes in *Leptosphaeria maculans* using Whole Genome Sequencing. *Functional & Integrative Genomics, 13*(3), 295-308. http://dx.doi.org/10.1007/s10142-013-0324-5

Appendices

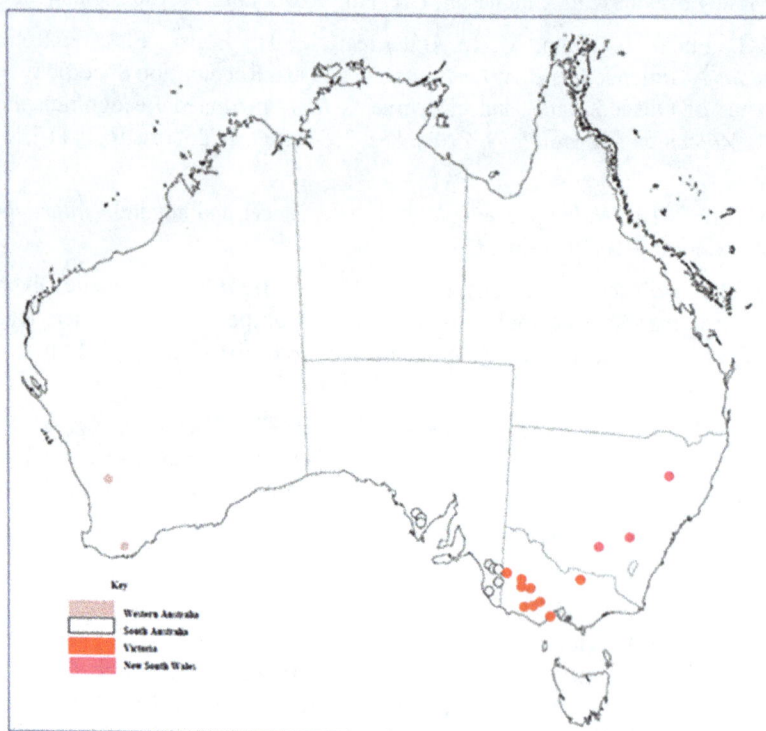

Appendix A. Site of collection of *L. maculans* in Australia

Appendix B. General statistics of 214 SNPs

SNP name	Total A	Total B	Total	NA	% A	% B	PIC
SNP1	35	56	91	5	38.46	61.54	0.47
SNP2	27	66	93	3	29.03	70.97	0.41
SNP3	35	57	92	4	38.04	61.96	0.47
SNP4	86	10	96	0	89.58	10.42	0.19
SNP5	23	62	85	11	27.06	72.94	0.39
SNP6	13	76	89	7	14.61	85.39	0.25
SNP7	89	5	94	2	94.68	5.32	0.10
SNP8	5	91	96	0	5.21	94.79	0.10
SNP9	36	54	90	6	40.00	60.00	0.48
SNP10	27	68	95	1	28.42	71.58	0.41
SNP11	82	12	94	2	87.23	12.77	0.22
SNP12	72	18	90	6	80.00	20.00	0.32
SNP13	32	54	86	10	37.21	62.79	0.47
SNP14	36	54	90	6	40.00	60.00	0.48
SNP15	8	64	72	24	11.11	88.89	0.20
SNP16	56	31	87	9	64.37	35.63	0.46
SNP17	40	47	87	9	45.98	54.02	0.50
SNP18	42	49	91	5	46.15	53.85	0.50
SNP19	23	67	90	6	25.56	74.44	0.38
SNP20	48	41	89	7	53.93	46.07	0.50
SNP21	33	55	88	8	37.50	62.50	0.47
SNP22	24	63	87	9	27.59	72.41	0.40
SNP23	37	55	92	4	40.22	59.78	0.48
SNP24	83	12	95	1	87.37	12.63	0.22
SNP25	34	58	92	4	36.96	63.04	0.47

SNP26	19	68	87	9	21.84	78.16	0.34
SNP27	30	56	86	10	34.88	65.12	0.45
SNP28	16	75	91	5	17.58	82.42	0.29
SNP29	5	82	87	9	5.75	94.25	0.11
SNP30	35	58	93	3	37.63	62.37	0.47
SNP31	90	5	95	1	94.74	5.26	0.10
SNP32	52	40	92	4	56.52	43.48	0.49
SNP33	37	48	85	11	43.53	56.47	0.49
SNP34	73	23	96	0	76.04	23.96	0.36
SNP35	23	64	87	9	26.44	73.56	0.39
SNP36	80	12	92	4	86.96	13.04	0.23
SNP37	57	33	90	6	63.33	36.67	0.46
SNP38	35	50	85	11	41.18	58.82	0.48
SNP39	32	60	92	4	34.78	65.22	0.45
SNP40	28	67	95	1	29.47	70.53	0.42
SNP41	31	60	91	5	34.07	65.93	0.45
SNP42	29	56	85	11	34.12	65.88	0.45
SNP43	65	28	93	3	69.89	30.11	0.42
SNP44	35	58	93	3	37.63	62.37	0.47
SNP45	45	44	89	7	50.56	49.44	0.50
SNP46	38	52	90	6	42.22	57.78	0.49
SNP47	7	83	90	6	7.78	92.22	0.14
SNP48	9	84	93	3	9.68	90.32	0.17
SNP49	72	20	92	4	78.26	21.74	0.34
SNP50	74	20	94	2	78.72	21.28	0.33
SNP51	20	71	91	5	21.98	78.02	0.34
SNP52	44	47	91	5	48.35	51.65	0.50
SNP53	30	65	95	1	31.58	68.42	0.43
SNP54	77	16	93	3	82.80	17.20	0.28
SNP55	93	3	96	0	96.88	3.13	0.06
SNP56	20	71	91	5	21.98	78.02	0.34
SNP57	23	71	94	2	24.47	75.53	0.37
SNP58	30	59	89	7	33.71	66.29	0.45
SNP59	68	25	93	3	73.12	26.88	0.39
SNP60	74	20	94	2	78.72	21.28	0.33
SNP61	10	75	85	11	11.76	88.24	0.21
SNP62	74	16	90	6	82.22	17.78	0.29
SNP63	63	28	91	5	69.23	30.77	0.43
SNP64	6	89	95	1	6.32	93.68	0.12
SNP65	7	84	91	5	7.69	92.31	0.14
SNP66	46	43	89	7	51.69	48.31	0.50
SNP67	90	6	96	0	93.75	6.25	0.12
SNP68	76	17	93	3	81.72	18.28	0.30
SNP69	33	61	94	2	35.11	64.89	0.46
SNP70	75	20	95	1	78.95	21.05	0.33
SNP71	61	32	93	3	65.59	34.41	0.45
SNP72	88	7	95	1	92.63	7.37	0.14
SNP73	10	84	94	2	10.64	89.36	0.19
SNP74	29	60	89	7	32.58	67.42	0.44
SNP75	15	67	82	14	18.29	81.71	0.30
SNP76	49	43	92	4	53.26	46.74	0.50
SNP77	22	66	88	8	25.00	75.00	0.38
SNP78	46	43	89	7	51.69	48.31	0.50
SNP79	30	59	89	7	33.71	66.29	0.45
SNP80	17	78	95	1	17.89	82.11	0.29
SNP81	3	93	96	0	3.13	96.88	0.06

SNP82	55	31	86	10	63.95	36.05	0.46
SNP83	45	38	83	13	54.22	45.78	0.50
SNP84	57	39	96	0	59.38	40.63	0.48
SNP85	3	81	84	12	3.57	96.43	0.07
SNP86	6	43	49	47	12.24	87.76	0.21
SNP87	47	44	91	5	51.65	48.35	0.50
SNP88	81	14	95	1	85.26	14.74	0.25
SNP89	66	28	94	2	70.21	29.79	0.42
SNP90	35	58	93	3	37.63	62.37	0.47
SNP91	27	59	86	10	31.40	68.60	0.43
SNP92	14	64	78	18	17.95	82.05	0.29
SNP93	16	80	96	0	16.67	83.33	0.28
SNP94	84	8	92	4	91.30	8.70	0.16
SNP95	48	40	88	8	54.55	45.45	0.50
SNP96	84	11	95	1	88.42	11.58	0.20
SNP97	88	7	95	1	92.63	7.37	0.14
SNP98	36	58	94	2	38.30	61.70	0.47
SNP99	23	69	92	4	25.00	75.00	0.38
SNP100	18	67	85	11	21.18	78.82	0.33
SNP101	36	49	85	11	42.35	57.65	0.49
SNP102	36	49	85	11	42.35	57.65	0.49
SNP103	10	85	95	1	10.53	89.47	0.19
SNP104	42	50	92	4	45.65	54.35	0.50
SNP105	67	24	91	5	73.63	26.37	0.39
SNP106	39	55	94	2	41.49	58.51	0.49
SNP107	47	41	88	8	53.41	46.59	0.50
SNP108	5	91	96	0	5.21	94.79	0.10
SNP109	25	66	91	5	27.47	72.53	0.40
SNP110	24	65	89	7	26.97	73.03	0.39
SNP111	42	49	91	5	46.15	53.85	0.50
SNP112	45	48	93	3	48.39	51.61	0.50
SNP113	64	29	93	3	68.82	31.18	0.43
SNP114	22	69	91	5	24.18	75.82	0.37
SNP115	90	4	94	2	95.74	4.26	0.08
SNP116	46	42	88	8	52.27	47.73	0.50
SNP117	5	89	94	2	5.32	94.68	0.10
SNP118	54	39	93	3	58.06	41.94	0.49
SNP119	90	5	95	1	94.74	5.26	0.10
SNP120	47	32	79	17	59.49	40.51	0.48
SNP121	53	38	91	5	58.24	41.76	0.49
SNP122	28	63	91	5	30.77	69.23	0.43
SNP123	37	57	94	2	39.36	60.64	0.48
SNP124	11	84	95	1	11.58	88.42	0.20
SNP125	4	92	96	0	4.17	95.83	0.08
SNP126	16	74	90	6	17.78	82.22	0.29
SNP127	14	76	90	6	15.56	84.44	0.26
SNP128	7	53	60	36	11.67	88.33	0.21
SNP129	80	11	91	5	87.91	12.09	0.21
SNP130	92	4	96	0	95.83	4.17	0.08
SNP131	4	79	83	13	4.82	95.18	0.09
SNP132	49	39	88	8	55.68	44.32	0.49
SNP133	23	70	93	3	24.73	75.27	0.37
SNP134	66	24	90	6	73.33	26.67	0.39
SNP135	3	83	86	10	3.49	96.51	0.07
SNP136	76	16	92	4	82.61	17.39	0.29
SNP137	4	92	96	0	4.17	95.83	0.08

SNP138	91	5	96	0	94.79	5.21	0.10
SNP139	85	11	96	0	88.54	11.46	0.20
SNP140	92	3	95	1	96.84	3.16	0.06
SNP141	3	89	92	4	3.26	96.74	0.06
SNP142	24	70	94	2	25.53	74.47	0.38
SNP143	51	36	87	9	58.62	41.38	0.49
SNP144	40	47	87	9	45.98	54.02	0.50
SNP145	74	21	95	1	77.89	22.11	0.34
SNP146	73	15	88	8	82.95	17.05	0.28
SNP147	92	4	96	0	95.83	4.17	0.08
SNP148	9	86	95	1	9.47	90.53	0.17
SNP149	38	50	88	8	43.18	56.82	0.49
SNP150	65	29	94	2	69.15	30.85	0.43
SNP151	4	88	92	4	4.35	95.65	0.08
SNP152	31	61	92	4	33.70	66.30	0.45
SNP153	3	89	92	4	3.26	96.74	0.06
SNP154	29	58	87	9	33.33	66.67	0.44
SNP155	45	42	87	9	51.72	48.28	0.50
SNP156	19	71	90	6	21.11	78.89	0.33
SNP157	86	7	93	3	92.47	7.53	0.14
SNP158	66	24	90	6	73.33	26.67	0.39
SNP159	70	23	93	3	75.27	24.73	0.37
SNP160	93	3	96	0	96.88	3.13	0.06
SNP161	3	60	63	33	4.76	95.24	0.09
SNP162	74	20	94	2	78.72	21.28	0.33
SNP163	91	5	96	0	94.79	5.21	0.10
SNP164	75	17	92	4	81.52	18.48	0.30
SNP165	69	25	94	2	73.40	26.60	0.39
SNP166	56	35	91	5	61.54	38.46	0.47
SNP167	61	27	88	8	69.32	30.68	0.43
SNP168	69	24	93	3	74.19	25.81	0.38
SNP169	29	63	92	4	31.52	68.48	0.43
SNP170	30	55	85	11	35.29	64.71	0.46
SNP171	39	50	89	7	43.82	56.18	0.49
SNP172	75	17	92	4	81.52	18.48	0.30
SNP173	17	76	93	3	18.28	81.72	0.30
SNP174	41	47	88	8	46.59	53.41	0.50
SNP175	4	88	92	4	4.35	95.65	0.08
SNP176	26	65	91	5	28.57	71.43	0.41
SNP177	17	76	93	3	18.28	81.72	0.30
SNP178	9	79	88	8	10.23	89.77	0.18
SNP179	59	34	93	3	63.44	36.56	0.46
SNP180	33	59	92	4	35.87	64.13	0.46
SNP181	3	69	72	24	4.17	95.83	0.08
SNP182	25	69	94	2	26.60	73.40	0.39
SNP183	71	17	88	8	80.68	19.32	0.31
SNP184	22	69	91	5	24.18	75.82	0.37
SNP185	48	42	90	6	53.33	46.67	0.50
SNP186	81	14	95	1	85.26	14.74	0.25
SNP187	22	63	85	11	25.88	74.12	0.38
SNP188	3	92	95	1	3.16	96.84	0.06
SNP189	54	35	89	7	60.67	39.33	0.48
SNP190	13	79	92	4	14.13	85.87	0.24
SNP191	15	76	91	5	16.48	83.52	0.28
SNP192	70	21	91	5	76.92	23.08	0.36
SNP193	7	85	92	4	7.61	92.39	0.14

SNP194	3	93	96	0	3.13	96.88	0.06
SNP195	39	49	88	8	44.32	55.68	0.49
SNP196	66	25	91	5	72.53	27.47	0.40
SNP197	7	80	87	9	8.05	91.95	0.15
SNP198	93	3	96	0	96.88	3.13	0.06
SNP199	41	50	91	5	45.05	54.95	0.50
SNP200	13	83	96	0	13.54	86.46	0.23
SNP201	53	39	92	4	57.61	42.39	0.49
SNP202	40	47	87	9	45.98	54.02	0.50
SNP203	3	93	96	0	3.13	96.88	0.06
SNP204	4	82	86	10	4.65	95.35	0.09
SNP205	92	3	95	1	96.84	3.16	0.06
SNP206	73	20	93	3	78.49	21.51	0.34
SNP207	75	18	93	3	80.65	19.35	0.31
SNP208	32	57	89	7	35.96	64.04	0.46
SNP209	60	33	93	3	64.52	35.48	0.46
SNP210	76	19	95	1	80.00	20.00	0.32
SNP211	45	46	91	5	49.45	50.55	0.50
SNP212	33	57	90	6	36.67	63.33	0.46
SNP213	19	72	91	5	20.88	79.12	0.33
SNP214	25	60	85	11	29.41	70.59	0.42

Note: PIC-Polymorphism Information Content (Totals excluding NAs).

Appendix C. Analysis of association between SNPs and state of collection of isolates

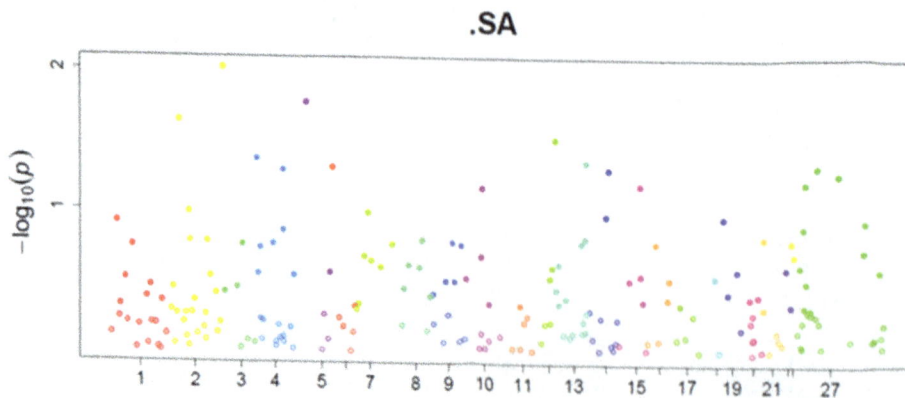

Figure C1: Manhattan plot of all isolates showing association between SNPs and South Australia
Note: x-axis: Supercontigs (27= SC0); y-xis $-\log_{10}$ values of association between stubble species and SNPs.

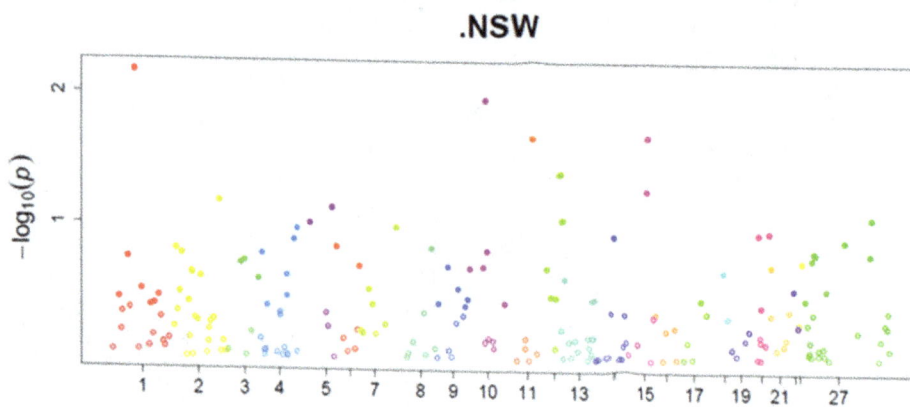

Figure C2: Manhattan plot of all isolates showing association between SNPs and New South Wales
Note: x-axis supercontigs (27= SC0); y-xis $-\log_{10}$ values of association between stubble species and SNPs.

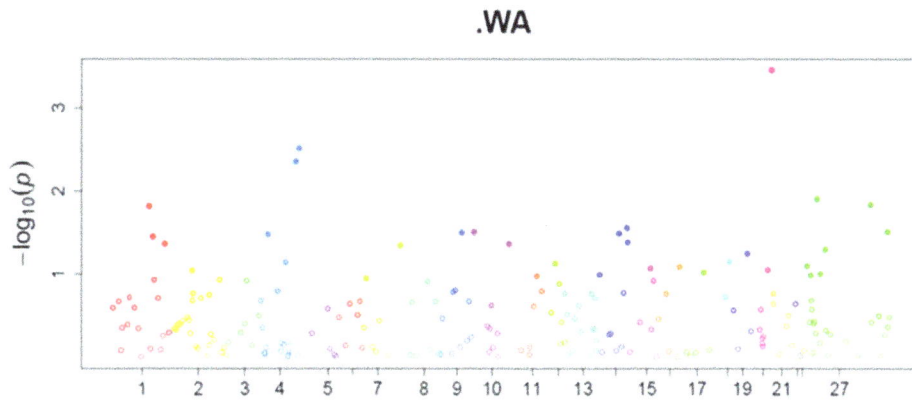

Figure C3 Manhattan plot of all isolates showing association between SNPs and Western Australia

Note: x-axis supercontigs (27= SC0); y-xis $-\log_{10}$ values of association between stubble species and SNPs.

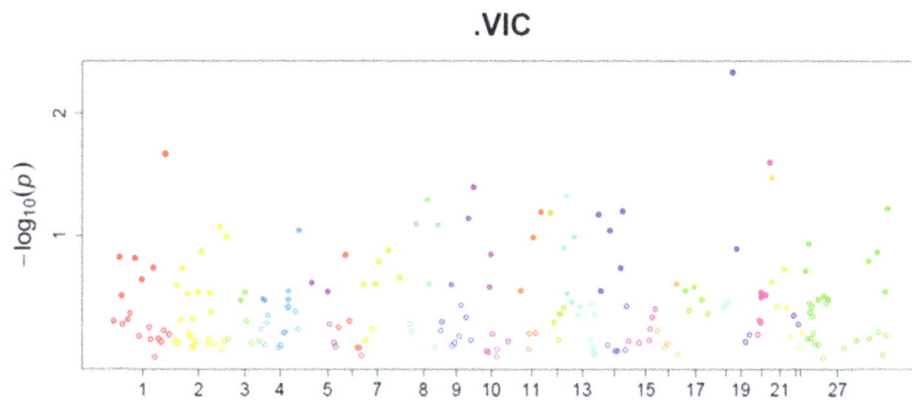

Figure C4 Manhattan plot of all isolates showing association between SNPs and Victoria

Note: x-axis supercontigs (27= SC0); y-xis $-\log_{10}$ values of association between stubble species and SNPs.

Permissions

List of Contributors

Uttamkumar S. Bagde
Applied Microbiology Laboratory, Department of Life Sciences, University of Mumbai, India
Amity Institute of Microbial Technology, Amity University-Uttar Pradesh, India

Ram Prasad
Amity Institute of Microbial Technology, Amity University-Uttar Pradesh, India

Ajit Varma
Amity Institute of Microbial Technology, Amity University-Uttar Pradesh, India

I. H. AL-Mishhadani Ibrahim
Biotechnology Research Center, AL-Nahrain University, P. O. Box 64074, Jadriah, Baghdad, Iraq

Bilal F. Zakariya
Biology Department, AL-Razi College of Education, University of Diyala, Iraq

N. Ismail Eman
Biotechnology Research Center, AL-Nahrain University, P. O. Box 64074, Jadriah, Baghdad, Iraq

M. Dawood Wisam
Biology Department, AL-Razi College of Education, University of Diyala, Iraq

Wesam Al Khateeb
Department of Biological Sciences, Yarmouk University, Jordan

A. A. Nwabueze
Department of Fisheries, Delta State University, Asaba Campus, Nigeria

Nadilia N. Gómez Raboteaux
Research Analyst – RIM, Pioneer Hi-Bred Seed Co., Johnston, IA, USA

Neil O. Anderson
Department of Horticultural Science, University of Minnesota, Saint Paul, MN, USA

Victoria Anatolyivna Tsygankova
Department Cell Signal System, Institute of Bioorganic Chemistry and Petrochemistry, National Academy of Sciences of Ukraine, Ukraine

Galyna Alexandrovna Iutynska
Department of General and Soil Microbiology, Zabolotny Institute of Microbiology and Virology, National Academy of Sciences of Ukraine, Ukraine

Anatoliy Pavlovych Galkin
Department Genomics and Molecular Biotechnology, Institute of Food Biotechnology and Genomics, National Academy of Sciences of Ukraine, Ukraine

Yaroslav Borisovych Blume
Department Genomics and Molecular Biotechnology, Institute of Food Biotechnology and Genomics, National Academy of Sciences of Ukraine, Ukraine

H Stambouli-Meziane
Laboratory of Ecology and Management of Natural Ecosystems, Tlemcen, Algeria

A Merzouk
Laboratory of Ecology and Management of Natural Ecosystems, Tlemcen, Algeria

M Bouazza
Laboratory of Ecology and Management of Natural Ecosystems, Tlemcen, Algeria

Jerome Y. Gaugris
Centre for Wildlife Management, University of Pretoria, Pretoria, South Africa
Flora Fauna & Man, Ecological Services Ltd., Road Town/ Tortola, British Virgin Island

Caroline A. Vasicek
Flora Fauna & Man, Ecological Services Ltd., Road Town/ Tortola, British Virgin Island

Margaretha W. van Rooyen
Ekotrust cc. 272 Thatcher's Field, Lynwood, Pretoria, South Africa

Raji A. Abdullateef
Department of Biotechnology, Kulliyyah of Science, International Islamic University, Malaysia
Sinwan Agricultural Research and Development Institute, Kwara State, Nigeria

Mohamad bin Osman
Faculty of Plantation and Agrotechnology, UniversitiTeknologi MARA (UiTM), Shah Alam, Selangor, Malaysia

Zarina bint Zainuddin
Department of Biotechnology, Kulliyyah of Science, International Islamic University, Malaysia

Hirokazu Fukunaga
Tokushima-cho, Tokushima city, Tokushima, Japan

Yutaka Sawa
Sawa Orchid Laboratory, Ikku, Kochi city, Kochi, Japan

Shinichiro Sawa
Kumamoto University, Graduate school of Science and Technology, Kumamoto, Japan

Heru Kuswantoro
Indonesian Legume and Tuber Crops Research Institute, Indonesian Agency for Agricultural Research and Development, Indonesia

Handan Çulal Kılıç
Department of Plant Protection, Faculty of Agriculture, Süleyman Demirel University, Turkey

Nejla Yardımcı
Department of Plant Protection, Faculty of Agriculture, Süleyman Demirel University, Turkey

Gözde Urgen
Department of Plant Protection, Faculty of Agriculture, Süleyman Demirel University, Turkey

Olutobi Otusanya
Botany Department, Obafemi Awolowo University, Ile-Ife, Nigeria

Olasupo Ilori
Biology Department Adeyemi College of Education, Ondo, Nigeria

Eduardo Costas
Genética, Facultad de Veterinaria, Universidad Complutense, Madrid, Spain

Emma Huertas
Instituto de Ciencias Marinas de Andalucía (CSIC), Cádiz, Spain

Beatriz Baselga-Cervera
Genética, Facultad de Veterinaria, Universidad Complutense, Madrid, Spain

Camino García-Balboa
Genética, Facultad de Veterinaria, Universidad Complutense, Madrid, Spain

Victoria López-Rodas
Genética, Facultad de Veterinaria, Universidad Complutense, Madrid, Spain

Reni Lestari
Center for Plant Conservation Bogor Botanical Gardens, Indonesian Institute of Sciences, Indonesia

Yiting Li
Key Laboratory of Food Nutrition and Safety, Tianjin University of Science and Technology, Ministry of Education, Tianjin 300457, China

Shili Meng
Graduate School of Life and Environmental Sciences, University of Tsukuba, Ibaraki 305-8577, Japan

Linbo Wang
Graduate School of Life and Environmental Sciences, University of Tsukuba, Ibaraki 305-8577, Japan

Zhenya Zhang
Graduate School of Life and Environmental Sciences, University of Tsukuba, Ibaraki 305-8577, Japan

Dhwani A. Patel
School of Agriculture and Food Sciences and Centre for Integrative Legume Research, University of Queensland, Brisbane, Australia

Manuel Zander
School of Agriculture and Food Sciences and Centre for Integrative Legume Research, University of Queensland, Brisbane, Australia

Angela P. Van de Wouw
School of BioSciences, University of Melbourne, Parkville, Australia

Annaliese S. Mason
Department of Plant Breeding, Land Use and Nutrition, Justus Liebig University, Giessen, Germany

David Edwards
School of Plant Biology, University of Western Australia, Perth, Australia

Jacqueline Batley
School of Plant Biology, University of Western Australia, Perth, Australia

www.ingramcontent.com/pod-product-compliance
Lightning Source LLC
Chambersburg PA
CBHW050442200326
41458CB00014B/5042